香港觀鳥全圖鑑

鏡頭下的247種雀鳥美態・鳥類保育分享

A PHOTOGRAPHIC GUIDE TO THE BIRDS OF HONG KONG

編著

萬里機構

香港觀鳥會《香港觀鳥全圖鑑》製作組

HKBWS "*A Photographic Guide to the Birds of Hong Kong*" Working Group

顧問	林超英
攝影	文超凡、孔思義、黃亞萍、文權溢、王煌容、王維萍、王學思、古愛婉、甘永樂、任永耀、任德政、伍昌齡、伍耀成、朱祖仁、朱詠兒、朱翠萍、朱錦滿、何志剛、何建業、何國海、何瑞章、何萬邦、何維俊、何錦榮、余日東、余柏維、吳思安、吳璉宥、呂德恒、宋亦希、李君哲、李佩玲、李炳偉、李振成、李偉仁、李啟康、李逸明、李雅婷．李鶴飛、杜偉倫、周家禮、林文華、林釗、林鳳兒、柯嘉敏、洪敦熹、胡明川、英克勁、夏敖天、容姨姨 / 李啟源、高俊雄、高偉琛、崔汝棠、張玉良、張勇、張振國、張浩輝、梁釗成、深藍、許光杰、許淑君、郭子祈、郭加祈、郭匯昌、陳土飛、陳巨輝、陳志光、陳志明、陳志雄、陳佩霞、陳佳瑋、陳俊兆、陳建中、陳家強、陳家華、陳詠琴、陳燕明、陳燕芳、陶偉意、陸一朝、勞浚暉、彭俊超、森美與雲妮、馮少萍、馮振威、馮啟文、蕭敏晶、馮漢城、黃才安、黃志俊、黃良熊、黃卓研、黃倫昌、黃理沛、江敏兒、黃瑞芝、黃福基、黃寶偉、楊加強、溫柏豪、葉紀江、詹玉明、賈知行、劉柱光、劉振鴻、劉健忠、劉匯文、劉劍明、蔡美蓮、馬志榮、鄧玉蓮、鄧仲欣、鄧詠詩、鄭兆文、鄭偉強、鄭諾銘、黎凱輝、盧嘉孟、霍棟豪、薛國華、謝鑑超、鍾潤德、簡志明、羅文凱、羅瑞華、羅錦文、譚業成、譚耀良、關朗曦、關子凱、關寶權、蘇毅雄、Aka Ho、Geoff Smith、Geoff Welch、John Clough、Michael Schmitz
撰文	區俊茵、陳慶麟、陳明明、張勇、周家禮、馮寶基、何文輝、孔思義、黃亞萍、洪維銘、江敏兒、林鳳兒、劉偉民、馬嘉慧、蘇毅雄、黃志俊、王學思、余日東、呂德恒
圖片搜集及挑選	譚業成、張浩輝、何萬邦、江敏兒、黃理沛、李佩玲、盧嘉孟、呂德恒、黃卓研、葉思敏、郭子祈
文稿編輯、翻譯及校對	陳慶麟、陳佳瑋、張嘉穎、張浩輝、孔思義、黃亞萍、林傲麟、李銘慧、勞浚暉、呂德恒、黃志俊、駱俊賓、馬嘉慧、溫翰芝、黃卓研、王學思、楊莉琪、余日東、劉偉民、陳燕明、梁嘉善、左治強、劉家雁、黎淑賢、陳翠楣、方海寧、李鍾海、郭子祈、羅偉仁、陳愷瑩、謝偉麟、簡瑋彤

Consultant	C.Y. Lam
Contributors (photos)	Geoff Carey, Allen Chan, Christina Chan, Daniel Chan, Gary Chan, Helen Chan, Isaac Chan, Jacky Chan, Chan Jun Siu Brian, Chan Kai Wai, Natalie Chan, Sam Chan, Simon Chan, Thomas Chan, Chan Wing Kam, Cheng Nok Ming, Raymond Cheng, Cheng Wai Keung, Andy Cheung, Cheung Ho Fai, Louis Cheung, Cheung Mok Jose Alberto, Owen Chiang, Jimmy Chim, Gary Chow, Chu Cho Yan, Chu Chui Ping, Doris Chu, Francis Chu, Frankie Chu, Chung Yun Tak, John Clough, Stanley Fok, Ken Fung, Maison Fung, Fung Siu Ping, Sonia & Kenneth Fung, Martin Hale, Aka Ho, George Ho, Jan Ho, Ho Kam Wing, Kinni Ho, Danny Ho, Marcus Ho, Pippen Ho, Wan Pak Ho, Jemi & John Holmes, Kami Hui, KK Hui, Tony Hung, Herman Ip, Kam Wing Lok, Kan Chi Ming, Koel Ko, Kitty Koo, Kou Chon Hong, Matthew Kwan & TH Kwan, Kwan Po Kuen, Andy Kwok, Kwok Ka Ki, Kwok Tsz Ki, Dick Lai, James Lam, Kenneth Lam, Shirley Lam, Angus Lau, Benson Lau, Lau Chu Kwong, Jac Lau, Jianzhong, Law Kam Man, Anita Lee, Eling Lee, Lee Hok Fei, Jasper Lee, KY Lee, Lee Kai Hong, Lee Yat Ming, Evans Leung, Geoffery Li, Harry Li, Li Wai Yan, Aaron Lo, Lo Chun Fai, Lo Kar Man, Roman Lo, Henry Lui, Mike Luk, Christine and Samuel Ma, Bill Man, Kennic Man, Felix Ng, Karl Ng, Ng Lin Yau, Ng Sze On, Or Ka Man, Pang Chun Chiu, Sammy Sam & Winnie Wong, Michael Schmitz, Leo Sit, Geoff Smith, Samson So, Sung Yik Hei, Tam Yip Shing, Tam Yiu Leung, Tang Chung Yan, Joyce Tang, Wing Tang, Allen To, Ann To, Wallace Tse, Wee Hock Kee, Geoff Welch, Captain Wong, Cherry Wong, Wong Choi On, Dickson Wong, Wong Hok Sze, Wong Leung Hung, Peter Wong & Michelle Kong, Wong Po Wai, Wong Shui Chi, Wong Wai Ping, Wong Wong Yung, Woo Ming Chuan, Kelvin Yam, Yam Wing Yiu, Yeung Ka Keung, Ying Hak King, Yu Yat Tung, Freeman Yue
Contributors (text)	Joanne Au, Alan Chan, Chan Ming Ming, Louis Cheung, Gary Chow, Robin Fung, Ho Man Fai, Jemi & John Holmes, Hung Wai Ming, Michelle Kong, Shirley Lam, Apache Lau, Carrie Ma, Samson So, Dickson Wong, Wong Hok Sze, Yat-tung Yu, Henry Lui
Photographic editors	Tam Yip Shing, Cheung Ho Fai, Marcus Ho, Michelle Kong, Peter Wong, Eling Lee, Lo Kar Man, Henry Lui, Cherry Wong, Cecily Yip, James Kwok
Text editing, translation & proofread	Alan Chan, Peter Chan, Fion Cheung, Cheung Ho Fai, Jemi & John Holmes, Alan Lam, Maggie Li, Lo Chun Fai, Henry Lui, Dickson Wong, Luk Tsun Pun, Carrie Ma, Judy Wan, Cherry Wong, Wong Hok Sze, Vicky Yeung, Yu Yat Tung, Apache Lau, Christina Chan, Katherine Leung, George Jor, Norma Lau, Lai Suk Yin, May Chan, Helen Fong, Tom Li, James Kwok, Lo Wai Yan, Zoey Chan, Ivan Tse, Vivienne Kan

序

香港山川秀美，城市邊緣不遠處，總有讓人感受自然悠閒之所，稍加留意，更可見到種類繁多的鳥類，與我們一起活在同一天空下。

勞碌匆忙的現代城市生活，往往把我們的視野局限在眼前的瑣碎。以物質享受為主導的價值觀念，使我們與自然生命的距離拉得越來越遠。自然雖近，心中卻無處安放。我們很容易在不經意之中陷入一種渾噩的機械狀態，日子一天一天過去，精神生活卻只剩下一片空白。

觀鳥表面上是一種消閒活動，但是多年經驗告訴我，觀鳥促使我們走近生命，認識生命的多元和微妙，感受生命的美麗和舒泰。我的觀鳥朋友全都開心樂觀，不論環境順逆，總是活在安祥自在之中。

在推廣觀鳥的過程中，我發現大家開始時都很重視分辨鳥種和很倚重照片的幫忙，因此一本囊括香港較多鳥種的照片圖鑑是十分需要的。2003年香港觀鳥會得到鳥友慷慨借出在野外多年努力才拍得的珍貴照片，結集出版了《香港鳥類攝影圖鑑》，是香港歷史上第一本以照片向廣大市民展示香港鳥類美麗一面的圖鑑，加上知識廣博的鳥友以文字提供相關知識，實在難得，更令人欣喜的是攝影圖鑑出版後受到市民熱烈歡迎，遠超想像。

在攝影圖鑑的鼓舞下，很多人愛上鳥類攝影，2009年吸納了大批精彩鳥類照片和增添豐富內容，改名《香港鳥類圖鑑》出版。十年來，鳥類攝影水平大幅提升，《香港鳥類圖鑑》因應重修，易名《香港觀鳥全圖鑑》，並更新照片和增添近年拍攝到的鳥種，使本書既精美又完備。

本圖鑑成書過程中，所有鳥友及攝影師都是義務投入的，實在是我們香港人的福分，謹向他們致以誠心的謝意。漁農自然護理署香港濕地公園於2009年的支持促成本圖鑑廣泛流通，我們銘記心中，是次2020年新版得以實現，有賴本會紅耳鵯俱樂部成員無名氏慷慨解囊襄助，我們衷心感激。

願望此書促進越來越多人親近大自然，並在觀賞鳥類中感受到生命的喜悅。

香港觀鳥會榮譽會長
林超英
二零二零年五月

Preface

Hong Kong is a land blessed with hills and streams. In spite of urbanization, one is never far away from a place for unwinding oneself in the embrace of Nature. For people who care to look around, they would find a great variety of feathered friends sharing the same sky with us.

While Nature lies virtually at our doorsteps, it escapes our attention most of the time. The hustle and bustle of the modern city confine our sight to the immediate and the trivial. Values centred on material pleasure distance us from Life itself. It is easy for us to slip obliviously into a form of muddled mechanical existence, in which time flows by while we remain spiritually impoverished.

Bird watching might seem like just another hobby. But years of experience tells me that it provides a path leading to our communion with Nature. It lets us see the diversity and wonders of Life, and enables us to appreciate its beauty and serenity. All my bird watching friends are happy and optimistic. In both times good or bad, their mind stays calm and peaceful.

While promoting bird watching, I realize that beginners are generally keen on the identification of species and need a lot of help from photographs. A photographic guide that covers a good number of the bird species in Hong Kong is therefore badly needed. This book has materialized thanks to the generosity of members of the Hong Kong Bird Watching Society donating amazing photographs derived from many years of hard work in the field. The book also benefits from the text kindly written by members knowledgeable in the subject. Taken together, this guide is a most valuable work presenting to the general public the beauty of the birds in Hong Kong, in pictures and in words. We were all very pleased to discover that the book received an unexpectedly warm reception by the public on publication.

The photographic guide motivated many people to join the army of bird photographers. During the past decades, more and more superb photos comes up attracting more public's attention. This new edition has incorporated more bird species and photos after the 2009 edition which make this book much heavier.

All birdwatchers and photographers who have contributed towards the production of this book have done so as volunteers. They have done our community a great service indeed. I thank them most sincerely. We are most grateful for the support from Hong Kong Wetland Park of Agriculture, Fisheries and Conservation Department, which has

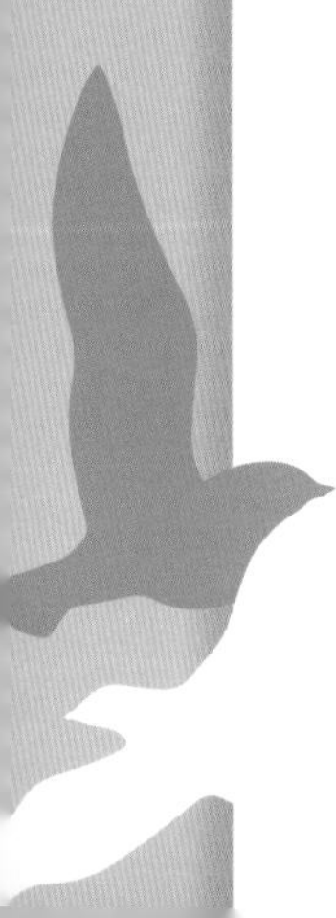

enabled the early publication of the 2009 edition. We are also grateful for the support of an anonymous member of the Crested Bulbul Club of the Hong Kong Bird Watching Society make this new edition in 2020 come true.

I truly hope that this book will bring people closer to Nature and help them appreciate the joy of Life.

CY Lam
Honorary President, Hong Kong Bird Watching Society
2020.05

Shenzhen Bay
深圳灣
（后海灣）
Deep Bay

尖鼻咀
Tsim Bei Tsui

塱原
Long Valley

米埔
Mai Po

南生圍
Nam Sang Wai

香港
濕地公園
Hong Kong Wetland Park

大帽山
Tai Mo Shan

城門水塘
Shing Mun Re

香港觀鳥熱點
Hot Spots for Birdwatching in Hong Kong

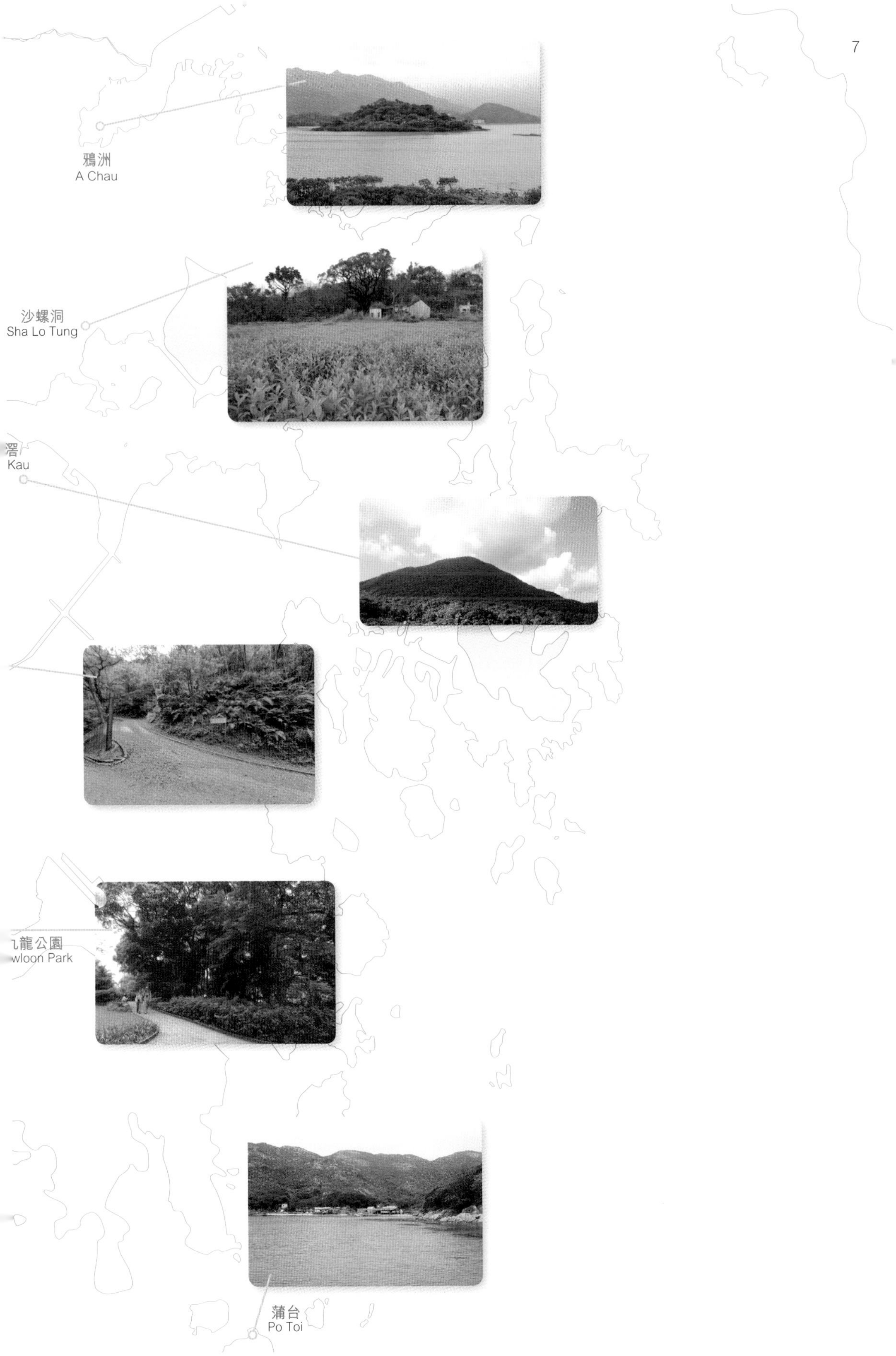

鴉洲
A Chau
沙螺洞
Sha Lo Tung
滘
Kau
龍公園
wloon Park
蒲台
Po Toi

目錄 Contents

* 本書的鳥類分類及排序主要根據「國際鳥類學大會」(International Ornithological Congress) 的分類方法。

The classification and ranking of birds in this book are basically according to the classification method of "International Ornithological Congress".

香港觀鳥會

願景

人鳥和諧　自然長存

使命

香港觀鳥會致力教育、科研、生境管理與保育政策倡議，啟發及鼓勵公眾一起欣賞與保護野生雀鳥及其生境。

香港觀鳥會成立於1957年，一直以推廣欣賞及保育鳥類及其生境為宗旨。2002年獲認可為公共性質慈善機構。2013年成為國際鳥盟 (BirdLife International)的成員(Partner)，國際鳥盟是一個關注鳥類的國際性非政府組織聯盟機構，致力保育鳥類以及牠們的生境、全球的生物多樣性，促進自然資源的永續利用。

我們的工作

鳥類生態研究

掌握鳥類和有關生態環境的數據及資料是制訂保育政策及實施相關措施的重要依據，因此，本會從不間斷地進行各項與鳥類有關的調查及研究工作。

環境監察及政策倡議

作為一個提供專業意見的綠色團體，本會持續地對各項自然保育政策、環境影響評估及大型發展的諮詢提供意見，亦同時監察各類型不同規模的發展項目對自然生態特別是鳥類及其生境的影響。

自然教育

本會深信推動公眾欣賞野生鳥類，令市民親身體驗鳥類的美麗和珍貴，大家自然會由衷地保護牠們。因此，我們一直致力舉辦各項觀鳥及自然教育活動，把鳥類、大自然和市民連繫起來。

生境管理 — 塱原與魚塘

◗「塱原自然保育管理計劃」

自2005年起，本會與長春社在上水塱原濕地開展香港首個農業式濕地的管理項目，保育塱原的生物多樣性、文化及景觀。計劃連結農友及土地擁有人，實踐一些有利鳥類和生物多樣性的濕地管理措施，共同建立有利鳥類的生活環境。

◗「香港魚塘生態保育計劃」

自2012年開始，本會於新界西北魚塘區開展 香港魚塘生態保育計劃」，約130多位漁民參與，覆蓋超過600公頃的魚塘。計劃主要目的是鼓勵漁民每年把魚塘的水位降低一次，讓水鳥可以在魚塘內捕食無經濟價值的雜魚及小蝦，改善及提高魚塘的生態價值。

跨地合作

本會與國際鳥盟於2005年合作展開中國項目。項目初期工作以能力建設為主，以鳥類監測、環境教育、觀鳥組織發展等主題舉辦培訓。其後專注鳥類的保育工作，包括受脅的中華鳳頭燕鷗、勺嘴鷸、靛冠噪鶥，以及華南地區非法捕鳥的問題。自1999年開始，本會成立「香港觀鳥會中國自然保育基金」，近年更擴展基金的範圍至亞洲各地，因此更名為「香港觀鳥會亞洲自然保育基金」。以上工作促進內地觀鳥機構的各方面發展及組織能力、增強各地鳥會之間的溝通和交流，推動鳥類保育工作。

支持香港觀鳥會

- 成為會員
- 捐款支持
- 訂閱香港觀鳥會電子報，緊貼鳥類和保育最新資訊！

電　話：(852) 2377 4387　　電　郵：info@hkbws.org.hk
網　頁：www.hkbws.org.hk
地　址：香港九龍荔枝角青山道532號偉基大廈7樓C室

The Hong Kong Bird Watching Society

Vision

People and birds living in harmony as nature continues to thrive.

Mission

HKBWS promotes appreciation and protection of birds and their habitats through education, research, habitat management and conservation advocacy.

The Hong Kong Bird Watching Society (HKBWS) was founded in 1957. The objective of the organization is to promote the appreciation and the conservation of birds as well as their habitats. It became a public charitable organization in 2002 and BirdLife International Partner in 2013. BirdLife International is a global partnership of conservation organizations that strives to conserve birds, their habitats and global biodiversity, working with people towards sustainably use of natural resources.

Our Work:

Bird Ecology and Research

Data and knowledge of bird species and their habitat ecology are key factors in supporting the formulation and execution of conservation policy and measures. Researches and surveys on birds and their habitats have been conducting throughout the past six decades.

Environmental Monitoring and Policy Advocacy

HKBWS acts as a green group providing professional views. We keep providing comments on government conservation policies, environmental impact assessments and large development projects. We also monitor different kinds and size of development projects to avoid any adverse impact towards the natural environment, especially to birds and their living environment.

Nature Education

We believe once the general public can directly experience the beauty and precious of birds, they will be more committed to the protection of birds. Therefore, we organize different kinds of bird-related educational programmes to link up among birds, nature and the people.

Habitat Management

- **"Nature Conservation Management for Long Valley"**

Since 2005, HKBWS and The Conservancy Association began an agricultural wetland conservation project in Long Valley in order to conserve the biodiversity, culture and landscape of Long Valley. With the engagement of local farmers and landowners, wetland management measures favouring birds and other wetland-dependent wildlife were implemented

- **"Hong Kong Fishpond Conservation Scheme"**

In 2012, a conservation scheme named "Hong Kong Fishpond Conservation Scheme" has been launched in the Northwest New Territories. This scheme covered more than 600 hectares of fishponds and more than 130 fishermen engaged. The purpose of the scheme is to encourage fishermen lowering the water level to allow waterbirds feeding on the trash fishes and shrimps left in the ponds. As a whole, the conservation value of fishpond could be maintained and even strengthened.

Transboundary Cooperation Programme

HKBWS and BirdLife International jointly established the China Programme in 2005. Initial work of the China programme focused on capacity building, which includes holding training workshops on bird monitoring, environmental education and organization development. Later, the Programme works more on bird conservation, including threated species like Chinese Crested Tern, Spoon-billed Sandpiper, Blue-crowned Laughingthrush, as well as problems of illegal hunting in South China. In 1999, HKBWS set up the "HKBWS China Conservation Fund", and later changed as "HKBWS Asia Conservation Fund" to extend the cooperation to the Asian countries. All these works enhance the capacity and development of bird watching societies in mainland China, encourage communication and exchange of information, as well as support and promote bird conservation work in different area.

Support HKBWS

- Be a Member
- Donate to HKBWS
- Subscribe E-news - Stick with the latest news of birds and our work!

Tel: (852) 2377 4387 E-mail: info@hkbws.org.hk
Website: www.hkbws.org.hk
Address: 7C, 532 Castle Peak Road, Lai Chi Kok, Kowloon, Hong Kong

觀鳥在香港
Birdwatching in Hong Kong

城市雀鳥
Birds in the Cities

城市的興起大約有二百多年歷史，而全球大規模城市化則只有數十年，在悠久的自然歷史上，城市只是新近出現的一種生境。一般來說，城市發展會將原本的植被剷除，然後鋪上混凝土，再栽種一些新植物，是破壞自然環境的人工生境，不利野生動植物。然而，有一些適應力強的生物，能夠利用城市的資源生活，甚至部分能大量繁衍，成為城市中佔優勢的物種。

在香港，1990年代有博士生研究市區的雀鳥生態，在考察的三年期間，共記錄111個鳥種，佔當時全港鳥種的四分之一；當中最多是留鳥，佔超過四成，其次是近三成的冬候鳥。研究發現雀鳥的多寡明顯與植被有關，有些鳥種偏好樹木，另有些偏好草地，因此公園中的雀鳥明顯比住宅範圍多。在市區的環境中，在地上撿拾種子穀物為食糧的鳥種最為適應，例如樹麻雀和珠頸斑鳩，在市區較郊外更為常見。

※ 呂德恆 Henry Lui

※ 郭子祈 Kwok Tsz Ki

公園可以説是動物的「城市綠洲」，內裏有用於綠化及美化環境的植物，為動物及昆蟲提供豐富的食物和安穩的棲息地。每當植物開花，太陽鳥、啄花鳥和繡眼鳥等雀鳥常於花間吸食花蜜，也間接為植物傳播花粉。秋冬時份，各式各樣的果實正好成為留鳥和候鳥度過寒冬的必需品。到了春夏之間，數量龐大的昆蟲為剛孵化的雛鳥提供成長所需的營養。

植物可算是雀鳥能留在城市生活的關鍵，植物是食物鏈中的生產者，除了花蜜和果實可供動物直接攝取外，不少原生樹種更能吸引大量昆蟲，成為雀鳥覓食的寶庫。種於路旁及屋苑的樹木，除了為人們抵擋刺眼的陽光，也為鳥類提供晚間棲息的場所。一些沿路種植的樹木，更可能成為雀鳥穿梭城市的「綠色通道」。

由於城市歷史不長，環境的變化亦較一般的自然生境快，城市中的生物群落(community)亦會隨時間演變。例如從一些非正式的日常觀察所見，八哥和黑領椋鳥的生態位(niche)相近，近數十年黑領椋鳥似乎漸漸較八哥更有優勢；以往市區公園常見大群白腰文鳥，近年則少見得多；近年紅嘴藍鵲進佔市區情況變得普遍，其營巢時具侵略性的驅趕行為，有時更與市民產生磨擦。

此外，可能由於鳥類競爭始終較自然野外生境低，一些外來鳥種會在市區建立穩定種群(population)，鳥友最熟悉的小葵花鳳頭鸚鵡、紅領綠鸚鵡、亞歷山大鸚鵡可説是香港市區最有代表性的幾種，而原生地在印尼的小葵花鳳頭鸚鵡，其香港種群更可能是印尼以外最大的野生族群。不少外來物種對環境生態帶來負面影響，例如鸚鵡咬斷樹枝的習慣也對樹木有一定影響，幸而影響不算十分嚴重，而對市民騷擾較大的家鴉則有着不同的命運，漁農自然護理署自2004年起開始監控家鴉的計劃，採取不同措施如使用藥餌及取走蛋和幼鳥等，以減少家鴉數目。

市區交通方便，城市觀鳥成為很好的自然教育活動。香港觀鳥會近年便推行一個「發現香港城市自然生態」的教育計劃，當中有中、小學及幼稚園的活動，其中的「城市生態大使」項目，除了訓練及戶外考察外，生態大使需要於自己學校所處的社區進行生態調查，認識社區的自然生態狀況。

樹麻雀是香港城市中最常見、市民最熟悉的鳥種。香港觀鳥會由2016年開始進行「全港麻雀普查」，最新估算(2019年)香港現時約有30萬隻樹麻雀。此項公民科學活動讓公眾有機會參與科學研究，不少更是一家大小參加，在專家統籌和訓練下，有系統地進行鳥類監測。

如欲進一步認識香港觀鳥會「發現香港城市自然生態」的教育計劃，請瀏覽以下網址：https://sparrow795.wixsite.com/hkbws-sparrow

The history of cities dates back approximately more than 200 years, with large-scale global urbanisation beginning to occur only decades ago. Compared to the long history of nature, cities are a habitat which has come into being only recently. Usually, as a city develops, the existing vegetation will be removed and replaced with concrete, and new plants will be planted. A city is therefore an artificial habitat which is damaging to the natural environment and unfavorable to wild plants and animals. However, certain highly adaptable creatures are able to make use of the resources which a city has to offer, some are even capable of multiplying in great numbers and becoming a dominant species in the city.

In the 1990's, a PhD student in Hong Kong conducted a study on urban bird ecology. During the three years covered by the study, 111 bird species were recorded, accounting for one-fourth of the total number of bird species then recorded in Hong Kong. Most of these species are residents, which accounts for over 40%, followed by winter visitors, which accounts for close to 30%. The study finds that bird numbers are obviously related to vegetation. Some species prefer trees, and some species prefer lawns. Therefore, there is a remarkably greater number of birds in a park than in a residential area. Birds which forage on the ground for seeds and grains, such as Eurasian Tree Sparrow and Spotted Dove, are most adaptable to urban environments and are more commonly found in the city than in the countryside.

Parks can be regarded as an "urban oasis" for animals. The plants planted in a park for greening and beautifying the environment serve as a plentiful food source and a secure home for animals and insects. When the plants bloom, sunbirds, flowerpeckers, white-eyes and other birds are often seen feeding on the nectar, inadvertently helping to pollenate the plants. The large variety of fruits that are available in autumn and winter are exactly what the residents and migrants need for surviving the cold winter. The abundant number of insects that appear between spring and summer provide hatchlings with the nutrition they need for their growth.

We may say plants are the key to birds' survival in cities. Plants are producers in the food chain. In addition to producing nectar and fruits for animals to feed on, many native tree species also attract a large number of insects, thus becoming a precious feeding ground for birds. Apart from shading pedestrians from the scorching sun, trees planted alongside pavements and around housing estates also provide a place for birds to roost at night. Trees planted alongside pavements may even serve as a "green corridor" for birds to move around the city.

The short history of cities and the more rapid changes in the urban environment compared to natural habitats in general means that the community of living organisms in cities also changes with time. For example, based on casual observations, Crested Myna and Black-collared Starling have similar niches, but Black-collared Starling seems to be gradually gaining dominance over Crested Myna in the recent decades. White-rumped Munia, which used to flock in large numbers in urban parks in the past, has become much less common in recent years. Also in recent years, it has become more common to find Red-billed Blue Magpie in urban areas. When nesting, Red-billed Blue Magpie is very aggressive and will chase away other birds or even come into conflict with people from time to time.

Certain alien bird species will build up a stable population in cities, probably because for birds, competition in cities is less intense than in the wild after all. Of these alien species, Yellow-crested Cockatoo, Rose-ringed Parakeet and Alexandrine Parakeet are probably the most representative species found in urban Hong Kong which birds watchers are most familiar with. The Yellow-crested Cockatoo population in Hong Kong may even be the largest wild population of this species outside its native Indonesia. Many alien species have a negative impact on the local ecology. The habit of parrots to bite off branches also has a certain impact on trees, luckily the impact is not very serious. By comparison, House Crow is a bigger nuisance to city dwellers, and this species has met with a different fate. Since 2004, the Agriculture, Fisheries and Conservation Department has been monitoring the number of House Crow and has taken various measures to control its number, such as using baits and removing its eggs and young birds.

Easy accessibility of the urban area has made bird watching in the city a great environmental education activity. In recent years, Hong Kong Bird Watching Society (HKBWS) has implemented an education programme called "Discover the Urban Nature of Hong Kong". The programme offers activities for secondary school, primary school and kindergarten students. One of the activities, called "Urban

Eco-ambassador", does not only provide training and organise field trips for the participants, but also requires the eco-ambassadors to conduct eco-surveys in the community in which their school is located so as familiarise themselves with the ecological conditions of the community.

Eurasian Tree Sparrow is the most common bird species in Hong Kong's urban areas. It is also the species most well known to Hong Kong people. HKBWS has been organising a territory-wide Sparrow Census in Hong Kong since 2016. According to the latest Sparrow Census (2019), there are estimated to be around 300,000 Eurasian Tree Sparrows in Hong Kong. This citizen science activity gives the public a chance to take part in scientific research. Many participants join the activity with their entire family. Under the coordination and training by experts, participants conduct a bird survey in a systematic manner.

For more information on the "Discover the Urban Nature of Hong Kong" education programme run by HKBWS, please visit the following website: https://sparrow795.wixsite.com/hkbws-sparrow

※ 郭子祈 Kwok Tsz Ki

塱原農田及新界西北魚塘
Farmland in Long Valley and Fishponds in NW New Territories

要數雀鳥的重要棲息地，除了天然生境，也有一些與人類活動關係密切的地方，例如農田和魚塘。在香港這個高度發展的大都會，漁農業已式微多年，但部分仍留存下來、面積較大片的農田和魚塘，其生態價值絕不能小覷！

位於香港新界北區的塱原，面積約37公頃，是香港少有大片完整的農地。塱原的生境可分為乾農地、濕農地、荷花池、紅蟲水滋塘和荒地等等，塱原大部分面積為菜田，但因栽種不同的蔬菜，所提供的環境亦有所不同，如種植通菜和西洋菜的水耕方法會形成很多不同的濕地小生境，包括水浸的渠道、儲水池、水浸的農地等。田邊很多時種有果樹或一些大樹，提供另一些樹木小生境，荷花池和池塘亦是一些有特色的生境。即使是廢棄耕地、剛犁過而未長有作物的農田、灌溉水道等，亦能提供各式各樣的生境。

塱原多樣化的生境吸引不同種類的雀鳥，香港觀鳥會自1993年至2019年在塱原錄得超過310種鳥類，超過香港鳥種的一半，當中還包括20種國際性瀕危物種，還有多種本地受關注鳥種，可算是香港另一片雀鳥天堂。另外，塱原共錄得10種兩棲類動物，佔香港兩棲類的42%，包括中國國家二級保護動物俗稱田雞的虎紋蛙，是繼鳥類之後第二豐富的物種。這些都突顯了農業式濕地作為補充本土天然濕地，以提供合適水鳥和其他野生生物另類棲息、覓食和繁殖地的重要性。

※ 梁巧晴 Della Leung

香港的傳統魚塘區位於新界西北部，在最高峰的1970及1980年代，魚塘面積超過二千公頃，漁獲年產量達七千公噸。後來內地輸入大量淡水魚，令本地魚價大幅下降，相反養殖成本卻不斷上漲，加上漁業人口老化，年青一代不願加入，漁業日漸式微。在2016年，魚塘佔地約1,135公頃，年產量2,543公噸。

魚塘屬於人工濕地，平常大片水體會吸引鸊鷉科、鸕鷀科、鴨科鳥類覓食；水邊則有鶺鴒、鷺科、鷸科、鶚等等；長滿植物的塘壆為秧雞科雀鳥提供藏身之所，而各種翠鳥、椋鳥、猛禽也是魚塘的常客。除雀鳥外，魚塘也為眾多動物提供棲息及繁殖的地方，包括兩棲類、爬行類、哺乳類、昆蟲(如蜻蜓、螢火蟲等)，生物多樣性極之豐富，具很高的生態價值。

養魚戶在「刮魚」(行內術語，即收獲塘魚)時會將水位降低，淺水的魚塘會吸引大量過境、越冬及居留本地的水鳥，如鷺科、鴴科、鷸科及鸛科等來捕捉刮魚後剩下經濟價值較低的雜魚。乾塘及曬塘時，鷸科及鴴科雀鳥可在露出的泥地覓食。現時大部分魚塘位於后海灣「拉姆薩爾濕地」範圍內，或是屬於其外圍的「濕地保育區」或「濕地緩衝區」。每年后海灣濕地吸引超過5萬隻、累計超過160多種水鳥，養魚戶進行清塘時正好為大量候鳥提供食物。

香港的農田和魚塘雖然有一定的經濟生產，不過相對於發展房地產的龐大經濟利益，農地和魚塘始終受一定程度都市發展的威脅。農田和魚塘亦要配合生態友善的作業模式，才能發揮其生態價值。有見及此，香港觀鳥會近年分別在塱原和后海灣，開展了管理協議計劃，保育這兩片雀鳥的重要生境。

塱原項目是與另一環保組織長春社合作，自2005年開展，是香港首個農業式濕地的管理項目，目標保育塱原的文化景觀及生物多樣性。此管理項目邀請農夫及土地擁有人合作，制定並實行有利水鳥和其他濕地生物多樣性的水田及濕地管理措施，包括建立更多元的水田濕地生境、減低化學農藥及肥料的投入、季節性淹水、人工鳥島、繁殖期保護措施等等。項目亦重新引入一些傳統的水田農作物，如水稻、荸薺(即馬蹄)及慈菇，以增加生境異質性。其中水稻復育的工作尤其成功，一些從前常見但隨着稻田消失而變得罕見的冬候鳥及過境遷徙鳥，如鵐科雀鳥，現在重新在塱原記錄到穩定的數量，水稻復育更帶來了其他意料之外的社區及教育效果，引起了相當的社會關注，更重要是鄉村社區的注目，對推動香港農業濕地的保育工作有非常正面的作用。

塱原管理項目其中一項重要工作就是舉辦不同類型的教育活動，活動主題主要圍繞濕地生態、雀鳥和農耕，而活動類型多樣，有生態導賞、講座及展覽、農耕體驗、清除外來物種等等，希望讓學生、親子及普羅大眾從不同角度認識塱原、體驗塱原。近年，管理項目亦積極推廣生態友善農業和本地農業，透過選購本地農產品，支持本地農業可持續發展。

魚塘項目則在2012年開展，「香港魚塘生態保育計劃」透過與當地養魚戶合作，在超過600公頃的魚塘進行生境管理工作，改善及提高魚塘的生態價值，維持對野生動物的吸引力，特別是提供更多棲息地及食物供水鳥使用。

另外，計劃更進行多項生態調查及多樣化的教育宣傳活動，更以藝術形式切入環境保育工作，加強公眾對魚塘和濕地保育的參與和認識，令活動更多元化。

如欲進一步認識香港觀鳥會塱原項目和魚塘項目，請瀏覽以下網址：

塱原自然保育管理計劃 —— https://cms.hkbws.org.hk/cms/work-tw/habitat-mgmt-tw/long-valley-tw

香港魚塘生態保育計劃 —— https://cms.hkbws.org.hk/cms/work-tw/habitat-mgmt-tw/fishpond-scheme-tw

In addition to the natural environment, certain areas closely linked to human activities, such as farmland and fishponds, can also be important habitats for birds. Hong Kong being a well-developed metropolitan, its agriculture and fisheries industries have declined for many years, yet some relatively large patches of farmland and fishponds still remain. The ecological value of these farmland and fishponds is absolutely not to be underestimated!

Covering an area of around 37 ha in North New Territories, Hong Kong, Long Valley is a stretch of intact agricultural land rarely seen in Hong Kong. The habitats it holds can be classified into wet farmland, dry farmland, lotus pond, water flea ponds and abandoned farmland, etc. While most of the area is dedicated to vegetable cultivation, different habitats are created according to the types of vegetables being planted. For example, the planting of aquatic plants such as water spinach and watercress will create different small wetland habitats, such as water channels and flooded farmland. The fruit trees or big trees often planted by the side of the fields offer another small habitat, so do the lotus ponds and water ponds. Even the abandoned farmland, farmland recently ploughed and yet to be planted with crops, and irrigation channels can add to the variety of habitats.

The diverse habitats offered by Long Valley attract a diverse assemblage of bird species. From 1993 to 2019, Hong Kong Bird Watching Society (HKBWS) recorded over 310 bird species in Long Valley, accounting for more than 50% of the total bird species recorded in Hong Kong. Of the bird species recorded in Long Valley, 20 are globally threatened species and many are species of local concern, making Long Valley one of Hong Kong's bird paradises. Long Valley is also home to 10 amphibians. This number represents 42% of all amphibians recorded in Hong Kong, making amphibians the second most abundant group of wildlife found in this

※ 梁釗成 Evans Leung

area, second only to birds. Among amphibians recorded in Long Valley is Chinese Bullfrog, a national Grade II protected species in China. All these have highlighted the importance of agricultural wetlands as a supplement to local natural wetlands in providing alternative roosting, foraging and breeding grounds for shorebirds and other forms of wildlife.

Traditionally fishponds in Hong Kong were located in North West New Territories. In the 1970's and 1980's, when Hong Kong's pond fish culture industry was at its peak, there were over 2,000 ha of fishponds, producing as much as 7,000 tonnes of catch each year. The pond fish culture industry subsequently declined as a result of the significant drop in local fish prices due to the import of large volumes of freshwater fish from mainland China, the continuous increase in culture cost, the aging of fishpond operators and the reluctance of young people joining the industry. In 2016, fishponds covered an area of approximately 1,135 ha, with an annual output of 2,543 tonnes.

Fishponds are artificial wetlands. Birds of the families Podicipedidae, Phalacrocoracidae and Anatidae usually like to forage in the large bodies of water, while birds of the families Motacillidae, Ardeidae and Scolopacidae as well as pipits prefer water edges. Pond bunds covered with vegetation provide hiding places for rails. Different types of kingfishers, starlings and raptors are also frequent visitors to fishponds. Apart from birds, fishponds also provide dwelling and breeding grounds for many different kinds of living organisms, such as amphibians, reptiles, mammals and insects (e.g.

dragon flies and fireflies). This great biodiversity gives fishponds a high ecological value.

To harvest the fish, fishpond operators will partially drain the ponds. A large number of passage migrant, wintering and resident shorebirds, such as birds of the families Ardeidae, Charadriidae, Scolopacidae and Threskiornithidae, will come and feed on the "trash fish" of low commercial value which are left in the ponds after harvesting. When the ponds are drained down and allowed to dry out by sunlight, birds of the families Scolopacidae and Charadriidae are able to forage on the exposed pond base. Currently most of the fishponds are located either within the Ramsar Site in the Deep Bay area or in the Wetland Conservation Area or Wetland Buffer Area outside the Ramsar Site. Over 50,000 shorebirds of over 160 species visit the Deep Bay wetlands each year. The pond maintenance work carried out by fishpond operators offers a perfect feeding opportunity for this huge group of migratory birds.

While farmland and fishponds in Hong Kong do support certain commercial activities, their financial return is minimal compared to the huge return that can be generated by real property developments, thus the survival of farmland and fishponds is always threatened by urban development to some extent. Furthermore, the ecological value of farmland and fishponds can only be realized if the farmland and fishponds are operated in an eco-friendly manner. Therefore, HKBWS launched a management agreement project in Long Valley and Deep Bay respectively several years ago in order to conserve these two crucial habitats for birds.

Nature Conservation Management for Long Valley is a project jointly run by HKBWS and the Conservancy Association since 2005 and the first agricultural wetland management project in Hong Kong. The aim is to preserve the cultural landscape and biodiversity of Long Valley. The project mainly involves co-operation with farmers in devising and implementing wet agricultural field and wetland management measures that are favorable to the diversity of shorebirds and other wetland-dependent wildlife. These measures include enhancing the diversity of paddy field and wetland habitats, reducing the use of chemical pesticides and fertilizers, seasonal flooding of the fields, building artificial bird islands and taking protective measures for the breeding season. Under the project, some traditional crops that are grown in irrigated fields, such as paddy rice, water chestnut and Chinese arrowhead, are also re-introduced to increase habitat heterogeneity. The re-introduction of paddy rice is particularly successful. Some species of winter migrants and passage migrants which were once common but have become scarce as paddy fields disappear, such as birds of the Emberizidae family, are once again recorded in Long Valley in steady numbers. The re-introduction of paddy rice has also had some unexpected effect on the community and education. It has captured a fairly large amount of public attention, most importantly the attention of the rural community, which has a very positive effect on the promotion of agricultural wetland conservation in Hong Kong.

One important component of the Long Valley project is the holding of a series of educational activities under such themes as wetland ecology, birds and farming. The activities come in many different forms, including guided eco-tours, seminars, exhibitions, farming experience courses and clearing of invasive species, etc. The aim is to enable students, parents and their children as well as the general public to get to know and experience Long Valley from different angles. In recent years, the management project also actively promotes eco-friendly agriculture and local agriculture and supports the sustainable development of local agriculture through the purchase of local agricultural produce.

The Fishpond Conservation Scheme was launched in 2012 to engage local fishpond operators in conducting habitat management work in over 600 ha of fishponds in order to improve and enhance the ecological value of the fishponds, maintain their appeal to wildlife and, in particular, provide more roosting grounds and food for shorebirds.

The Scheme also comprises ecological surveys, education and publicity. More environment conservation work will be conducted in artistic form, so as to enhance public participation in fishpond and wetland conservation and raise their awareness of the issue through a large variety of activities.

For further information on the Long Valley and fishpond projects of HKBWS, please visit the following websites:

Nature Conservation Management for Long Valley —— https://cms.hkbws.org.hk/cms/work-tw/habitat-mgmt-tw/long-valley-tw

Hong Kong Fishpond Conservation Scheme —— https://cms.hkbws.org.hk/cms/work-tw/habitat-mgmt-tw/fishpond-scheme-tw

遇到雀鳥受傷或死亡應如何處理
What to Do when You Find an Injured or Dead Bird

雀鳥一般都怕人，不會與人類近距離接觸，不過日常生活中，有時都會近距離遇上受傷的雀鳥或需要協助的幼鳥，甚至需要處理雀鳥屍體。由於鳥類受驚時可能會攻擊人類，加上雀鳥身體有機會懷有病菌或病毒，為安全起見，建議大家參考以下守則才接觸或處理這些雀鳥。

受傷成鳥的處理方法

如果發現雀鳥受傷，首先要注意不要接觸傷鳥，以免弄傷雀鳥或受雀鳥攻擊。如傷鳥是被霧網或繫有魚鈎的魚絲纏住，由於很可能有人觸犯相關法例，請立即致電政府熱線1823向漁農自然護理署報告。如果傷鳥是躺在地上，為避免傷鳥受驚，應盡快尋找一個體積較雀鳥大的紙箱、藤籃或竹籮，並在紙箱底部開數個小孔作通氣用，再覆蓋傷鳥。傷鳥在黑暗環境中會較安靜，可減少因胡亂掙扎而導致傷勢加重。

※ 胡明川 Woo Ming Chuan

醫治受傷雀鳥需要專業獸醫處理，因此要向相關機構求助，可致電以下機構：

1. 愛護動物協會24小時緊急拯救熱線2711 1000；或
2. 漁農自然護理署1823（政府熱線）；或
3. 嘉道理農場野生動物拯救中心2483 7200（但需自行運送受傷動物到農場。嘉道理農場暨植物園地址：新界大埔林錦公路）

致電相關機構時，請提供以下資料：

1. 「發現受傷雀鳥，欲要求派員前來帶走傷鳥。」
2. 數目
3. 詳細位置
4. 你的姓名及手提電話號碼

如你未能在現場等候相關人士到場處理，請嘗試尋找在附近工作的人士協助。

受傷幼鳥的處理方法

假如你遇到沒有親鳥照顧的幼鳥停留在地上，情況就有點不一樣。假如是剛出世或只有少許羽毛的雛鳥，你可以嘗試確認牠的鳥巢，把牠放回巢中。如幼鳥已長出羽毛，具有成鳥的雛型，請先遠離幼鳥最少10米，觀察約十分鐘。親鳥通常也會站在幼鳥附近，當你站遠了，親鳥便有機會盡快引領幼鳥離開。

假若十分鐘後親鳥仍未出現，可嘗試用一枝粗幼相等於手指尾的棒狀物，緊貼在幼鳥腳前，以圖引導牠抓緊棒狀物，然後把牠放在附近一些安全而略高的地方，避免幼鳥受貓、狗或蛇的襲擊。若以上步驟均不能成功完成，請參考「受傷成鳥的處理方法」，致電相關機構交由專人處理。

切記不要把幼鳥帶回家飼養，主要因為幼鳥需要具足夠訓練或經驗的人士特別照料，食物也有特別要求，不恰當的護理反而會令牠們承受不必要的痛苦。此外，藏有野鳥亦屬違法行為。

雀鳥屍體處理方法

1997年，香港出現市民感染禽流感致病的個案，政府高度重視雀鳥屍體的處理，因此會派專人負責相關工作。若你發現雀鳥屍體，應致電政府熱線1823，告訴接線員發現雀鳥屍體，需要漁農自然護理署派員清理，並告之屍體數量、詳細位置，並留下姓名及手提電話號碼。如未能在現場等候職員到場，請嘗試尋找在附近工作的人士協助。

除了通報相關部門，亦建議通知香港觀鳥會(info@hkbws.org.hk)發現日期、環境資料及懷疑死因，如可以請拍下照片，並通知香港觀鳥會作記錄及有關跟進工作。雀鳥常見的死亡原因包括撞擊大廈幕牆或隔音屏障、遷徙途中體力不支、雨燕科雀鳥意外墮地、被霧網或魚絲纏住、或是放生儀式等等。

所有野生雀鳥均受香港法例第170章《野生動物保護條例》保護，而瀕危雀鳥(例如鸚鵡)是受香港法例第586章《保護瀕危動植物物種條例》保護。切勿於討論區、社交媒體、或向傳媒披露鳥巢的位置，以免對雀鳥造成不必要的干擾。如遇到任何干擾野生雀鳥、鳥巢及蛋的行為，若情況嚴重或緊急，請即致電999報警，最好同時致電1823(政府熱線)通知漁農自然護理署及香港觀鳥會。

售賣野生雀鳥或瀕危雀鳥亦是違法行為，如有發現，請以電郵或致電政府熱線1823通知漁農自然護理署，請告訴接線員所涉及的物種是瀕危或是野生雀鳥。此外，亦可參考漁農自然護理署建議的舉報方法(http://www.afcd.gov.hk/tc_chi/conservation/con_end/con_end_rew/con_end_rew.html)，也同時用電郵或電話通知香港觀鳥會有關個案。為免影響調查進度，請勿於討論區或社交媒體向香港觀鳥會查詢或張貼線索。

Birds are generally scared of and tend to stay away from humans. However, from time to time, one may come into close contact with an injured bird or a young bird that needs help, or may even need to remove or handle a dead bird. As birds may attack humans when scared and may carry bacteria or viruses, so for safety, we recommend that you follow the guidelines below when touching or handling a bird.

What to do with an injured adult bird

If you find an injured bird, the first thing to note is not to touch it. This is to avoid the possibility of hurting the bird or being attacked by it. If the bird is entangled by a mist net or a fishing line attached with a hook, please call Government hotline 1823 immediately to report the case to the Agriculture, Fisheries and Conservation Department (AFCD) as someone may have probably broken the law. If the injured bird is lying on the ground, in order not to scare it, you should find a cardboard box, rattan basket or bamboo basket which is big enough to cover the bird as soon as possible. If it is a cardboard box, cut several holes at its bottom for ventilation. Cover the bird with the box or basket to keep it in the dark and quiet, so as to reduce the chances of the bird struggling and aggravating its injury.

Injured birds require the professional care of a veterinarian. You may contact any of the following organizations for assistance by calling:

1. The 24-hour emergency hotline of the Society for the Prevention of Cruelty to Animals: 2711 1000; or
2. Agriculture, Fisheries and Conservation Department: 1823 (Government hotline); or
3. Wild Animal Rescue Centre of Kadoorie Farm and Botanic Garden (KFBG): 2483 7200 (you are required to make your own arrangements for delivering the injured animal to KFBG. Address: Lam Kam Road, Tai Po, New Territories)

Please provide the following information when you call the relevant organization:

1. "I have found an injured bird/some injured birds, please send someone here to pick it/them up."
2. Number of injured birds found
3. Specific location
4. Your name and mobile phone number

If you are unable to stay there to wait for the relevant person to arrive, please try to find someone who is working nearby to help.

What to do with an injured young bird

The approach is a little bit different if you find a young bird on the ground seemingly abandoned by its parents. If it is a baby bird newly hatched or having only a few feathers, you may try to look for its nest and if you can find it, put it back to the nest. If it is a young bird with feathers that is beginning to look like an adult bird, you should first withdraw to a distance of at least 10 metres from the bird and observe it for about 10 minutes. The young bird's parents are usually staying nearby and by stepping back, you will give the parents a chance to lead the young bird away as soon as possible.

If neither of the parents show up after 10 minutes, you may try to find a stick-like article about the same thickness as the pinkie and put it immediately in front of the feet of the young bird to induce it to hold on to the stick so that you can put the bird in a safe, somewhat elevated place to protect it from cats, dogs and snakes. If you are unable to complete the above steps, please refer to the procedures for handling injured adult birds and leave the matter with the specialists by calling a relevant organization.

One important thing to remember is that you must not bring the young bird home and keep it as a pet. The main reasons being that only someone with sufficient training or experience knows how to take care of young birds, and birds have special requirements for diet. Caring for the birds in an inappropriate way will cause them to suffer unnecessarily. It is also illegal to keep a wild bird.

What to do with a dead bird

In 1997, a number of Hong Kong residents were infected by avian flu. Since then the Government has taken the handling of dead birds very seriously and will assign the task to a dedicated group of people. If you find a dead bird, you should call the Government hotline 1823 and tell the telephone operator that you have found a dead bird/some dead birds and would like the AFCD Agriculture, Fisheries and Conservation Department to send someone to pick it up. You should specify the number of dead bird found, the specific location where it is/they are found, and leave your name and mobile phone number. If you are unable to stay there to wait for the AFCD staff to arrive, please try to find someone who is working nearby to help.

Besides reporting to the relevant department, we also suggest that you inform the Hong Kong Bird Watching Society (HKBWS) (info@hkbws.org.hk) of the date you found the dead bird(s), the information about the surroundings and the suspected cause of death. If possible, please take a photo and send it to the Hong Kong Bird Watching SocietyHKBWS both for its record and for it to follow up with the case. Common causes of bird deaths include collision with a tall glass curtain wall of buildings or noise barrier, exhaustion during migration, accidental landing on the ground (in the case of birds of the swift family), entanglement by a mist net or fishing line, and "mercy" release, etc.

※ 呂德恒 Henry Lui

All wild birds are protected under the Wild Animal Protection Ordinance (Cap. 170), while endangered birds (such as parrots) are protected under the Protection of Endangered Species of Animals and Plants Ordinance (Cap. 586). To avoid causing unnecessary disturbance to the birds, you must not disclose the position of a bird nest in an Internet forum, on Facebooksocial media, or to the media. If you find someone tampering with a wild bird, a bird nest or bird eggs, and the situation is grave or urgent, please call 999 to report to the police immediately. We also recommend that you call the Agriculture, Fisheries and Conservation DepartmentFCD at 1823 (Government hotline) and inform the HKBWSHong Kong Bird Watching Society as well.

It is illegal to sell wild birds or endangered birds. If you find someone selling these birds, please inform the AFCD Agriculture, Fisheries and Conservation Department by email or by calling the Government hotline 1823. Please tell the telephone operator that the species involved is an endangered bird or a wild bird. You may also follow the reporting method recommended by the AFCD Agriculture, Fisheries and Conservation Department (https://www.afcd.gov.hk/english/conservation/con_end/con_end_rew/con_end_rew.html), and at the same time inform the HKBWS Hong Kong Bird Watching Society of the case by email or telephone. To avoid causing any delay to the investigations, please do not make any enquiry with the HKBWS Hong Kong Bird Watching Society or pose any clue in a public forum or on social media Facebook.

保育瀕危鳥類
Conservation of Threatened Birds

保護瀕危鳥類，過去的做法可能會將其捕捉，放進動物園中，在安全的環境下飼養和繁殖。但是過百年的經驗告訴我們，這不是一個理想的保育方案，動物始終要在天然情況和環境下，才較健康地成長和繁衍，尤其是大多會飛行，甚至跨越地域的鳥類，部分鳥種根本不可能困在動物園或保育中心中生活。近年，隨着普羅大眾的自然保育意識日漸提升，現時保育瀕危鳥類的方向，已主要針對保護其棲息地，這樣既可保護目標物種，其生境內其他各式各樣、已知或未知的生物也一併受到保護。

鳥類是活動範圍極大的動物，要保育個別鳥種，需要了解其整個生命周期所到過的生境和需要，很多時候更要不同國家的合作，才能有效制定保育方案，這涉及多方面深入的研究，每種鳥種的保育方案亦不盡相同。

香港觀鳥會一直參與保育瀕危鳥類的工作，其中包括以下四個重點鳥種，保育工作包括研究、調查、教育、生境管理等，當中亦不乏與其他地區合作。

※ 郭子祈 Kwok Tsz Ki

黃胸鵐

黃胸鵐，學名*Emberiza aureola*，俗稱「禾花雀」，曾是一種極度普遍並廣泛分布於歐洲和亞洲的鳥種，但根據研究顯示，黃胸鵐的數量不但出現災難性的劇減，數量在1980至2013年間下跌了9成，其分布範圍更縮減了5,000公里，主要原因是過度捕殺。國際自然保護聯盟（IUCN）於2017年12月5日公布最新修訂的「瀕危物種紅色名錄」（俗稱紅皮書），黃胸鵐更由「瀕危」再調升至「極度瀕危」。

塱原是黃胸鵐在香港最重要的棲息和覓食地點，自2005年開始，香港觀鳥會與長春社合作推行「塱原自然保育管理計劃」，保育塱原的文化景觀及生物多樣性。其中的水稻復育工作，當水稻種植面積擴大後，黃胸鵐的數量亦慢慢提升。該計劃亦舉辦不同類型的教育和推廣活動，當中包括黃胸鵐的保育。

香港觀鳥會把2018年訂為「國際禾花雀關注年」，在內地、香港進行保育行動，並與國際合作，呼籲各地民眾一起參與保育黃胸鵐，承諾「不吃禾花雀」。除了在香港開展一連串的公眾教育工作外，香港觀鳥會更派出研究人員遠赴蒙古Khurhk Valley，到黃胸鵐已知的繁殖點了解牠們繁殖的情況、數量及分佈，及進行環誌與安裝跟蹤器的工作。

黑臉琵鷺

黑臉琵鷺（*Platalea minor*）主要棲息在亞洲東部沿潮間帶泥灘的水鳥，1990年代初期只在東亞的數個地點記錄到不多於300隻，因此被國際自然保護聯盟列為「瀕危」。但隨着最近二十多年各地的保育措施不斷改善，特別是地區間的相互合作及調查研究不斷加強，而各地的保育團體紛紛以黑臉琵鷺作為推廣保育雀鳥或濕地的明星，令普羅市民對保育濕地的意識大大提高，令保護黑臉琵鷺得到重要的支持，成為亞洲拯救瀕危鳥類的最佳典範。

早於1997年冬季，港台地區學者聯同日本的研究員首次於香港和台灣地區為黑臉琵鷺戴上衛星追蹤器，團隊於1999年成功追蹤黑臉琵鷺的遷徙路線，是揭開黑臉琵鷺神秘面紗重要的一步。

香港觀鳥會於2003年起統籌黑臉琵鷺全球同步普查，普查自1994年起舉行至今，調查地點約有100個。過去二十年，黑臉琵鷺的數字保持着上升的趨勢，比九十年代的二、三百隻增加了十多倍，2020年的數目是歷史新高的4,864隻，這是各參與保育工作的個人及機構的最佳鼓勵。

勺嘴鷸

勺嘴鷸(*Calidris pygmaea*)是香港春季稀少過境鳥，本身在西伯利亞東北部繁殖，於亞洲東南部越冬。2009/2010年度勺嘴鷸的全球數量是360-600隻，其中120-200對屬有繁殖力的成鳥。2008年國際自然保護聯盟把勺嘴鷸列入「極度瀕危」的級別，其面對的最大威脅包括棲息地喪失及退化，以及捕獵。

香港觀鳥會自2011年起加入保育勺嘴鷸的工作，開展了一系列教育活動，包括在2013年發動500位來自勺嘴鷸遷徙路線上8個國家、12個地區的小孩參與製作短片；在2015年發起明信片交換計劃，宣揚保育勺嘴鷸和遷徙航道上其他水鳥的訊息。同時，香港觀鳥會發現勺嘴鷸在南中國一帶的分佈資料不足，在南中國沿岸進行初步調查後，發現了勺嘴鷸在南中國的越冬地，亦同時發現當地的非法捕獵嚴重威脅勺嘴鷸的生存。由2014年起，香港觀鳥會招募當地義工成立保育小組，並為義工提供鳥類調查和環境教育的培訓。同時與政府當局建立通報和合作機制，以提升執法效力。隨着保育小組成員增加，更多保育工作得以進行。現時非法捕獵的威脅已大大減少。香港觀鳥會成員亦與英國、新西蘭、南京師範大學的人員，於2018年展開在江蘇省的勺嘴鷸考察，調查全球勺嘴鷸的數量，了解牠們的移動情況、生活史及存活率，並環誌勺嘴鷸和其他涉禽，以研究牠們的遷徙。

香港觀鳥會亦與日本野鳥會及本地插畫師合作，於2018年出版《認識勺嘴鷸教材套》，內容包括勺嘴鷸的狀況及濕地的資訊。教材套也提供多種室內室外活動方案與教材，方便及協助導師設計課程，令年輕人了解保育濕地的重要，提高他們的保育意識。教材套分別以英文、繁體中文、簡體中文、日文和緬甸語出版。

中華鳳頭燕鷗

中華鳳頭燕鷗(*Thalasseus bernsteini*)又名黑嘴端鳳頭燕鷗，是全球最稀有的海鳥之一，甚至曾被認為已經滅絕，因為自1937年在中國出現之後，往後60年再無任何目擊記錄。直至2000年，終於在台灣馬祖列島發現四對中華鳳頭燕鷗。目前為止，全球只有不多於100隻中華鳳頭燕鷗，也自然被國際自然保護聯盟列為「極度瀕危」的鳥類。中華鳳頭燕鷗屬候鳥一種，經常混在大鳳頭燕鷗群中，分享相近的生境及棲息地，行為也相類似。

2012年起，香港觀鳥會與內地多個機構、國際鳥盟、香港海洋公園保育基金、俄勒崗州立大學及多位海外專家先後進行研究及招引計劃。2019年更聯同美國、日本和印尼的燕鷗專家和海鳥研究員到印尼，把衛星追蹤器放在其中兩隻大鳳頭燕鷗身上進行衛星追蹤。是項研究可間接地了解中華鳳頭燕鷗的遷徙路線，有助進一步與相關地區的政府部門協商保育其生態環境。

How do we protect endangered birds? In the past, people used to catch the birds and raise and breed them in the safety of a zoo. Yet we have learnt from more than 100 years of experience that this is not an ideal way to conserve the birds, as animals that grow up and breed in the natural environment are healthier. This is particularly true for most birds that can fly or even migrate across vast regions. It is impossible for most bird species to live in confinement, such as in a zoo or a conservation center. With the gradual increase in public awareness about nature conservation in recent years, the main focus of endangered bird conservation has shifted to the protection of bird habitats. This does not only protect the target species, but also protects many other species of wildlife, both known and unknown, that share the same habitat with the target species.

Birds' activities span across an extremely large area. To conserve a particular bird species, we need to find out all the habitats it visits and its needs during its entire life cycle, and different countries also need to co-operate closely with each other in order to come up with effective conservation plans. This involves in-depth research in many areas and the conservation plan will vary from one bird species to another.

The Hong Kong Bird Watching Society (HKBWS) has long been engaged in the conservation of endangered birds. In recent years, it has focused its conservation efforts on the four key species below, with work including research, survey, education and habitat management, etc. Some of these tasks require co-operation with other regions.

Yellow-breasted Bunting

The Yellow-breasted Bunting (*Emberiza aureola*), commonly known as the "rice bird", was once abundant and widely found in Europe and Asia. However, research has found that not only did its population suffer a catastrophic decline by 90% between 1980 and 2013, its area of distribution also shrank by 5,000 kilometres, mainly due to overhunting. According to the latest updates to the Red List of Threatened Species (also known as the "Red List") announced by the International Union of Conservation of Nature (IUCN) on 5 December 2017, the conservation status of the Yellow-breasted Bunting was further uplisted from "Endangered" to "Critically Endangered".

Long Valley is the most important resting and feeding site for Yellow-breasted Buntings when they stop over Hong Kong during migration. Starting from 2005, HKBWS and the Conservancy Association co-organized the "Nature Conservation Management for Long Valley", a project designed to preserve the cultural landscape and biodiversity of Long Valley. Following the resumption of paddy-rice growing and the expansion of paddy-rice cultivation area, the number of the Yellow-breasted Bunting started to increase slowly. The project also entails various types of educational and promotional activities, with the conservation of the Yellow-breasted Bunting being one of the themes.

HKBWS designated 2018 the "Year of International Concern for the Yellow-breasted Bunting". Besides taking conservation actions in mainland China and Hong Kong, it also joined forces with other countries in appealing for the public to participate in the conservation of the Yellow-breasted Bunting and to take the "no rice bird" pledge. In addition to organizing a series of activities for public education in Hong Kong, HKBWS also sent their researchers to the remote Khurhk Valley of Mongolia to study the breeding status, population and distribution of the Yellow-breasted Bunting in its known breeding sites, as well as to carry out such work as banding and tracker mounting.

Black-faced Spoonbill

The Black-faced Spoonbill (*Platalea minor*) mainly stages in intertidal mudflats in East Asia. In early 1990's, not more than 300 Black-faced Spoonbills were recorded across several sites in East Asia. It is currently listed as "Endangered" by the IUCN. With the implementation of measures for its protection in the past 20 years, especially the cooperation in research among the regional areas, plus using the bird as an icon in many promotion activities, the public awareness on wetland conservation has been raised rapidly. It also drew a great support from the general public in protecting this endangered bird. This successful story has brought saving Black-faced Spoonbill campaign the best demonstration among all Asian countries.

The first cooperation among the researchers from Japan and cities like Hong Kong, Taiwan started early in the 1997 winter. They attached satellite tracking devices in the Black-faced Spoonbill in Hong Kong as well as in Taiwan. Finally, the mysterious migrating route of Black-faced Spoonbill discovered in 1999.

HKBWS began coordinating the simultaneous global survey of the Black-faced Spoonbill in 2003. The survey has been held continuously since 1994, covering around 100 sites. Judging by the data collected over a period of more than twenty years, the Black-faced Spoonbill population is showing a gradual and steady increase. During the 2020 census, 4,864 Black-faced Spoonbills were recorded. This is a new record high and almost 10 times of the amount in the 1990s. This result is the best encouragement to all the people and organizations who have participated in the conservation of Black-faced Spoonbill.

Spoon-billed Sandpiper

The Spoon-billed Sandpiper (*Calidris pygmaea*) is a scarce spring passage migrant in Hong Kong which breeds in Northeast Siberia and winters in Southeast Asia. In the year 2009/2010, there were 360-600 Spoon-billed Sandpipers worldwide, including 120-200 pairs of adult birds capable of breeding. The Spoon-billed Sandpiper was listed as "Critically Endangered" by IUCN in 2008. The largest threats to this species include habitat loss and degradation as well as hunting.

HKBWS started taking part in Spoon-billed Sandpiper conservation in 2011 by launching a series of educational campaigns. These included an initiative in 2013 for 500 children from 12 districts in 8 countries located along the migration route of the Spoon-billed Sandpiper to take part in making a short film, and an initiative in 2015 for postcard exchange to advocate conservation of the Spoon-billed Sandpiper and other water birds using the same migration route. In the course of doing so, HKBWS became aware of the scarcity of information on the distribution of the Spoon-billed Sandpiper in South China. A preliminary survey along the South China coast identified some wintering sites of the Spoon-billed Sandpiper, as well as the serious threat posed by local poaching to their survival. Starting from 2014, HKBWS has teamed up with local conservation organizations to recruit local volunteers into conservation teams and provide training to the volunteers on bird survey and

※ 呂德恒 Henry Lui

environmental education. It has also set up reporting and co-operation mechanisms with government authorities to enhance law enforcement. As the size of the conservation teams increases, more conservation work is getting done. The threat from poaching has been reduced significantly. In 2018, HKBWS members launched a joint survey of the Spoon-billed Sandpiper in Jiangsu Province with researchers from the UK, New Zealand and Nanjing Normal University. The purpose of the survey was to find out about the global population of the Spoon-billed Sandpiper, its migratory pattern, life history and survival rate, and to study the migration of the Spoon-billed Sandpiper and other waders through banding.

In 2018, HKBWS, the Wild Bird Society of Japan and a local illustrator jointly published a Spoon-billed Sandpiper education kit. Besides providing information on the status of the Spoon-billed Sandpiper and wetland, the kit also offers a variety of indoor and outdoor activity proposals and teaching materials to assist teachers in designing lessons aimed at educating the younger generation about the importance of wetland conservation and raising their awareness about conservation. The education kit is available in English, Traditional Chinese, Simplified Chinese, Japanese and Burmese.

Chinese Crested Tern

The Chinese Crested Tern (*Thalasseus bernsteini*) is one of the rarest sea birds on earth. It was not seen for 60 years after appearing in China in 1937 and was once thought to be extinct. It was not until 2000 when four pairs of Chinese Crested Tern were sighted in the Matsu Islands, Taiwan. To date, there are no more than 100 Chinese Crested Terns worldwide. As a result, it is listed as "Critically Endangered" by IUCN. The Chinese Crested Tern is a migratory bird. It often lives in mixed groups with the Greater Crested Tern. The two species have similar behavior and share similar sites and habitats.

HKBWS has been working with a number of mainland Chinese organizations, BirdLife International, Hong Kong Ocean Park Conservation Foundation, Oregon State University and a group of overseas experts since 2012 on research and recruitment projects. A HKBWS team went to Indonesia in 2019 with tern experts and sea bird researchers from the United States, Japan and Indonesia to mount satellite trackers on two Greater Crested Terns. This study will provide indirect information on the migratory route of the Chinese Crested Tern and facilitate further discussions with the government authorities of the relevant regions regarding conservation of the ecological environment in which the Chinese Crested Tern is found.

觀鳥地點
Places to Visit

觀鳥地點 Birdwatching spots	交通 Transportation	目標鳥種 Target species	預計觀鳥時間 Time required at the site
		濕地 Wetlands	
尖鼻咀 Tsim Bei Tsui	在元朗泰豐街乘35號專線小巴於尖鼻咀下車。 Take the green minibus no. 35 from Tai Fung Street, Yuen Long. Get off at terminal at Tsim Bei Tsui.	多種猛禽和濕地鳥類如黑臉琵鷺；傍晚時份有機會看到鵰鴞。 A wide variety of raptors and wetland birds, e.g. Black-faced Spoonbill. Possibility of Eurasian Eagle-Owl in the evening.	4小時 4 hours
南生圍 Nam Sang Wai	元朗乘往上水的76K巴士或17號紅色小巴，在南生圍路口下車。 Take the bus no. 76K or red minibus no. 17 from Yuen Long to Sheung Shui. Get off at the main road into Nam Sang Wai, then walk.	濕地鳥類 Wetland birds	2小時 2 hours
米埔自然護理區 Mai Po Marshes Nature Reserve	在元朗或上水乘76K巴士或17號小巴(上水新發街或元朗水車館街)，在米埔村下車，沿担竿洲路步行20分鐘便可到達(需持有由漁農自然護理署發出的「進入米埔沼澤區許可證」才可進入保護區)。 Take the bus no. 76K in Sheung Shui or Yuen Long, or red minibus no. 17 from Sheung Shui (San Fat Street) or Yuen Long (Shui Che Kun Street). Get off at Mai Po Village. Follow Tam Kon Chau Road and walk for about 20 minutes. (It is a legal requirement for all visitor to have a "Mai Po Entry Permit" issued by the AFCD.)	多種濕地鳥類，如黃嘴白鷺、黑臉琵鷺、黑嘴鷗、勺嘴鷸、小青腳鷸等；以及猛禽，如白肩鵰和烏鵰。 A wide variety of wetland birds, such as Chinese Egret, Black-faced Spoonbill, Saunders's Gull, Spoon-billed Sandpiper and Nordmann's Greenshank, as well as raptors such as Eastern Imperial Eagle and Greater Spotted Eagle.	5小時 5 hours
米埔新村/担竿洲 Mai Po San Tsuen / Tam Kon Chau	米埔附近的魚塘，交通同上。在米埔村下車，然後沿担捍洲路步行，沿途觀鳥。 Fishponds around Mai Po. Transport as above, then explore the fishponds along the access road.	濕地鳥類及多種鵐 Wetland birds and buntings	2小時 2 hours

觀鳥地點 Birdwatching spots	交通 Transportation	目標鳥種 Target species	預計觀鳥時間 Time required at the site
塱原 Long Valley	上水乘76K巴士；或於上水港鐵站鄰近的小巴站乘搭綠色專線小巴50K、51K或55K；在燕崗村下車(進入青山公路後第一條行人天橋)，沿右面小路進入。 Take the bus no. 76K or green minibus no. 50K, 51K or 55K from Sheung Shui MTR Station. Get off at the first footbridge at Yin Kong Tsuen, not long after the bus turns right to Castle Peak Road. Walk along the path at the right hand side.	彩鷸、扇尾沙錐、棕扇尾鶯、田鷚、藍喉歌鴝和絲光椋鳥 Greater Painted-snipe, Common Snipe, Zitting Cisticola, Richard's Pipit, Bluethroat and Red-billed Starling	2-3小時 2-3 hours
鹿頸、南涌 Luk Keng, Nam Chung	粉嶺港鐵站乘56K往鹿頸的專線小巴，總站下車。夏季時，沿鹿頸路途中可眺望鴉洲鷺林。 Take the green minibus no. 56K from Fanling MTR station to Luk Keng. A Chau (with an egretry in summer) is visible along Luk Keng.	普通翠鳥、白胸翡翠、藍翡翠、斑魚狗、黑冠鵑隼和多種鷺鳥 Common Kingfisher, White-throated Kingfisher, Black-capped Kingfisher, Pied Kingfisher, Black Baza, egrets and herons	2小時 2 hours
荔枝窩 Lai Chi Wo	大埔墟港鐵站乘綠色專線小巴往烏蛟騰，然後步行經上苗田、下苗田、三椏涌往荔枝窩村。 Take the green minibus from Tai Po Market MTR station to Wu Kau Tang. Walk though Sheung Miu Tin, Ha Miu Tin, Sham A Chung to Lai Chi Wo. A long hike.	冠魚狗、蛇鵰和多種鷺鳥 Crested Kingfisher, Crested Serpent Eagle, egrets and herons	8小時 8 hours
香港濕地公園 Hong Kong Wetland Park	天水圍鐵路站轉乘輕鐵705或706號，於濕地公園站下車。 Take the Light Rail no. 705 or 706. Get off at Wetland Park Station.	多種水鳥如黑臉琵鷺、水雉及彩鷸 A wide variety of wetland species such as Black-faced Spoonbill, Pheasant-tailed Jacana and Greater Painted-snipe	2小時 2 hours

觀鳥地點 Birdwatching spots	交通 Transportation	目標鳥種 Target species	預計觀鳥時間 Time required at the site
船灣、洞梓、汀角 Shuen Wan, Tung Tsz, Ting Kok	大埔港鐵站乘75K往大美督巴士在三門仔、洞梓或汀角下車。 Bus no. 75K from Tai Po Market MTR Station to Sam Mun Tsai, Tung Tsz or Ting Kok.	濕地鳥類 Wetland birds	3小時 3 hours
高地 Uplands			
飛鵝山 Fei Ngo Shan (Kowloon Peak)	彩虹港鐵站乘計程車經飛鵝山道往基維爾營，在往基維爾營的標誌下車，然後步行下山回彩虹或沿衛奕信徑往西貢蠔涌方向步行下山。 Best to take a taxi from Choi Hung MTR Station to Gilwell Campsite. Watch birds around the hillside, walk along the road back to Choi Hung or down the other side to Ho Chung.	山鷚和中華鷓鴣 Upland Pipit and Chinese Francolin	2小時 2 hours
大帽山 Tai Mo Shan	荃灣港鐵站乘51號巴士，在大帽山郊野公園路口下車，步行上山；或乘計程車前往大帽山近山頂的閘口，再往山上觀鳥或步行下山。 Take the bus no. 51 from Tsuen Wan MTR Station. Get off at the sign at the entrance to Tai Mo Shan Country Park. Alternatively, take a taxi to the last barrier of the Park, then walk along the concrete road to the summit.	山鷚、鷓鴣、大草鶯和棕頭鴉雀 Upland Pipit, Chinese Francolin, Chinese Grassbird and Vinous-throated Parrotbill	2小時 2 hours
開闊原野 Open Country			
榕樹澳 Yung Shue O	沙田市中心乘299號往西貢的巴士，在水浪窩下車，沿往榕樹澳的車路步行約30分鐘。 Take the bus no. 299 from Sha Tin Town Centre to Sai Kung. Get off at Shui Long Wo, there is a road leading to Yung Shue O. It might take half an hour to walk there.	中華鷓鴣、松鴉、小鴉鵑、褐魚鴞、普通夜鷹和領角鴞 Chinese Francolin, Eurasian Jay, Lesser Coucal, Brown Fish Owl, Grey Nightjar and Collared Scops Owl	2小時 2 hours

觀鳥地點 Birdwatching spots	交通 Transportation	目標鳥種 Target species	預計觀鳥時間 Time required at the site
沙螺洞 Sha Lo Tung	大埔墟港鐵站乘巴士74K，在鳳園下車，沿路上山；或可在大埔墟乘計程車直接前往沙螺洞。山上可通往鶴藪水塘。 Take the bus no. 74K from Tai Po Market MTR Station. Get off at Fung Yuen and walk up the hill. Alternatively, take a taxi to Sha Lo Tung directly. A trail from there leads to Hok Tau Reservoir.	金頭扇尾鶯和多種鵐 Golden-headed Cisticola and buntings	3小時 3 hours
洲頭 Chau Tau	元朗或上水乘76K巴士，或上水新發街或元朗水車館街乘17號小巴，在洲頭下車，沿路上山。 Take the bus no. 76K or red minibus no. 17 from Sheung Shui (San Fat Street) or Yuen Long (Shui Che Kun Street). Get off at Chau Tau and walk towards the hills.	白腹隼鵰、中華鷓鴣和小鴉鵑（小毛雞）；傍晚時份有機會看到林夜鷹和鵰鴞 Bonelli's Eagle, Chinese Francolin and Lesser Coucal. Possibility of Savanna Nightjar and Eurasian Eagle-owl in the evening.	2小時 2 hours

※ 郭子祈 Kwok Tsz Ki

觀鳥地點 Birdwatching spots	交通 Transportation	目標鳥種 Target species	預計觀鳥時間 Time required at the site
林地、溪流 Woodlands and Streams			
香港仔水塘 Aberdeen Reservoir	香港仔市中心沿石排灣道步行上山，另一方便的方法是乘坐計程車直接前往香港仔郊野公園或任何往石排灣邨的交通工具。 From the Central Aberdeen, walk uphill along Shek Pai Wan Road to reach the entrance of Aberdeen Country Park. Alternatively, take a taxi to the Country Park directly or any available transport to Shek Pai Wan Estate.	大嘴烏鴉、夜鷺和黑鳶（麻鷹） Large-billed Crow, Black-crowned Night Heron and Black Kite	2小時 2 hours
摩星嶺 Mount Davis	往堅尼地城乘5B或47號巴士到摩星嶺總站，從山腳上山；或乘計程車往山頂青年旅舍，步行下山。 Take the bus no. 5B or 47 to Kennedy Town. Get off at the terminal, and walk up Mount Davis Road. Alternatively, take a taxi to the Youth Hostel near the summit of of Mount Davis and walk down.	髮冠卷尾、紅嘴藍鵲以及多種過境遷徙鳥 Hair-crested Drongo, Red-billed Blue Magpie and various passage migrants	2 小時 2 hours
龍虎山和柯士甸山 Lung Fu Shan and Mount Austin	穿過香港大學校園，沿克頓道往山頂，或到達龍虎亭後，沿小山路下山。 From the University of Hong Kong, walk up Hatton Road towards the Peak, or from Mount Austin walk downhill and follow the narrow tracks to the left after reaching Lung Fu Pavilion.	鳳頭鷹、黑喉噪鶥、紅嘴藍鵲、烏鶇、烏灰鶇和灰背鶇 Crested Goshawk, Black-throated Laughingthrush, Red-billed Blue Magpie, Chinese Blackbird, Japanese Thrush and Grey-backed Thrush	2小時 2 hours
馬己仙峽 Magazine Gap	中環交易廣場乘15號往山頂巴士在舊山頂道和僑福道交界的車站下車。 Take the bus no. 15 from Exchange Square, Central. Get off at the stop near the junction of Old Peak Road and Guildford Road.	傍晚時份觀賞晚棲黑鳶（麻鷹） Roosting Black Kites at sunset	1小時 1 hour

觀鳥地點 Birdwatching spots	交通 Transportation	目標鳥種 Target species	預計觀鳥時間 Time required at the site
大埔滘自然護理區 Tai Po Kau Nature Reserve	大埔墟港鐵站乘72、72A、73A或74A巴士或乘計程車前往。 Take the bus no. 72, 72A, 73A or 74A or taxi from Tai Po Market MTR Station.	多種林區鳥類，如海南藍仙鶲、白眉鵐、黑領噪鶥、黃頰山雀、山鶺鴒、橙腹葉鵯、白腹鳳鶥、絨額鳾、紫綬帶、灰喉山椒鳥、赤紅山椒鳥、藍翅希鶥和叉尾太陽鳥 Many woodland species such as Hainan Blue Flycatcher, Tristram's Bunting, Greater Necklaced Laughingthrush, Yellow-cheeked Tit, Forest Wagtail, Orange-bellied Leafbird, White-bellied Erpornis, Velvet-fronted Nuthatch, Japanese Paradise Flycatcher, Grey-chinned Minivet, Scarlet Minivet, Blue-winged Minla and Fork-tailed Sunbird	4小時 4 hours
城門水塘 Shing Mun Reservoir	荃灣港鐵站附近兆和街乘82號專線小巴前往城門郊野公園。 Take the green minibus no. 82 from Siu Wo Street near Tsuen Wan MTR Station terminating at Shing Mun Country Park.	多種林區鳥類，如海南藍仙鶲、紅頭穗鶥和棕頸鉤嘴鶥 Many woodland species such as Hainan Blue Flycatcher, Rufous-capped Babbler and Streak-breasted Scimitar Babbler	3小時 3 hours
梧桐寨 Ng Tung Chai	太和港鐵站乘64K往元朗巴士，在梧桐寨站下車，沿路入梧桐寨村，經萬德寺上山，沿路經過多個瀑布。沿路可步行上大帽山頂。 Take the bus no. 64K from Tai Wo MTR Station. Get off at Ng Tung Chai bus stop. Follow path across farmland uphill to Ng Tung Chai Village, from where a road leads to a temple and further to the waterfalls and eventually to Tai Mo Shan.	灰鶺鴒、灰樹鵲、日本歌鴝和棕腹大仙鶲 Grey Wagtail, Grey Treepie, Japanese Robin and Fujian Niltava	3小時 3 hours

觀鳥地點 Birdwatching spots	交通 Transportation	目標鳥種 Target species	預計觀鳥時間 Time required at the site
碗窰 Wun Yiu	大埔墟港鐵站步行，穿過運頭塘村，到達運路，再沿路步行至碗窰。 Walk through Wan Tau Tong Estate from Tai Po Market MTR Station. On reaching Tat Wan Road, walk uphill to Wun Yiu.	紅尾水鴝、銅藍鶲、三寶鳥、蛇鵰、赤紅山椒鳥和灰喉山椒鳥 Plumbeous Water Redstart, Verditer Flycatcher, Oriental Dollarbird, Crested Serpent Eagle, Scarlet Minivet and Grey-chinned Minivet	1小時 1 hour
大蠔河 Tai Ho River	大嶼山東涌港鐵站乘計程車到白芒村，沿路經白芒學校、牛牯塱村到達大蠔灣。沿路上山步行1.5小時可到達梅窩，乘坐小輪返市區。 Take a taxi from Tung Chung MTR Station to Pak Mong Village. Walk along village trail to Tai Ho Wan passing Pak Mong School and Ngau Kwu Long. Then walk for about 1.5 hours uphill and down the other side to Mui Wo. There is a regular ferry back to Central.	三寶鳥、岩鷺及多種冬候鳥為主的林區鳥類 Oriental Dollarbird, Pacific Reef Heron and various woodland species in particular, winter visitors	4小時 4 hours

※ 郭子祈 Kwok Tsz Ki

觀鳥地點 Birdwatching spots	交通 Transportation	目標鳥種 Target species	預計觀鳥時間 Time required at the site
 海洋、沿岸和海島 Ocean, Coastal Areas and Islands			
蒲台 Po Toi	可在香港仔碼頭或赤柱乘船前往。可預先查詢最新班次。 Take a ferry at Aberdeen Pier or Stanley. You may check the updated ferry schedule.	多種海鳥，包括黑枕燕鷗、褐翅燕鷗和紅頸瓣蹼鷸，還有罕見的過境遷徙鳥，例如黑鳽、黑枕黃鸝、黃眉姬鶲、灰山椒鳥等 Seabirds such as Black-naped Tern, Bridled Tern and Red-necked Phalarope, as well as rare passage migrants, such as Black Bittern, Black-naped Oriole, Narcissus Flycatcher, Ashy Minivet.	3小時 3 hours
塔門 Tap Mun	大學港鐵站步行15分鐘至馬料水碼頭，乘小輪前往（最早開8:30）。或由西貢黃石碼頭乘搭小輪前往。可預先查詢最新班次。 Take the earliest ferry from Ma Liu Shui Pier (15-minute walk from University MTR Station) at 8:30am daily. There is also a regular ferry from Wong Shek Pier in Sai Kung East Country Park. You may check the updated ferry schedule.	於船上觀察白腹海鵰、岩鷺，以及海鳥和燕鷗等鳥類 White-bellied Sea Eagle, Pacific Reef Heron and terns from the ferry	3小時 3 hours
東平洲 Tung Ping Chau	大學港鐵站步行15分鐘至馬料水碼頭，乘小輪前往，早上9:00開船，只於週六、日及公眾假期開出，當日有一班於下午5:15開出的回程船。可預先查詢最新班次。 Take a ferry leaving from Ma Liu Shui Pier (15-minute walk from University MTR Station). Ferries are only available on Saturdays, Sundays and public holidays, leaving at 9:00am. There is only one retrun boat at 5:15pm on the same day.	燕鷗、白腹海鵰和春秋過境遷徙鳥，如海鳥和棕尾褐鶲 Terns, White-bellied Sea Eagle and passage migrants such as seabirds and Ferruginous Flycatcher	8小時 8 hours

觀鳥地點 Birdwatching spots	交通 Transportation	目標鳥種 Target species	預計觀鳥時間 Time required at the site
市區 Urban Area			
香港公園 Hong Kong Park	金鐘港鐵站太古廣場後面。 Easily accessed from Admiralty MTR Station, or by other means; next to Pacific Place shopping centre.	小葵花鳳頭鸚鵡、紅領綠鸚鵡、黃眉柳鶯、叉尾太陽鳥和紅嘴藍鵲 Yellow-crested Cockatoo, Rose-ringed Parakeet, Yellow-browed Warbler, Fork-tailed Sunbird and Red-billed Blue Magpie	1小時 1 hour
香港動植物公園 Hong Kong Zoological and Botanical Gardens	中環港鐵站步行到花園道。公園入口位於花園道上山方向的右邊。 From Central MTR Station, walk up Garden Road. The Gardens are on the right.	小葵花鳳頭鸚鵡、黃眉柳鶯、叉尾太陽鳥和紅嘴藍鵲 Yellow-crested Cockatoo, Yellow-browed Warbler, Fork-tailed Sunbird and Red-billed Blue Magpie	1小時 1 hour
九龍公園 Kowloon Park	入口位於尖沙咀港鐵站A1出口。 Enter from Tsim Sha Tsui MTR Station Exit A1.	夜鷺、白胸翡翠，亞歷山大鸚鵡、黃眉柳鶯、紫綬帶、叉尾太陽鳥、灰背鶇、紅嘴藍鵲和黑領椋鳥 Black-crowned Night Heron, White-throated Kingfisher, Alexandrine Parakeet, Yellow-browed Warbler, Japanese Paradise Flycatcher, Fork-tailed Sunbird, Grey-backed Thrush, Red-billed Blue Magpie and Black-collared Starling	2小時 2 hours
彭福公園 Penfold Park	沙田火炭港鐵站附近。 Near Fo Tan MTR Station.	常見市區鳥類及多種鷺鳥 Many common urban bird species, and also egrets and herons	2小時 2 hours

※ 郭子祈 Kwok Tsz Ki

香港觀鳥地圖
Hong Kong Map for Birdwatching
后海灣
Deep Bay
米埔自然護理區
Mai Po Marshes Nature Reserve
米埔新村 / 担桿洲
Mai Po San Tsuen / Tam Kon Chau
塱原
Long Valley
洲頭
Chau Tau
尖鼻咀
Tsim Bei Tsui
南生圍
Nam Sang Wai
林村郊野公園
Lam Tsuen Country Park
元朗
Yuen Long
錦田
Kam Tin
大欖郊野公園
Tai Lam Country Park
港珠澳大橋香港接線
大蠔河
Tai Ho River
北大嶼郊野公園
Lantau North Country Park
大嶼山
Lantau Island
南大嶼郊野公園
Lantau South Country Park

沙頭角海
Starling Inlet
荔枝窩
Lai Chi Wo
東平洲
Tung Ping Chau
鹿頸、南涌
Nam Chung, Luk Keng
船灣郊野公園
Plover Cove Country Park
八仙嶺郊野公園
Pat Sin Leng Country Park
塔門
Tap Mun
沙螺洞
Sha Lo Tung
船灣、洞梓、汀角
Shuen Wan, Tung Tse, Ting Kok
大埔
Tai Po
吐露港
Tolo Harbour
大埔滘自然護理區
Po Kau Nature Reserve
榕樹澳
Yung Shue O
新界
Territories
西貢西郊野公園
Sai Kung West Country Park
西貢東郊野公園
Sai Kung East Country Park
彭福公園
Penfold Park
馬鞍山郊野公園
Ma On Shan Country Park
沙田
Sha Tin
萬宜水庫
High Island Reservoir
獅子山郊野公園
Lion Rock Country Park
飛鵝山
Fei Ngo Shan
(Kowloon Peak)
九龍
Kowloon
九龍公園
清水灣郊野公園
Clear Water Bay Country Park
香港公園
HK Park
香港島
Hong Kong Island
大潭郊野公園
Tai Tam Country Park
藍塘海峽
Tathong Channel
蒲台
Po Toi

鳥種分類
Family Description

鷓鴣、鵪鶉和雉雞
Francolins, Quails and Pheasants

※ 陳佳瑋 Chan Kai Wai

雉科 Phasianidae（雞形目 Galliformes）

外形笨重，頭部細小，大部分時間在陸地行走。不易觀察，但叫聲明顯易認。

Heavy-looking birds with small heads. Spend a lot of time on the ground. Difficult to see but easy to identify by their call.

鴨、雁和天鵝 Ducks, Geese and Swans

※ 江敏兒、黃理沛 Michelle and Peter Wong

鴨科 Anatidae（雁形目 Anseriformes）

嘴寬而扁平，腳短，多位於體後部，前三趾間有蹼，喜歡漂浮於水面，善泳。兩性外貌同色或異色，異色者雄鴨較雌鴨大，色彩亦較鮮艷。翅上多有翼鏡，呈金屬光澤。食物種類多樣，大部分為雜食性，食物有雜草種子、水生植物、昆蟲、貝類、魚類和兩棲類等。飛行時拍翼急速。

Aquatic birds characterized by broad flat bills. Their legs are short with webs between the first 3 toes. Good dabblers or divers. Sexes alike or differ. For those sexes differ, male ducks generally are brighter in colours. Most species have glossy speculum on wings. Mostly omnivorous, feed on grass seeds, aquatic plants, insects, shellfish and amphibians. Fly with rapid wing beats.

夜鷹 Nightjars

※ 江敏兒、黃理沛 Michelle and Peter Wong

夜鷹科 Caprimulgidae（夜鷹目 Caprimulgiformes）

與鴞形目一樣屬夜行性鳥類，但嘴部較弱，基部寬闊，捕食飛行中的昆蟲。嘴鬚長而發達。飛羽較長而闊，翅尖亦長。喜棲息於空曠草地，能靜止於地上或樹枝上，保護色極佳。

Nocturnal. Wide, weak bill with long bristles for catching insects in flight. Long, broad and pointed wings. Favour open grassland. Well camouflaged on ground or in trees.

*此鳥種分類是按兩冊介紹的雀鳥而定。
Family description is according to the birds mentioned in the guide book.

雨燕 Swifts

雨燕科 Apodidae（雨燕目 Apodiformes）

中至小型空中覓食雀鳥，身體短小，頸短。翼長而尖。尾短，呈方形或長叉尾狀。嘴短而闊，腿十分短而有毛，腳小而強壯，腳趾直而尖。

Small to medium-sized aerial insect feeders. Bodies compact, necks short. Wings long and pointed. Tails vary from short and square to long and forked. Bills short and broad, with wide gape. Legs extremely short and feathered; feet small and strong, with sharp, pointed toes.

※ 深藍 Owen Chiang

杜鵑和鴉鵑 Cuckoos and Coucals

杜鵑科 Cuculidae（鵑形目 Cuculiformes）

體型和鴿子相近但較修長。嘴尖稍向下彎，尾長而闊成楔形。羽色多為灰色和褐色，常躲在茂密的樹冠中，要靠獨特的叫聲辨認。杜鵑科鳥類自己不築巢，多在其他鳥類巢中產卵寄生(巢寄生)。

Size comparable to doves but slimmer. Bill slightly decurved. Long and broad tail forming wedge shape. Plumage mostly grey or brown, Cuckoos are often hidden in dense canopy. Identified by distinctive calls. They lay eggs in other birds' nests (Brood parasitism).

※ 江敏兒、黃理沛 Michelle and Peter Wong

鳩和鴿 Doves and Pigeons

鳩鴿科 Columbidae（鴿形目 Columbiformes）

體型較圓碩，腳短而強壯，適合在地上行走。羽色多樣，有些鳥種顏色鮮艷。棲息地廣泛分佈，包括樹木、灌叢、以至市區公園。主要以以種子及果實為食，築巢於樹、灌叢、岩崖或建築物上，以樹枝組成一個盤狀巢。

Stout bodies and short, strong legs, adapted to walking on the ground. Often colourful. They favour a wide range of habitats, including woodland, scrubland and urban parks. Feed mainly on seeds and fruits. Build platform-shaped nests of twigs on trees, scrubland, rock cliffs or buildings.

※ 何志剛 Pippen Ho

秧雞 Rails, Crakes and Coots

秧雞科 Rallidae（鶴形目 Gruiformes）

體型較小的雞，頭小而頸長，嘴短而強，四趾及跗蹠部較長。不善飛行，逃避敵人時會拔足狂奔。喜歡棲息於沼澤及密集的灌叢中，以小魚、水生昆蟲和植物嫩草為食。

Small head and long neck, short and strong bill. Toes and tarsus are longer. Poor flyers, runing to escape predators. Favour marshes and dense shrubland. Diet of small fish, aquatic insects and seedlings.

※ 郭匯昌 Andy Kwok

鶴 Cranes

鶴科 Gruidae（Gruiformes 鶴形目）

※ 孔思義、黃亞萍 Jemi and John Holmes

儀態優雅的大型涉禽，嘴、頸和腳都很長。雌雄同色，飛行時頸部伸直，組成人字形行列。

Large but elegant wading birds. Long bills, necks and legs. Sexes alike. Fly in V-formation with outstretched necks.

鸊鷉 Grebes

鸊鷉科 Podicipedidae（鸊鷉目 Podicipediformes）

※ 江敏兒、黃理沛 Michelle and Peter Wong

體型似鴨但較扁平，頸部長而細，嘴直而尖，趾間有瓣狀蹼，善於潛泳。起飛時會在水面奔跑，留下一列浪花。在蘆葦間的水面築巢，剛孵出的幼鳥喜停在成鳥的背上。主要棲息於濕地，以水生植物為食。

Diving birds resembling ducks but look much flatter, with small long necks and pointed sharp bill. Toes, with webs around toes for diving. Run on water surface for a short distance on take-off. Build nests on water surface between reeds. When hatched, juveniles often sit on the back of adults. Favour wetlands and feed on aquatic plants.

三趾鶉 Buttonquails

三趾鶉科 Turnicidae（鴴形目 Charadriiformes）

※ 孔思義、黃亞萍 John and Jemi Holmes

體型似鵪鶉，但較細小。大多只有三趾，沒有後趾。雌雄異色，雌鳥較大而顏色鮮艷。生活在開闊的沼澤或耕地，非常怕人，受驚時會突然飛出。

Resemble Japanese Quail but smaller. Three-toed, no hind claw. Sexes differ, females are more colourful. Inhabit open country including marshes and farmland. They stay well-hidden in vegetation, bursting into flight when flushed.

石鴴 Stone-curlews

石鴴科 Burhinidae（鴴形目 Charadriiformes）

※ 江敏兒、黃理沛 Michelle and Peter Wong

體型較大的涉禽，腿長，色彩暗淡，棲於開闊多石地帶或海灘。

Larger waders with long legs and cryptic coloration. Live in open, stony land or on beaches.

蠣鷸 Oystercatchers

蠣鷸科 Haematopodidae（鴴形目 Charadriiformes）

※ 陳志雄 Allen Chan

軀體粗壯的涉禽，腳短，嘴粗長。

Bulky shorebirds with short legs and stout bill.

長腳鷸和反嘴鷸 Stilts and Avocets

反嘴鷸科 Recurvirostridae（鴴形目 Charadriiformes）

體型較大的涉禽，嘴細長而上翹。脛裸露而跗蹠較長，尾短而方。常棲息於淺水區如泥灘。

Larger waders. Fine upcurved bills. Long, bare tarus and tibia. Tails short and square. Favour shallow water including mudflats.

※ 孔思義、黃亞萍 Jemi and John Holmes

鴴和麥雞 Plovers and Lapwings

鴴科 Charadriidae（鴴形目 Charadriiformes）

小、中型涉禽，嘴短而直，尖端較寬。翼形尖短，尾羽短小。主要羽色為棕色、黑色和白色。棲息於水邊，以無脊椎動物為食。

Small to medium-sized waders. Their bills are short and straight but thicker at tip. Wings are short, usually pointed. Most of them have brown, black and white colour pattern. Usually found at water edge and feed on small invertebrates.

※ 深藍 Owen Chiang

彩鷸 Painted-snipes

彩鷸科 Rostratulidae（鴴形目 Charadriiformes）

嘴直而長，尾羽較短。雌鳥比雄鳥大，且色澤鮮艷。常棲於低窪沼澤，躲藏於矮叢中，生性怕人。清晨或黃昏到耕地上覓食，以昆蟲和無脊椎動物為食。

Bills straight and long, short tails. Females are larger and more colourful. Inhabit lowland marshes. Secretive. Feed on amphibians and invertebrates in farmland at dawn and dusk.

※ 林文華 James Lam

水雉 Jacanas

水雉科 Jacanidae（鴴形目 Charadriiformes）

本科尾羽一般較短，唯水雉特長。四趾特長，有堅硬的爪。喜棲息於淡水沼澤，喜吃昆蟲、兩棲類和無脊椎動物。

Long toes and strong claws facilitate walking on floating vegetation. Habitat freshwater marshes, feeding on insects, amphibians and invertebrates.

※ 江敏兒、黃理沛 Michelle and Peter Wong

鷸、沙錐、瓣蹼鷸等 Sandpipers, Snipes, Phalaropes and allies

鷸科 Scolopacidae（鴴形目 Charadriiformes）

本科涉禽體型大小不一，由細小的濱鷸到大型的白腰杓鷸。嘴的形狀多樣，有的向上翹，有的向下彎，有的甚至呈勺形。腳和頸都較長。常成群於泥灘濕地覓食，以軟體動物、甲殼類為主要食物。

Waders with a wide range of sizes, from small Great Knot to large Eurasian Curlew. They have different sizes and bill shapes, including upcurved, decurved and some spoonbill shapes. Legs and neck are long. Feed on molluscs and shellfish on mudflat.

※ 夏敖天 Martin Hale

燕鴴 Pratincoles

※ 林文華 James Lam

燕鴴科 Glareolidae（鴴形目 Charadriiformes）

嘴短闊而尖端稍曲，初級飛羽長度超過尾端，尾呈叉狀，飛行時似燕子。棲於海岸沼澤濕地，以昆蟲為主食，是唯一可在飛行中捕食的涉禽。

Bill short and broad. Primaries are longer than tail tip. Forked-tailed, resembling Swallow in flight. Inhabit coastal wetlands, feed on insects. It is the only waders that feed on insects in flight.

鷗和燕鷗 Gulls and Terns

※ 黃卓研 Cherry Wong

鷗科 Laridae（鴴形目 Charadriiformes）

體型小至中等，羽色較單調，以黑、白、灰和褐色為主。翅長而尖，尾長而呈開叉或扇形。本科多屬海洋性鳥類，但也會在濕地泥灘等地方出現。

Medium to large-sized. Simple colour combination, mainly black, white, grey and brown. Wings long and pointed. Tails may be fork or fan-shaped. At sea or along mudflat and shoreline habitats.

賊鷗 Jaegers

※ 江敏兒、黃理沛 Michelle and Peter Wong

賊鷗科 Stercorariidae（鴴形目 Charadriiformes）

大小如鷗類，靠搶掠維生，追逐飛行中的鷗和燕鷗。賊鷗在接近極圈緯度繁殖，長途跋涉遷徙。雌雄同色，但毛色變化多端。

Jaeger are gull-sized, predatory seabirds that harass gulls and terns in flight to force them to jettison their catch. They breed in sub-polar latitudes and most undertake long migrations. Sexes alike, but plumage patterns are variable.

海雀 Murrelets

※ Geoff Welch

海雀科 Alcidae（鴴形目 Charadriiformes）

主要生活在海洋中。身體長、頸短、翼短和窄。飛行呈直線方式，快速而有力地拍翼。

Highly aquatic and exclusively marine. Bodies elongate; necks short, wings short and narrow. Flight direct and strong with rapid whirring wing-beats.

鸏 Tropicbirds

※ 宋亦希 Sung Yik Hei

鸏科 Phaethontidae（鸏形目 Phaethontiformes）

熱帶海洋上中等大小和非常善於飛行的海鳥，飛行姿態優雅似鳩鴿，盤旋升至空中。俯衝入海覓食，經常在海面浮游，並把長尾豎起。身體主要白色，成鳥以嘴和長尾的顏色配搭來辨別鳥種。

Medium-sized highly aerial seabirds of tropical seas. Graceful pigeon-like flight soaring high into the air. Plunge-dive and often sit on the sea with their long tails cocked. Mostly white and adults specially identified by a combination of bill and tail-streamer colour.

潛鳥 Loons

潛鳥科 Gaviidae（潛鳥目 Gaviiformes）

※ 江敏兒、黃理沛 Michelle and Peter Wong

體型大，吃魚為主。用腳推動游泳及潛水，只在開闊水面出現，冬天一般在海上。雌雄同色，辨別鳥種不易，香港非常罕見。

Large, mainly fish-eating, foot-propelled swimming/diving birds. Only found on open water, usually at sea in winter. Sexes alike. Identification is difficult. Very rare in Hong Kong.

鸌 Petrels and Shearwaters

鸌科 Procellariidae（鸌形目 Procellariiformes）

※ 何志剛 Pippen Ho

長途遷徙的遠洋海鳥，只在築巢或受風暴影響時才靠近陸地。翼尖長，不常拍動，常見於中國南海。

Oceanic seabirds which undertake long migrations. Only seen close to land when nesting or storm-driven. Long narrow rigidly held wings. Often seen in the South China Sea.

鸛 Storks

鸛科 Ciconiidae（鸛形目 Ciconiiformes）

※ 譚耀良 Tam Yiu Leung

大型水鳥，體形粗壯，嘴鈍而長，腿甚長，脛部半裸露。翅長而寬，飛行時頸部伸直，尾甚短。

Very large waterbirds. Stout body, long and thick bill. Long legs, tibia partly feathered. Long and broad wings, fly with neck extended, tail relatively short.

軍艦鳥 Frigatebirds

軍艦鳥科 Fregatidae（鰹鳥目 Suliformes）

※ 黃卓研 Cherry Wong

大型海鳥。嘴長成鈎形。翼形長而窄，末端尖，展翅可長達二米。尾長開叉。羽色以黑和深褐為主。常靠海面的氣流滑翔。喜歡搶奪其他海鳥的漁獲，亦是捕魚能手。

Large-sized oceanic birds. Long and strongly hooked bill. Wings very long, narrow and pointed, with wingspan over 2 metres. Long and distinctive forked tail. Black and dark brown feather. Superb gliders, soaring effortlessly on thermals. Renowned for their piratical behaviour on other seabirds catches. They are also skillful fishers.

鰹鳥 Boobies

鰹鳥科 Sulidae（鰹鳥目 Suliformes）

※ 譚耀良 Tam Yiu Leung

大型海鳥，軀體雪茄形，翼尖長，尾楔狀，嘴粗長而尖。振翅有力，沿直線前進，在岩山島嶼上集體築巢。

Large seabirds with cigar-shaped bodies. Long pointed wings, wedge-shaped tails and stout pointed bills. Strong direct flight. Nest colonially on rocky islands.

鸕鷀 Cormorants

鸕鷀科 Phalacrocoracidae（鰹鳥目 Suliformes）

※ 夏敖天 Martin Hale

體羽以黑色為主，嘴長而尖端有鈎，潛水能力極高，是捕魚能手。由於體羽缺防水油脂，常展翅待乾。喜歡成群聚集於樹枝上，排泄物會將全株植物染成白色。

Plumage mainly black in colour. Bills long, pointed and hooked at tip. Good at diving and fishing. Often seen with wings extended for drying due to lack of oily waterproof secretion. Favour gathering in groups on tree branches, which are often stained white by their droppings.

琵鷺和鹮 Spoonbills and Ibises

鹮科 Threskiornithidae（鵜形目 Pelecaniformes）

※ 盧嘉孟 Lo Kar Man

鹮的嘴長，嘴尖向下彎；琵鷺的嘴尖端扁平呈琵琶狀。飛行時，嘴和頸部伸直。

Ibises have long and decurved bills. Spoonbills have bills with a spoon-shaped end. Bill and neck are outstretched in flight.

鷺鳥和鳽 Egrets, Herons and Bitterns

鷺科 Ardeidae（鵜形目 Pelecaniformes）

※ 李雅婷 Anita Lee

體型多樣，由較小的鳽到較大的蒼鷺都有。嘴長而直，腿和頸也很長，脛下部常裸露。善於在淺水區覓食，是濕地如魚塘、溪流、海岸及沼澤等常見的水鳥。飛行時頸向後縮，腳向後伸直。繁殖期常大群在樹林中營巢，聚集成鷺鳥林。食物主要為魚類、兩棲動物和昆蟲等。

Sizes range from small bitterns to large grey herons. Bills long and straight, legs and neck are very long. Naked tibia adapted for feeding in shallow water. Often found at fishponds, coast and wetland areas. In flight, neck pulled back and legs stretched behind body. Breed in large groups in woodland, forming egretries. Feed mainly on fish, amphibians and insects.

鵜鶘 Pelicans

鵜鶘科 Pelecanidae（鵜形目 Pelecaniformes）

※ 夏敖天 Martin Hale

體型巨大的水鳥，可長達1.8米。嘴長而扁平，下嘴有大型喉囊。善於游泳和飛翔，常結小群於空中翱翔。俗名「塘鵝」。

Large waterbirds with wingspan up to 1.8 metres. Characterised by a long bill with conspicuous pouch underneath. Pelicans are good at hunting fish. Often flies and fishes in small groups.

魚鷹 Ospreys

鶚科 Pandionidae（鷹形目 Accipitriformes）

廣泛分佈。魚鷹的外趾能反轉，趾底多刺突，常衝入水中捕捉魚類。

Worldwide distribution. Reversible outer toe and sharp spicules on the underside of the foot used to catch and hold their prey, which is caught by diving feet-first into water.

※ 黃卓研 Cherry Wong

鵰和鷹 Eagles and Hawks

鷹科 Accipitridae（鷹形目 Accipitriformes）

本科的鳥種由細小的雀鷹到大型的禿鷲，體型各異，雌鳥比雄鳥大。嘴尖、向下彎曲成鈎狀，翅大，尾羽形狀不一，善於飛行。具強而有力的腳和趾，尖銳的鈎爪有助獵食。食性複雜，由小型的兩棲類、爬蟲類至較大的哺乳類；海岸濕地的猛禽則以魚類、海蛇為食糧。棲息於不同生境，通常在空中飛行時見到。

Large and varied family, from small Besra to large Black Vulture. Females are generally larger than males. Broad wings and tails with various shapes support powerful flight. Strong legs, toes and sharp talons facilitate hunting. Feed on a wide variety of food, from small reptiles, amphibians to larger mammals. Species that depend on wetlands feed on fish and water snakes. Found in a wide variety of habitats. Many are easy to see in flight.

※ 黃卓研 Cherry Wong

鴞（貓頭鷹）Owls

鴟鴞科 Strigidae（鴞形目 Strigiformes）

體型大小不一，頭形寬大，眼圓大，眼周圍的羽毛組成面盤，部分鳥種頭兩側有角羽。嘴強而有力，下彎成鈎狀。為夜行性猛禽，視覺和聽覺敏銳，飛行時無聲。以昆蟲和蜥蜴為食。

Wide range of sizes. Broad heads with round and big forward-looking eyes, with feathers around the eyes forming a facial disk. Some species have long eartufts. Strong hook-shaped bill. Nocturnal predators with good eye-sight and hearing, silent flight. Prey includes insects and lizards.

※ 江敏兒、黃理沛 Michelle and Peter Wong

佛法僧 Rollers

佛法僧科 Coraciidae（佛法僧目 Coraciiformes）

體型較粗壯，嘴大而嘴基闊，飛行時翼長而闊，棲息於開闊環境，以昆蟲、爬蟲及兩棲動物為食。

Stout body. Large bill with wide bill-base. In flight they show long and broad wings, favouring open country. Feed on insects, reptiles and amphibians.

※ 郭匯昌 Andy Kwok

戴勝 Hoopoes

戴勝科 Upupidae（犀鳥目 Bucerotiformes）

※ 李佩玲 Eling Lee

冠羽明顯突出，嘴細長而微向下彎。波浪形飛行。棲息於開闊生境，喜在地上找尋昆蟲和小型無脊椎動物為食。

Unmistakable crown feathers. Bills slender and slightly decurved. Undulating flight. Prefer open country. Feed on insects and small invertebrates.

翠鳥 Kingfishers

翠鳥科 Alcedinidae（佛法僧目 Coraciiformes）

※ 陳家強 Isaac Chan

體型小，頭大，嘴長而粗直。羽色多變，常具金屬光澤。棲於林地的翠鳥以昆蟲或小型動物為主食，棲於濕地的翠鳥以魚、蝦等為食。多在河岸或土崖挖洞營巢。

Birds with small body and big head. Bill long, thick and straight. Colourful plumage, often with metallic gloss. Open country-dwelling kingfishers favour insects and small animals. Wetland species favour fish, shrimps, etc. In breeding season, they build nests at river banks.

蜂虎 Bee-eaters

蜂虎科 Meropidae（佛法僧目 Coraciiformes）

※ 崔汝棠 Francis Chu

體型小，嘴細長而尖，由基部開始稍微向下彎。羽色以藍、褐為主，鮮艷美麗。棲息於開闊環境，常停在電線上，在空中捕捉昆蟲為食。

Small size. Long and sharp, decurved bill. Beautiful blue and brown plumage. Favour open country, often perching on electric wires. Feed on insects in flight.

亞洲擬鴷 Asian Barbets

鬚鴷科 Megalaimidae（鴷形目Piciformes）

※ 黃卓研 Cherry Wong

體型粗壯，嘴大而粗壯，上嘴略長於下嘴。棲於成熟森林中，以種子、果實及昆蟲為食。

Stout body, bill strong and large. Upper bill slightly longer than lower bill. Favour mature woodland. Feed on seeds, fruits and insects.

啄木鳥 Woodpeckers

啄木鳥科 Picidae（鴷形目Piciformes）

※ 黃才安 Wong Choi On

中至小型雀鳥，趾爪有力和尾短，特別適應爬樹，大部分有粗壯的嘴，飛行呈波浪形，叫聲粗糙。

Medium-small size birds. Powerful feet and short stiff tails are specially adapted for climbing trees. Most have strong bills. Undulating flight. Harsh and screaming call notes.

隼 Falcons

隼科 Falconidae（隼形目 Falconiformes）

體型較小的猛禽，雌鳥比雄鳥大。飛行時翅膀長而尖，尾短，頸短。能在空中停留，以高速飛行捕獵食物。生性兇猛，常以較小的雀鳥為食。

Small raptors. Females larger than males. In flight they show long and pointed wings, short tail and neck. They hover, and dive at high speed to catch prey, Which can include small birds.

※ 江敏兒、黃理沛 Michelle and Peter Wong

鳳頭鸚鵡 Cockatoos

鳳頭鸚鵡科 Cacatuidae（鸚形目 Psittaciformes）

體型多變，具強而有力的嘴，尖端明顯鈎曲。翼形強而闊大，尾長。羽色多變，鮮艷奪目。頭頂具明顯冠羽。叫聲多變響亮，常模仿其他鳥類聲音。原居地為熱帶雨林，以硬殼果實為主要食糧。

Varied in sizes. Strong and powerful hooked bills. Wings strong and broad with long tail. Colourful plumage and conspicuous crested feather. Very loud voice and often mimic other sounds. Their native habitat is tropical rainforest where they feed mainly on hard shell nuts.

※ 陳志雄 Allen Chan

鸚鵡 Parakeets

鸚鵡科 Psittacidae（鸚形目 Psittaciformes）

嘴鈎曲如猛禽，能打開硬殼、咬破果實和樹葉。腳短而強壯，善爬樹。羽色較艷麗，多為綠色和紅色。喜群居，叫聲響亮。

Hook-shaped bill resembles that of raptors. Able to open nut shells and demolish fruit and foliage. Legs short and strong, good at climbing trees. Colourful plumage, mostly red and green. Often in noisy groups.

※ 黃卓研 Cherry Wong

八色鶇 Pittas

八色鶇科 Pittidae（雀形目Passeriformes）

羽色鮮艷但隱蔽難見的林鳥，隱藏於濃密的叢林下層。叫聲響亮，飛行時拍翼呼呼作響。

Enigmatic and elusive, brightly coloured forest birds that keep to thick undergrowth. Loud calls and whirring flight.

※ Geoff Welch

山椒鳥和鵑鵙 Minivets and Cuckooshrikes

山椒鳥科 Campephagidae（雀形目 Passeriformes）

體型大小不一，身體瘦長，腳弱小。嘴短而壯，基部略闊，嘴端向下彎。山椒鳥多成群活動，並不時鳴叫。鵑鵙則單獨或數隻出沒。棲於茂密林中，以昆蟲、果實為食。

Sizes vary. Slim body, small legs. Short and strong bills with wide base and decurved tips. Minivets often fly and call in groups. Cuckoo-shrikes usually appear as singles, and sometimes in small groups. Favour dense woodland. Feed on insects and fruits.

※ 江敏兒、黃理沛 Michelle and Peter Wong

伯勞 Shrikes

伯勞科 Laniidae（雀形目 Passeriformes）

※ 江敏兒、黃理沛 Michelle and Peter Wong

身體中至小型，嘴強壯，有利鈎。尾長，腳強健，有鈎爪，大多有黑色過眼紋。雌雄羽色相似，幼鳥下體有橫斑。棲息於開闊平原至林區邊緣，有很強的領域行為。捕食昆蟲、蛙、蜥蜴、小鳥或鼠類，會將獵物插在樹枝上儲存。

Small to medium-sized. Bill strong and tipped with sharp hook. Long tail, strong legs, hooked claws. Dark eye-stripe on most species. Sexes alike, juveniles have barred underparts. Habitats include open country to edge of woodland. Strong territorial behaviour. Prey includes insects, frog, lizards, small birds and ratsk. Prey "stored" by impaling on spikes, hence they are named "butcher bird".

白腹鳳鶥 White-bellied Erpornis

鶯雀科 Vireonidae（雀形目 Passeriformes）

※ 江敏兒、黃理沛 Michelle and Peter Wong

此科大部分鳥種見於美洲，只有兩個屬於亞洲出現。中小型雀鳥，以林鳥為主，喜於林區中上層活動。食性廣泛，由細小的脊椎動物、昆蟲以至果實均是牠們的食物。

This family of birds mainly occur in the Americans and only two genera occur in Asia. Medium small-sized birds mainly inhabit mid-upper level in woodland. They have a wide-range of diet including small vertebrates, insects and fruits.

黃鸝 Orioles

黃鸝科 Oriolidae（雀形目 Passeriformes）

※ 呂德恒 Henry Lui

體型中等，嘴尖略向下彎，翅尖長，尾甚短呈扇狀。雄鳥羽色較雌鳥鮮艷。棲於樹林間，喜在樹冠活動。以昆蟲和果實為食。

Medium-sized. Bill tip slightly decurved. Wings are pointed and long, tails short and slightly fan-shaped. Male is more colourful than females. Active in woodland and tree canopies. Feed on insects and fruits.

卷尾 Drongos

卷尾科 Dicruridae（雀形目 Passeriformes）

※ 江敏兒、黃理沛 Michelle and Peter Wong

體型中等，嘴稍向下彎，嘴尖有鈎。尾長呈叉狀，有些種類尾羽向上卷曲，體羽多為灰或黑色，雌雄相似。常停在樹梢上，以昆蟲包括蜻蜓為食。

Medium-sized. Bill tip slightly decurved and hooked. Long forked tails and black plumage, sexes alike. Often perch on exposed branches. Prey on insects including dragonflies.

王鶲和綬帶
Monarchs and Paradise Flycatchers

王鶲科 Monarchinae（雀形目 Passeriformes）

毛色鮮艷、外型似鶲的大型鳥類，嘴強壯。

Large, brightly plumaged flycatcher-like birds with strong bills.

※ 郭匯昌 Andy Kwok

烏鴉、喜鵲等 Crows, Magpies and allies

鴉科 Corvidae（雀形目 Passeriformes）

大型鳥，嘴和腳都很粗壯，喜在開闊地上行走。大多為雜食性的鳥類，吃種子、果實、小動物和屍體。

Large perching birds. Bills and legs thick and strong. Often active on the ground. Omnivorous diet of seeds, fruits and small animals.

※ 江敏兒、黃理沛 Michelle and Peter Wong

太平鳥 Waxwings

太平鳥科 Bombycillidae（雀形目 Passeriformes）

高遷徙性群居鳥，數目時有間歇性爆發。身體渾圓，頸短，嘴短而嘴基寬闊。翼尖長，尾短呈方形。羽毛柔軟、濃密和呈絲質光澤，有濃密短毛的冠羽。

Migratory and nomadic, with periodic irruptions. Body plump-looking; neck short. Bill short, thick and broad at base. Wings long and pointed. Tail short and square. Plumage soft, dense and silky. Crest with short dense feathers.

※ 霍棟豪 Stanley Fok

方尾鶲 Grey-headed Canary-flycatcher

玉鶲科 Stenostiridae（雀形目 Passeriformes）

此科鳥種全都是活躍的食蟲鳥種；會於樹枝上下飛躍捕食飛行中的昆蟲，或於樹枝上跳躍捕捉獵物，亦習慣經常拍動及伸展翅膀。

Birds in this family are all active insectivores. They hunt for flying insects by leaping from tree branches or hopping among branches for other preys. Often seen beating or stretching their wings.

※ 李啟康 Lee Kai Hong

山雀 Tits

山雀科Paridae（雀形目 Passeriformes）

體型大小不一，外形健碩。翅近圓形，尾圓而微叉。喜於樹林及高大灌叢間出沒，捕捉昆蟲為食。

Variable sizes. Stout-shaped, round wings, round and slightly forked tail. Favour woodland and shrubland with large trees. Feed on insects.

※ 郭匯昌 Andy Kwok

攀雀 Penduline Tits

攀雀科Remizidae（雀形目 Passeriformes）

※ 黃卓研 Cherry Wong

小型雀鳥，嘴錐狀，翅短圓而尖，尾方形。常見於蘆葦叢中，善於攀緣，有時會腹部朝天倒懸。以昆蟲為食，有時也吃種子。

Small-sized. Cone-shaped bills, round and pointed wings, square tails. Favour reedbed, good at climbing and sometimes hang themselves upside-down. Feed on insects and sometimes seeds.

百靈 Larks

百靈科 Alaudidae（雀形目 Passeriformes）

※ 陳家強 Isaac Chan

體型細小如麻雀，頭有冠羽，嘴較細小而呈圓錐形，後爪長而直。常見於平坦開闊環境，如草地、耕地、沼澤地等。主要以雜草種子、嫩芽為食，也吃昆蟲。常在空中邊飛邊唱，叫聲悅耳。

As small as Tree Sparrows, with small crown and small cone-shaped bill. Hind claws long and straight. Often found in flat open country, e.g. grassland, farmland and marshes. Favour grass seeds, buds and insects. Often call sweetly in flight.

鵯 Bulbuls

鵯科 Pycnonotidae（雀形目 Passeriformes）

※ 羅錦文 Law Kam Man

中小型鳥類，嘴形多變，如雀嘴鵯屬的嘴短厚而向下彎，鵯屬的則細尖。大多成群活動，生境由市區公園、灌叢林地以至濕地紅樹林等。喜食果實和種子，也吃昆蟲。

Small to medium size. Bills vary, for example Collared Finchbill has thick decurved bill, Bulbuls have small and pointed bills. Most of them are gregarious. Favours urban parks, shrubland or mangroves. Feeds on seeds and fruits, sometimes insects.

燕和沙燕 Swallows and Martins

燕科 Hirundinidae（雀形目 Passeriformes）

※ 江敏兒、黃理沛 Michelle and Peter Wong

體型細小而輕巧，嘴形短闊，略呈三角形。翅狹長而尾羽呈叉狀。常成群在開闊地方出現，亦喜站在電線上。飛行速度快，在空中捕食昆蟲。巢以泥土、雜草築成半碗形，置於屋簷之下或峭壁懸岩之間。

Small size. Bills are broad and triangular. Wings are long and narrow, tail forked. Often appear in flocks in open country, perching on electric wires. In flight, they fly at high speed and feed on flying insects. Their nests are half-bowl shaped, made of mud and grass and built beneath overhangs, rocks and cliffs.

鷦鶥 Wren-babblers

鱗胸鷦鶥科 Pnoepygidae（雀形目 Passeriformes）

此科鳥種為亞洲東南部特有的小型鳥種，尾巴極短至近乎無尾。活躍於林區下層，生性隱蔽，會發出獨特易認的尖鋭叫聲，要於林中尋找此科雀鳥建議先尋找其叫聲。

This family of small birds are endemic to South-east part of Asia. Their extremely short tail almost appeared missing. Active but secretive at the low level of woodland. Often give a distinctive sharp call, which is the key to find this bird in the undergrowth.

※ 何國海 Danny Ho

樹鶯 Bush Warblers

樹鶯科 Cettiidae（雀形目 Passeriformes）

此科鳥種大部分雀鳥都是行蹤隱蔽的褐色鳥種，於密集的樹叢中活動。大部分鳥種於地面活動，警戒時亦不會驚飛，會像老鼠般匆忙逃走；另有部分鳥種亦會於植被的中上層活動。

Most species in this family are secretive brown birds inhabiting dense bushes. Rather than fly away, most of them remain on the ground of undergrowth and escape like a mouse when frightened. Some species inhabit the mid-upper level of vegetations.

※ 關朗曦 Matthew Kwan

長尾山雀 Long-tailed Tits

長尾山雀科 Aegithalidae（雀形目 Passeriformes）

體型細小，嘴短而粗厚，尾較長，翅短而圓，體羽蓬鬆，雌雄羽色相似。

Small size. Short and thick bills with long tails. Short and broad wing with fluffy plumage. Sexes alike.

※ 黃卓研 Cherry Wong

柳鶯 Leaf Warblers

柳鶯科 Phylloscopidae（雀形目 Passeriformes）

體型細小的鳥種，於植被的中上層活動，以昆蟲為食。成員以綠色及褐色為主色調，部分有眉紋及翼帶。部分鳥種的外型差異極少，難以分辨；但是各種柳鶯繁殖時都有其獨特的歌聲，足以分辨出不同鳥種，可惜香港並非柳鶯繁殖地故牠們極少鳴唱。

Small-sized birds inhabit mid-upper level of vegetation. Mainly feed on insects. Most member of this family are green or brown, some of them have supercilia and wing bars. Several of them are indistinguishable due to their highly similar appearances. Each of them have a distinctive call during breeding season and these calls are diagnostic for species identification. Unfortunately, Hong Kong is not their breeding range and they are rarely heard singing.

※ 關朗曦 Matthew Kwan

葦鶯 Reed Warblers

葦鶯科 Acrocephalidae（雀形目 Passeriformes）

小至中型鳥種，於多種生境均有分佈，以褐色及橄欖色為主色調，不少成員背上都有條紋。此科很多鳥種都是太平洋小島上的特有種。於香港出沒的成員常見於濕地生境，並以蘆葦床為甚。

Small to medium-sized birds occurring in various habitats. Mainly brown or olive-green. Some of them have streaks on their back. Many species in this family are Pacific Islands endemics. In Hong Kong, most species of this family occur in wetland, especially reedbed.

※ 深藍 Owen Chiang

短翅鶯和蝗鶯
Bush Warblers and Grasshopper Warblers

蝗鶯科 Locustellidae（雀形目 Passeriformes）

此科包含很多行蹤隱蔽、於濕地及灌叢出沒的鳥種。牠們善於在濃密的植被中行走穿梭，較少飛行，相對上較喜歡像老鼠般流竄。

This family includes many secretive species which occur in wetland and bushes. They are good at running among dense vegetations like mice, not often fly.

※ 文權溢 Bill Man

扇尾鶯、鷦鶯等
Cisticolas, Prinias and allies

扇尾鶯科 Cisticolidae（雀形目 Passeriformes）

與鶯科非常接近的草地鳥類。

Grassland birds closely related to warblers.

※ 林文華 James Lam

鈎嘴鶥和穗鶥
Scimitar Babblers and Babblers

林鶥科 Timaliidae（雀形目 Passeriformes）

中等大小陸棲鳥，嘴強直而稍向下彎，腳強壯，善於奔走和攀爬樹枝，不善飛行。大多於灌叢中或樹林的下層活動，以昆蟲、果實和其他植物為食。歌聲嘹亮悦耳。

Medium-sized, terrestrial or arboreal. Bills vary in this family. Strong legs adapted for walking on ground (Scimitar-babbler and Laughingthrush) or hopping from branch to branch (Minlas and Yuhinas). Favour shrubland or woodland, feeding on insects, fruits and other plants. Calls variable.

※ 郭匯昌 Andy Kwok

雀鶥和草鶯 Fulvettas and Grassbirds

幽鶥科 Pellorneidae（雀形目 Passeriformes）

此科鳥種多於林區低層或地面活動，經常鳴叫，觀鳥時往往先聞其聲後才見其蹤影。林區其他雀鳥漸趨寂靜時仍會聽到此科鳥種的叫聲。

Species in this family inhabit low or bottom level of woodland. Their call are often heard before the birds are located. They keep calling even when the woodland gone quiet without noise from other birds.

※ 陳志雄 Allen Chan

噪鶥、希鶥等 Laughingthrushes, Minlas and allies

噪鶥科 Leiothrichidae（雀形目 Passeriformes）

多為小至中等體型的鳥種，大部分都有較長的尾巴及強壯的身體。此科的亞洲鳥種多於林區植被棲息，經常小群出沒。

Mostly small to medium-sized species with relatively long tails and strong bodies. Asian species of this family often occur in groups in woodland.

※ 黃卓研 Cherry Wong

林鶯和鴉雀等 Sylviid Warblers, Parrotbills and allies

鶯鶥科 Sylviidae（雀形目 Passeriformes）

此科大部分成員分布於歐亞非大陸，只有一種居於美洲大陸。此科鳥種分布於多種生境，包括濕地、溫帶森林、熱帶雨林及竹林。

Most members of this family occur in Afro-Eurasia, only one species occur in Americans. They inhabit various habitats including wetland, temperate forest, tropical rain forest and bamboo forest.

※ 呂德恒 Henry Lui

繡眼鳥 White-eyes

繡眼鳥科 Zosteropidae（雀形目 Passeriformes）

體型細小，全身大致綠色，眼周有白圈，嘴小而尖。常成群活動，穿梭於樹林及其外圍，以昆蟲和果實為食。

Small and green. Conspicuous white eyerings, small and sharp bill. Often in flocks in woodland and forest edge. Feed on insects and fruits.

※ 羅錦文 Law Kam Man

鳾 Nuthatches

鳾科 Sittidae（雀形目 Passeriformes）

小型雀鳥，嘴強而直，腳短但很發達和強壯。常於樹幹上快速攀爬穿梭，找尋昆蟲為食。

Small size. Strong and straight bill, short and strong feet with long toes adapted to climbing on tree trunks to look for insects.

※ 李佩玲 Eling Lee

椋鳥和八哥 Starlings and Mynas

椋鳥科 Sturnidae（雀形目 Passeriformes）

體型大小適中，嘴直而尖，尾短呈方形。腿長而強健，多為陸棲，常於樹上活動。性合群，叫聲嘈雜。以果實、漿果為食，也吃昆蟲。

Medium-sized. Bills straight and pointed. Short and square tails. Legs long and strong. Mostly terrestrial and favour perching on trees. Gregarious and noisy. Feed on fruits, berries and sometimes insects.

※ 呂德恒 Henry Lui

鶇 Thrushes

鶇科 Turdidae（雀形目 Passeriformes）

中型雀鳥，嘴較短，翅長而平，尾外形不一。大多以昆蟲、幼蟲、漿果為食。棲息地多樣化，有林地、河流、灌叢、草地、沼澤、岩石等。

Medium sized. Short bill, long and flat wing, tails varied. Feed on insects, grubs and berries. Habitats include woodland, river, shrubland, grasslands, marshland and cliffs.

※ 夏敖天 Martin Hale

鶲和鴝 Flycatchers and Chats

鶲科 Muscicapidae（雀形目 Passeriformes）

體型略小，嘴扁平而基部較闊，嘴鬚發達，腳較弱。體羽大都是褐、灰、藍色，雌雄異色，雄鳥較鮮艷。常見於茂盛的樹林中，通常會在空中飛捕昆蟲。

Small-sized. Bills flat with broad base and bristles. Weak legs. Plumage is mainly brown, grey or blue. Sexes differ, males are more colourful. Often found in dense woodland. Feed on flying insects.

※ 郭匯昌 Andy Kwok

葉鵯 Leafbirds

葉鵯科 Chloropseidae（雀形目 Passeriformes）

體型較鵯大，嘴與頭長度相若，形狀細尖。雌雄羽色相異，但大多以綠色為主，雄性較鮮艷。常混在其他鳥群之間，在樹冠頂以花粉、果實、昆蟲為食。

Larger than bulbuls. Bill small and pointed, with similar length as the head. Sexes differ, both are mainly green but males are more colourful. Often flock with other birds. Feed on fruits and insects in woodland canopy.

※ 陳志雄 Allen Chan

啄花鳥 Flowerpeckers

啄花鳥科 Dicaeidae（雀形目 Passeriformes）

體型細小，嘴小翼小，尾短，常於樹頂槲寄生類植物間活動。雌雄異色，雄性色彩較艷麗。

Small size. Small bill, small wings and short tail. Often active between parasitic mistletoes growths on top of trees. Males are more colourful.

※ 李君哲 Jasper Lee

太陽鳥 Sunbirds

花蜜鳥科 Nectariniidae（雀形目 Passeriformes）

體型纖細，嘴細長而稍向下彎。雄性羽色華麗，有金屬光澤，雌鳥較暗淡。飛行時常發出尖鋭的叫聲，活躍於市區公園或林間的樹冠及花叢間。

Small and slender-bodied, long and slightly decurved bill. Males have colourful and glossy plumage, while females are mainly green. Call at high pitch in flight, often active on top of tree canopy, or at flowering plants in urban parks or woodland areas.

※ 陳志雄 Allen Chan

麻雀 Sparrows

雀科 Passeridae（雀形目 Passeriformes）

體型細小，短尾而矮胖鳥類，具備用來吃植物種子的厚嘴，性好合群。

Small-sized, plump greyish brown birds with short tails. Powerful bills for eating seeds. Gregarious.

※ 黃卓研 Cherry Wong

文鳥 Munias

梅花雀科 Estrildidae（雀形目 Passeriformes）

體型細小。嘴粗厚和圓錐形。翼短闊，翼尖圓和鈍。喜歡在開闊原野的草被生活，有時亦見於樹林和蘆葦。群居性。

Small size. Bill thick and conical-shaped. Wings short and broad with rounded or bluntly pointed tip. Typically birds of open country grasses, but also found in woodland and reedbeds. Gregarious.

※ 何志剛 Pippen Ho

織雀（織布鳥）Weavers

織雀科 Ploceidae（雀形目 Passeriformes）

體型中等。嘴粗厚呈圓雉形。多群居生活，在樹洞內或屋簷下築巢。以種子為食，繁殖時也吃昆蟲。

Medium-sized. Bill short, thick and conical. Mostly live in flocks. Nests in tree holes and under overhangs of buildings. Feed on seeds, and sometimes insects when breeding.

※ 關朗曦、關子凱 Matthew and TH Kwan

鶺鴒和鷚 Wagtails and Pipits

※ 江敏兒．黃理沛 Michelle and Peter Wong

鶺鴒科 Motacillidae（雀形目 Passeriformes）

體型纖小，嘴形細長，翅尖長，尾羽外側近乎白色。多棲息於草地、河或沼澤邊。善於在地上走動，停留時常上下或左右擺尾。飛行時成波浪形起伏，邊飛邊叫。多以昆蟲為食，也食植物種子。

Small. Bill slender and long, wings pointed. Outer tail feathers whitish. Inhabit grassland, edge of rivers or marshes. Often flick tail at rest. Undulating flight with calls. Feed on insects and sometimes seeds.

燕雀和金翅雀類 Brambling and Cardueline Finches

※ 李啟康 Lee Kai Hong

燕雀科 Fringillidae（雀形目 Passeriformes）

體型細小，嘴強厚，上下嘴喙互相緊接。生活於樹林、矮叢及蘆葦叢中，以果實和種子為食。

Small birds with powerful bills. They live in woodland, shrubland and sometimes reedbeds, feeding on fruits and seeds.

鵐 Buntings

※ 江敏兒、黃理沛 Michelle and Peter Wong

鵐科 Emberizidae（雀形目 Passeriformes）

嘴圓錐形而頗尖，嘴峰直，上下喙不完全切合而有縫隙。食物主要是種子，也吃昆蟲。

Small birds with straight, sharp conical bills. Upper and lower bills are not well fitted. Feed on seeds and sometimes insects.

香港有記錄的全球瀕危鳥類
Globally Threatened Birds in Hong Kong

中文鳥名 Chinese Name	英文鳥名 English Name	學名 Scientific Name	全球保育狀況 Global Conservation Status
青頭潛鴨	Baer's Pochard	*Aythya baeri*	極度瀕危 Critical Endangered (CR)
白鶴	Siberian Crane	*Leucogeranus leucogeranus*	
勺嘴鷸	Spoon-billed Sandpiper	*Calidris pygmaea*	
白腹軍艦鳥	Christmas Frigatebird	*Fregata andrewsi*	
小葵花鳳頭鸚鵡	Yellow-crested Cockatoo	*Cacatua sulphurea*	
黃胸鵐	Yellow-breasted Bunting	*Emberiza aureola*	
紅腰杓鷸	Far Eastern Curlew	*Numenius madagascariensis*	瀕危 Endangered (EN)
大濱鷸	Great Knot	*Calidris tenuirostris*	
小青腳鷸	Nordmann's Greenshank	*Tringa guttifer*	
東方白鸛	Oriental Stork	*Ciconia boyciana*	
黑臉琵鷺	Black-faced Spoonbill	*Platalea minor*	
栗鳽	Japanese Night Heron	*Gorsachius goisagi*	
草原鵰	Steppe Eagle	*Aquila nipalensis*	
鵲色鸝	Silver Oriole	*Oriolus mellianus*	
小白額雁	Lesser White-fronted Goose	*Anser erythropus*	易危 Vulnerable (VU)
棕頸鴨	Philippine Duck	*Anas luzonica*	
紅頭潛鴨	Common Pochard	*Aythya ferina*	
角鸊鷉	Horned Grebe	*Podiceps auritus*	
三趾鷗	Black-legged Kittiwake	*Rissa tridactyla*	
黑嘴鷗	Saunders's Gull	*Chroicocephalus saundersi*	
遺鷗	Relict Gull	*Ichthyaetus relictus*	
白腰燕鷗	Aleutian Tern	*Onychoprion aleuticus*	
冠海雀	Japanese Murrelet	*Synthliboramphus wumizusume*	
黃嘴白鷺	Chinese Egret	*Egretta eulophotes*	
卷羽鵜鶘	Dalmatian Pelican	*Pelecanus crispus*	
烏鵰	Greater Spotted Eagle	*Clanga clanga*	
白肩鵰	Eastern Imperial Eagle	*Aquila heliaca*	
仙八色鶇	Fairy Pitta	*Pitta nympha*	
白頸鴉	Collared Crow	*Corvus torquatus*	
飯島柳鶯	Ijima's Leaf Warbler	*Phylloscopus ijimae*	

中文鳥名 Chinese Name	英文鳥名 English Name	學名 Scientific Name	全球保育狀況 Global Conservation Status
遠東葦鶯	Manchurian Reed Warbler	*Acrocephalus tangorum*	易危 Vulnerable (VU)
史氏蝗鶯	Styan's Grasshopper Warbler	*Helopsaltes pleskei*	
大草鶯	Chinese Grassbird	*Graminicola striatus*	
白喉林鶲	Brown-chested Jungle Flycatcher	*Cyornis brunneatus*	
田鵐	Rustic Bunting	*Emberiza rustica*	
硫磺鵐	Japanese Yellow Bunting	*Emberiza sulphurata*	
鵪鶉	Japanese Quail	*Coturnix japonica*	近危 Near Threatened (NT)
羅紋鴨	Falcated Duck	*Mareca falcata*	
白眼潛鴨	Ferruginous Duck	*Aythya nyroca*	
黑海番鴨	Black Scoter	*Melanitta americana*	
紅頂綠鳩	Whistling Green Pigeon	*Treron formosae*	
斑脇田雞	Band-bellied Crake	*Porzana paykullii*	
大石鴴	Great Stone-Curlew	*Esacus recurvirostris*	
蠣鷸	Eurasian Oystercatcher	*Haematopus ostralegus*	
鳳頭麥雞	Northern Lapwing	*Vanellus vanellus*	
白腰杓鷸	Eurasian Curlew	*Numenius arquata*	
斑尾塍鷸	Bar-tailed Godwit	*Limosa lapponica*	
黑尾塍鷸	Black-tailed Godwit	*Limosa limosa*	
紅腹濱鷸	Red Knot	*Calidris canutus*	
彎嘴濱鷸	Curlew Sandpiper	*Calidris ferruginea*	
紅頸濱鷸	Red-necked Stint	*Calidris ruficollis*	
飾胸鷸	Buff-breasted Sandpiper	*Calidris subruficollis*	
半蹼鷸	Asian Dowitcher	*Limnodromus semipalmatus*	
灰尾漂鷸	Grey-tailed Tattler	*Tringa brevipes*	
黃嘴潛鳥	Yellow-billed Loon	*Gavia adamsii*	
白額鸌	Streaked Shearwater	*Calonectris leucomelas*	
黑頭白䴉	Black-headed Ibis	*Threskiornis melanocephalus*	
禿鷲	Cinereous Vulture	*Aegypius monachus*	
亞歷山大鸚鵡	Alexandrine Parakeet	*Psittacula eupatria*	
紫綬帶	Japanese Paradise Flycatcher	*Terpsiphone atrocaudata*	
小太平鳥	Japanese Waxwing	*Bombycilla japonica*	
斑背大尾鶯	Marsh Grassbird	*Helopsaltes pryeri*	
琉璃藍鶲	Zappey's Flycatcher	*Cyanoptila cumatilis*	
紅頸葦鵐	Japanese Reed Bunting	*Emberiza yessoensis*	

* 取自國際自然保護聯盟瀕危物種紅皮書 Retrieved from The International Unoin for Conservation of Nature (IUCN) Red List of Threatened Species

鏡頭下的247種香港鳥類

247 Kinds of Bird under the Lens

如何使用這本書
How to Use this Book

① 中、英文名稱
Chinese and English common names

② 學名
Scientific name

150 鵯科 Pycnonotidae

1 紅耳鵯 [普]hóng ěr bēi [粵]鵯：音卑 12

1 Red-whiskered Bulbul 2 *Pycnonotus jocosus*

3 體長 length：18-20.5cm 4

其他名稱 Other names：高髻冠, Crested Bulbul

13 14

5

9 10

6 全身偏褐，頭、嘴和腳黑色，有獨特的直立冠羽，耳羽紅色，面頰及喉白色，有明顯黑色頰紋。上體至尾部褐色，尾羽末端有白點。下體淡褐色，臀部橙紅色。幼鳥無紅色耳羽，臀部紅色較淡。常發出清脆的「bulbit…bulbit…」聲。

Dark brown upperparts. Black head, bill and legs, prominent erect crown feather. Red coverts, white cheeks and throat, with black moustachial stripe. Brown upperparts to tail tipped with white. Underparts pale brown with orange red vent. Juvenile lacks red cheeks and has paler red vent. Call a cheerful "bulbit…bulbit…".

8
[1] adult 成鳥：Siu Lek Yuen 小瀝源：Dec-08, 08 年 12 月：Ken Fung 馮漢城
[2] adult 成鳥：Hang Tau 坑頭：Mar-07, 07 年 3 月：Jemi and John Holmes 孔思義．黃亞萍
[3] adult 成鳥：Kowloon Park 九龍公園：Jul-08, 08 年 7 月：Lo Chun Fai 勞浚暉
[4] adult 成鳥：Cheung Chau 長洲：Dec-04, 04 年 12 月：Henry Lui 呂德恆
[5] adult 成鳥：Hong Kong Park 香港公園：8-Sep-07, 07 年 9 月 8 日：Anita Lee 李雅婷
[6] juvenile 幼鳥：Mai Po 米埔：Aug-07, 07 年 8 月：Sammy Sam and Winnie Wong 森美與雲妮
[7] adult 成鳥：Hong Kong Park 香港公園：Sep-07, 07 年 9 月：Bill Man 文權溢
[8] juvenile 幼鳥：Mai Po 米埔：Aug-07, 07 年 8 月：Cherry Wong 黃卓研

7

11 常見月份

春季過境遷徙鳥 Spring Passage Migrant			夏候鳥 Summer Visitor			秋季過境遷徙鳥 Autumn Passage Migrant			冬候鳥 Winter Visitor		
1	2	3	4	5	6	7	8	9	10	11	12
留鳥 Resident				迷鳥 Vagrant				偶見鳥 Occasional Visitor			

③ 體長
Body length

④ 體長和輪廓與本書的對比
Size compared to size of this book

⑤ 本地拍攝的照片
Photos taken in Hong Kong

⑥ 描述鳥種外形特徵和特別行為
Description of characteristics and behaviour

⑦ 外形特徵
Characteristics

⑧ 圖片説明
Caption

圖片説明的術語解釋：

幼鳥 juvenile	離巢後至第一次換羽之間的鳥。 The first immature plumage after the nestling stage.
成鳥 adult	鳥類的羽毛變化到了最後一個階段，即以後的羽色及模式不會再有變化。 The bird acquires its final or definitive plumage that is then repeated for life.
未成年鳥 immature	除了成鳥之外的所有階段。 Denotes all plumages phases except adult.
繁殖羽 breeding	鳥類在繁殖季節呈現異常鮮艷的顏色。 Plumage that turn more attractive during breeding season.
非繁殖羽雄鳥 male eclipse	部分雄性鳥類在繁殖期過後身上出現類似雌鳥的毛色，主要見於鴨類和太陽鳥。 Post-nuptical, female-like plumage that occurs in males of some groups of birds - notably ducks and sunbirds.
第一年冬天 1st winter	鳥類在幼鳥階段之後所披的羽色。 Plumage adopted by many species after juvenile plumage.
第一年夏天 1st summer	鳥類在第一年冬天之後轉變成的羽色。 The plumage following the first winter plumage in birds.

⑨ 地圖 Map

指出該鳥較常出現的地區、或曾經出現的地區，而並非是該鳥在全港的分佈圖。
Areas that the birds are usually found, or once found. It is not a distribution map of the bird in Hong Kong.

注：全粉紅色的地圖指在全港廣泛地區出現
Note: Pink coloured map indicates the bird is widely distributed in Hong Kong

⑩ 生態環境 Habitats

	濕地（淡水—魚塘、濕農地） Wetland (freshwater - fishponds, wet agricultural land)
	濕地（鹹淡水—蘆葦、紅樹林、基圍、泥灘） Wetland (brackish - reedbed, mangrove, *gei wai*, mudflat)
	溪流 Streams
	海洋、沿岸和海島 Ocean, Coastal Waters and Islands
	開闊原野（灌木叢、草地、仍有耕作或棄耕的農地） Open Country (shrubland, grassland, active and abandoned farmland)
	林地 Woodland
	高地 Upland / High Ground
	市區 Urban Area

⑪ 居留狀況
Status icon

常見 Common	不常見 Uncommon	稀少 Scarce	罕有 Rare	無記錄 No record

常見月份 Icon for abundance throughout the year

⑫ 普通話 普 拼音和廣東話 粵 難字讀音
Putonghua 普 and Cantonese 粵 pronunciation

注：這是拼音，並不是鳥種名字
Note: This is the pronunciation, not the name of the bird

⑬ 名稱讀音及鳥鳴聲 The pronunciation of the bird name and the sound of bird

⑭ 鳥種片段 The video of the bird

小葵花鳳頭鸚鵡

㊟ xiǎo kuí huā fèng tóu yīng wǔ

體長 length：33cm

Yellow-crested Cockatoo | *Cacatua sulphurea*

其他名稱 Other names：小葵花鸚鵡

1

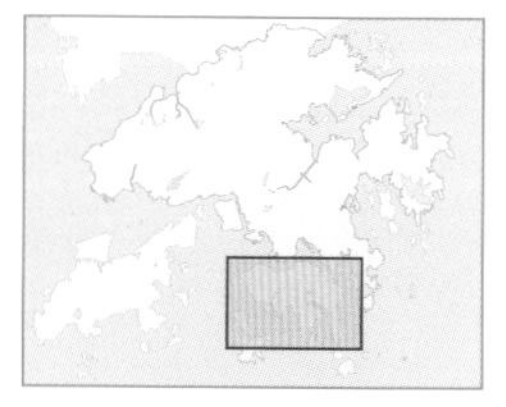

外地引入鳥種，全身白色，嘴和腳灰黑，經常豎起黃色或橙色冠羽。雄性眼黑色；雌性眼紅褐色。大多棲於香港島，常見棲息於香港公園和香港動植物公園內的樹洞。通常小群活動，叫聲沙啞嘈吵。

Introduced cockatoo, overall white in colour, bill and legs dark grey. Yellow or orange crest often erects. Male has black eye and female has red-brown eye. Mostly lives in Hong Kong Island, often observed living in tree holes at Hong Kong Park and Hong Kong Zoological and Botanical Gardens. Usually appears in small noisy groups.

1 adult 成鳥；Hong Kong Park 香港公園；Mar-08, 08 年 3 月；Allen Chan 陳志雄
2 adult 成鳥；Hong Kong Park 香港公園；Mar-08, 08 年 3 月；Allen Chan 陳志雄
3 adult 成鳥；Hong Kong Park 香港公園；Nov-08, 08 年 11 月；Eling Lee 李佩玲
4 adult 成鳥；Hong Kong Park 香港公園；Mar-03, 03 年 3 月；Henry Lui 呂德恆
5 adult 成鳥；Hong Kong Park 香港公園；Mar-08, 08 年 3 月；Allen Chan 陳志雄
6 adult 成鳥；Hong Kong Park 香港公園；Nov-04, 04 年 11 月；Henry Lui 呂德恆
7 adult 成鳥；Hong Kong Park 香港公園；Nov-04, 04 年 11 月；Henry Lui 呂德恆

	春季過境遷徙鳥 Spring Passage Migrant			夏候鳥 Summer Visitor			秋季過境遷徙鳥 Autumn Passage Migrant			冬候鳥 Winter Visitor		
常見月份	1	2	3	4	5	6	7	8	9	10	11	12
	留鳥 Resident				迷鳥 Vagrant				偶見鳥 Occasional Visitor			

7

亞歷山大鸚鵡

㊟yà lì shān dà yīng wǔ

體長 length：50-62cm

Alexandrine Parakeet | *Psittacula eupatria*

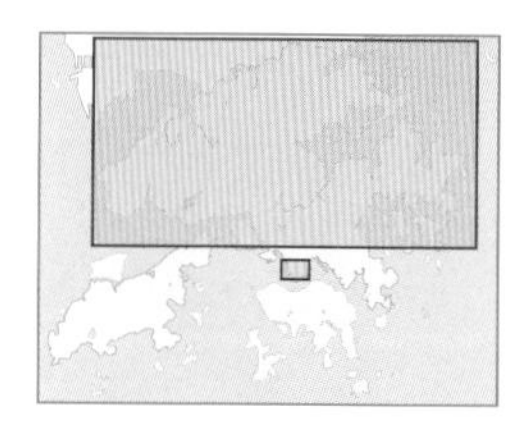

外地引入鳥種，全身綠色，嘴大紅色，眼圈黃色。頸上有粉紅領環，肩部紅色。飛行時形狀似「十」字，尾長而尖。常小群出現。

Introduced green parakeet with big red bill, yellow eye-rings, pink neck ring and shoulder patch. In flight, it shows a "flying cross" profile with long and pointed tail. Occurs in small parties.

1 male 雄鳥；Kowloon Park 九龍公園；Jan-06, 06 年 1 月；Cherry Wong 黃卓研
2 male and female 雄鳥和雌鳥；Kowloon Park 九龍公園；May-04, 04 年 5 月；Henry Lui 呂德恆
3 male 雄鳥；Kowloon Park 九龍公園；May-04, 04 年 5 月；Henry Lui 呂德恆
4 male 雄鳥；Kowloon Park 九龍公園；Feb-06, 06 年 2 月；Kami Hui 許淑君

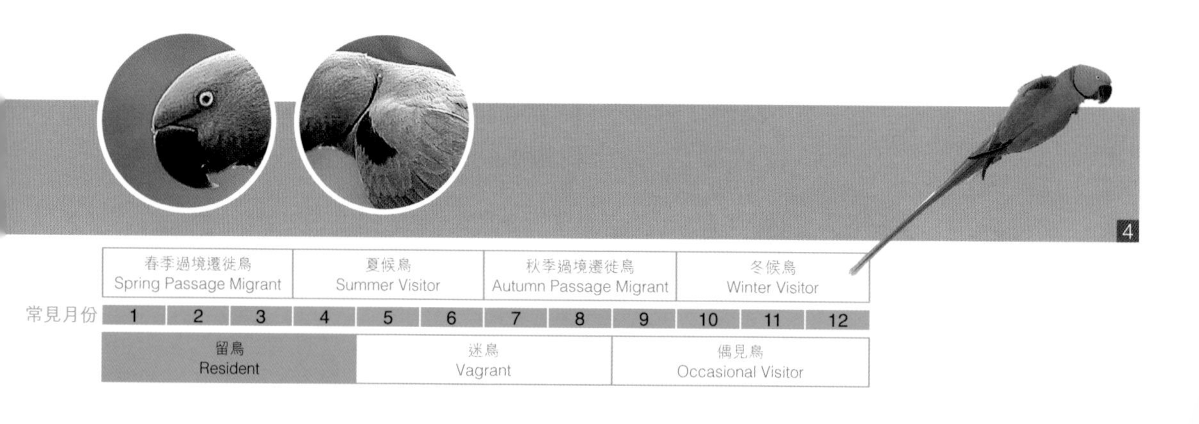

春季過境遷徙鳥 Spring Passage Migrant			夏候鳥 Summer Visitor			秋季過境遷徙鳥 Autumn Passage Migrant			冬候鳥 Winter Visitor		
1	2	3	4	5	6	7	8	9	10	11	12

常見月份

留鳥 Resident	迷鳥 Vagrant	偶見鳥 Occasional Visitor

紅領綠鸚鵡 [普]hóng lǐng lǜ yīng wǔ

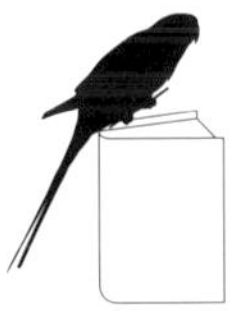

體長 length：37-43cm

Rose-ringed Parakeet | *Psittacula krameri*

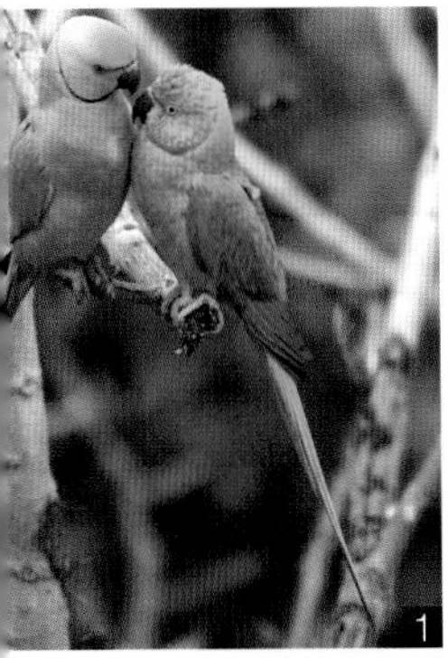
1

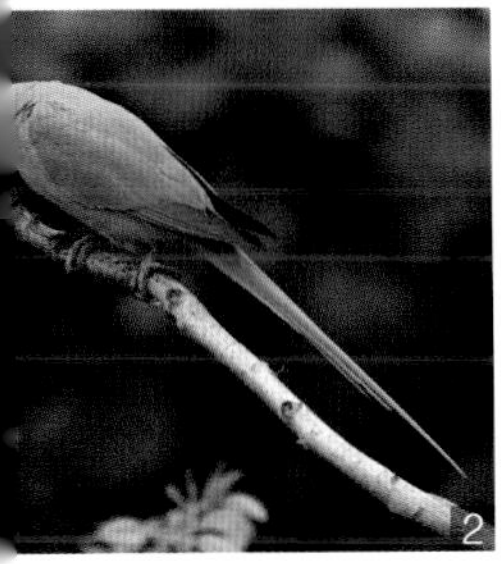
2

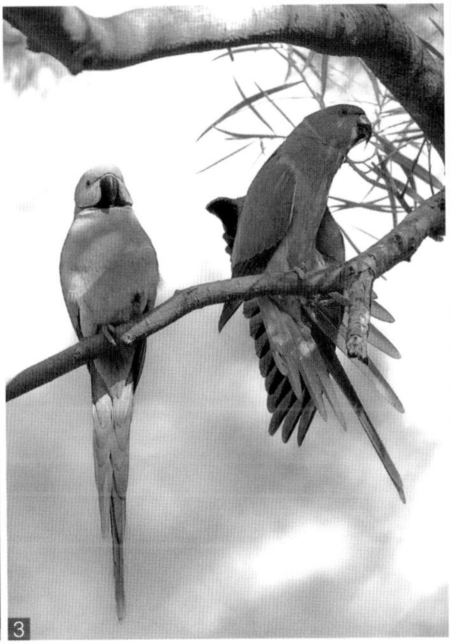
3

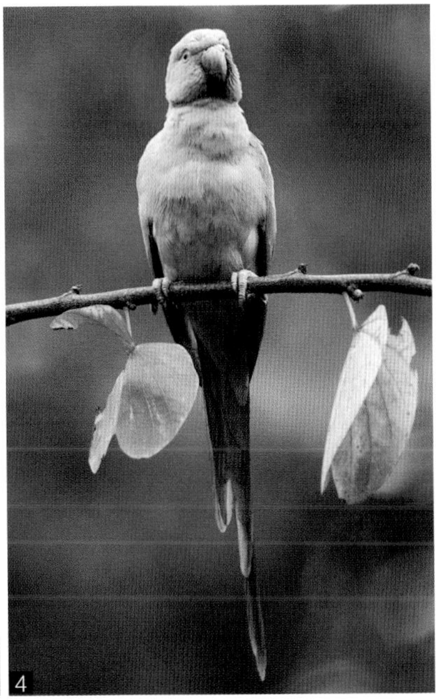
4

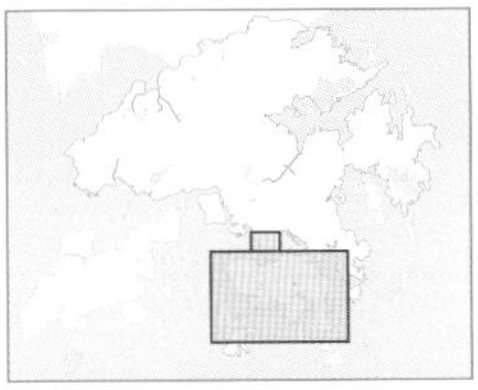

🏢

外地引入鳥種，全身綠色，嘴紅色，眼圈紅色。雄鳥頸部有粉紅色細領環，環上有黑邊，頸側至後頸沾藍。尾部藍色且很長而尖。飛行時形狀似「十」字，在市區內大型公園出現。

Introduced green parakeet, with red bill and eye-rings. Male bird has thin pink collar with black fringe. Side and hind neck bluish. Long and pointed tail blue in colour. In flight, it shows a "flying cross" profile. Occurs in large urban parks.

[1] adult male and female 雄成鳥和雌成鳥：Hong Kong Park 香港公園：Apr-07, 07 年 4 月：Cherry Wong 黃卓研
[2] adult male 雄成鳥：Hong Kong Park 香港公園：Apr-07, 07 年 4 月：Cherry Wong 黃卓研
[3] adult male and female 雄成鳥和雌成鳥：HK Zoological and Botanical Gardens 香港動植物公園：Jan-03, 03 年 1 月：Henry Lui 呂德恒
[4] adult female 雌成鳥：Hong Kong Park 香港公園：Jun-08, 08 年 6 月：Eling Lee 李佩玲
[5] Mar-05, 05年3月：Henry Lui 呂德恒

5

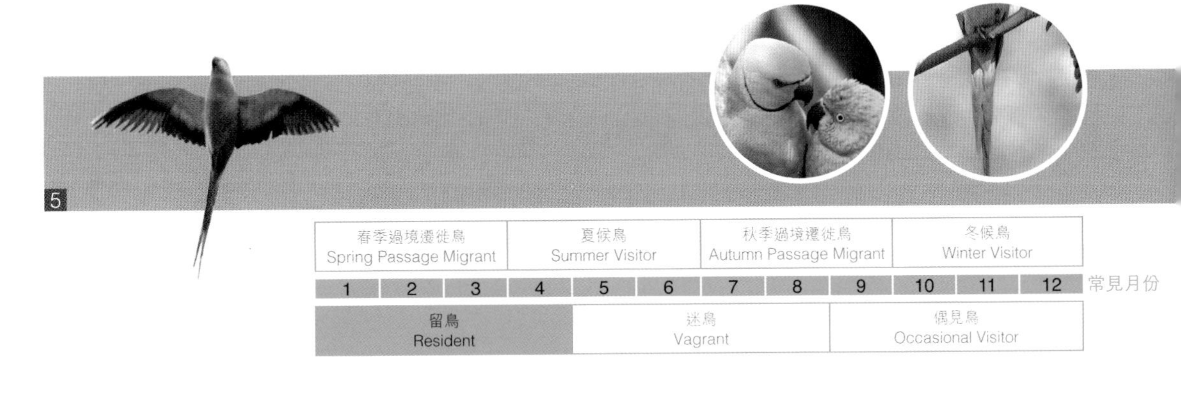

春季過境遷徙鳥 Spring Passage Migrant	夏候鳥 Summer Visitor	秋季過境遷徙鳥 Autumn Passage Migrant	冬候鳥 Winter Visitor

1	2	3	4	5	6	7	8	9	10	11	12	常見月份

留鳥 Resident	迷鳥 Vagrant	偶見鳥 Occasional Visitor

仙八色鶇

普 xiān bā sè dōng
粵 鶇：音東

體長 length：19cm

Fairy Pitta | *Pitta nympha*

其他名稱 Other names：藍翅八色鶇

1

2

色彩鮮艷，頭黑，冠紋黑色，冠兩旁淡褐色，白眼線，黑色貫眼紋。上體綠色，覆羽有小片藍斑，飛羽黑色，腰藍色，下體淺黃褐色。喉白色，腹部中央至尾下覆羽鮮紅色。飛行時可見翼上白斑。

Colourful pitta. Black head with black crown strip and brown lateral crown strip, buff supercilium and black eye-stripes. Green upperparts with small blue patch on wing coverts. Black flying feathers and blue rump. Underparts pale buff. White throat. Middle of belly to vent bright red in colour. In flight, it shows prominent white patch.

1 Po Toi 蒲台：3-May-08, 08 年 5 月 3 日：Wong Choi on 黃才安
2 Po Toi 蒲台：3-May-08, 08 年 5 月 3 日：Wong Choi on 黃才安

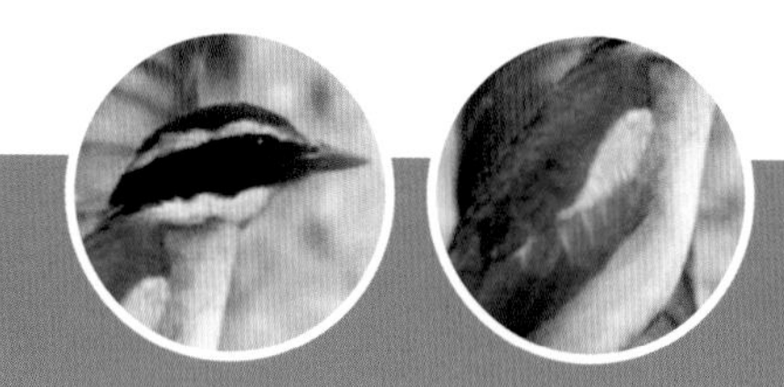

春季過境遷徙鳥 Spring Passage Migrant	夏候鳥 Summer Visitor	秋季過境遷徙鳥 Autumn Passage Migrant	冬候鳥 Winter Visitor

常見月份	1	2	3	4	5	6	7	8	9	10	11	12

留鳥 Resident	迷鳥 Vagrant	偶見鳥 Occasional Visitor

藍翅八色鶇

㊢ lán chì bā sè dōng
[粵] 鶇：音東

體長 length：18-20cm

Blue-winged Pitta | *Pitta moluccensis*

其他名稱 Other names：紫藍翅八色鶇

色彩鮮艷。頭黑色，冠紋黑色，冠紋兩旁淡褐色。上體綠色，覆羽有小片藍斑。喉白色，下體橙色，腹部中央至尾下覆羽鮮紅色。飛行時，翼鮮藍色，有大塊的白斑。幼鳥嘴基鮮紅色。

Colourful pitta. Black head with black crown strip and buff lateral crown strip. Upperparts green with blue patch on wing coverts. White throat. Underparts orange. Middle of belly to vent bright red. Bright blue wings with large white patches can be seen in flight.

1 adult 成鳥：Po Toi 蒲台：1-May-08, 08 年 5 月 1 日：Geoff Welch

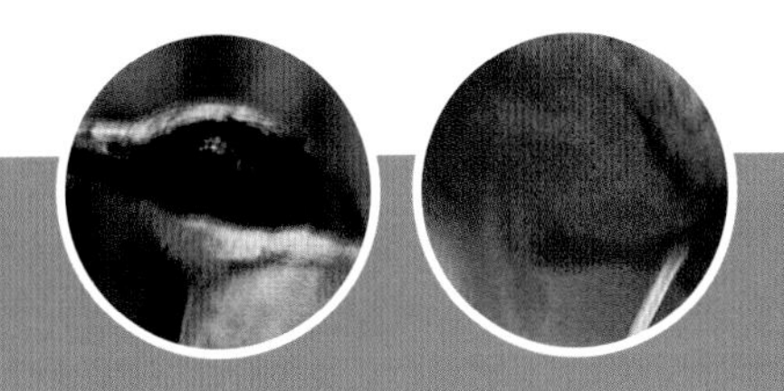

春季過境遷徙鳥 Spring Passage Migrant			夏候鳥 Summer Visitor			秋季過境遷徙鳥 Autumn Passage Migrant			冬候鳥 Winter Visitor			
1	2	3	4	5	6	7	8	9	10	11	12	常見月份
留鳥 Resident				迷鳥 Vagrant				偶見鳥 Occasional Visitor				

灰喉山椒鳥

[普]huī hóu shān jiāo niǎo

體長length：17-19cm

Grey-chinned Minivet | *Pericrocotus solaris*

1

2

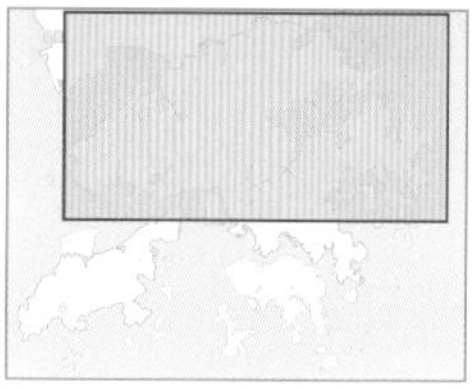

顏色鮮艷奪目。頭、面頰和喉部灰色，頭頂較深色，嘴和腳黑色。雄性上背和翼黑色，有倒轉「7」字型的橙紅色翼斑，腰及下體均為橙紅色，尾上黑色，尾下紅色。雌鳥似雄鳥，但身上橙紅色均代以黃色。幼鳥下體淺色，但有黃色或紅色翼斑。經常與赤紅山椒鳥混在一起，活躍於成熟樹林中。飛行時發出輕柔的「tsee-sip」叫聲。

Colourful and eye-catching. Grey head, cheeks and throat, darker crown. Black bill and legs. Male has black mantle and wings, with a reddish orange "inverted tick" mark on each wing. Reddish orange rump and underparts. Tail is black above and reddish orange below. Female resembles male but the red colour is replaced by yellow. Juvenile has paler underparts, wing marks can be yellow or red. Often flocks with Scarlet Minivets in mature woodland. Makes repeated soft "tsee-sip" calls in flight.

1 female 雌鳥：Tai Po Kau 大埔滘：Nov-05, 05 年 11 月：Michelle and Peter Wong 江敏兒、黃理沛
2 male 雄鳥：Tai Po Kau 大埔滘：Apr-05, 05 年 4 月：Cherry Wong 黃卓研
3 female 雌鳥：Kowloon Hills Catchment 九龍水塘引水道：Jan-04, 04 年 1 月：Henry Lui 呂德恆
4 female 雌鳥：Tai Po Kau 大埔滘：Nov-05, 05 年 11 月：Michelle and Peter Wong 江敏兒、黃理沛
5 female 雌鳥：Tai Po Kau 大埔滘：Dec-08, 08 年 12 月：Ann To 陶偉意
6 male 雄鳥：Tai Po Kau 大埔滘：Nov-06, 06 年 11 月：Andy Kwok 郭匯昌
7 male 雄鳥：Tai Po Kau 大埔滘：Nov-05, 05 年 11 月：Michelle and Peter Wong 江敏兒、黃理沛
8 male 雄鳥：Tai Po Kau 大埔滘：Mar-04, 04 年 3 月：Henry Lui 呂德恆
9 female 雌鳥：Tai Po Kau 大埔滘：Nov-04, 04 年 11 月：Jemi and John Holmes 孔思義、黃亞萍

春季過境遷徙鳥 Spring Passage Migrant			夏候鳥 Summer Visitor			秋季過境遷徙鳥 Autumn Passage Migrant			冬候鳥 Winter Visitor		
常見月份 1	2	3	4	5	6	7	8	9	10	11	12
留鳥 Resident				迷鳥 Vagrant				偶見鳥 Occasional Visitor			

3
4
5
6
7
8

9

赤紅山椒鳥

[普] chì hóng shān jiāo niǎo

體長 length：17-22cm

Scarlet Minivet | *Pericrocotus speciosus*

顏色鮮艷奪目，體型較灰喉山椒鳥大，嘴和腳黑色。雄性頭部全黑，上背和翼黑色，有「倒轉『7』字加一點」的紅色翼斑，腰及下體均為紅色，尾上黑色，尾下紅色。雌鳥頭至上背灰色，腰及下體黃色。前額、面頰、喉部至下體為黃色，翼斑黃色。雄性幼鳥偏黃但羽色沾紅。經常與灰喉山椒鳥混在一起，活躍於成熟樹林中。飛行時發出輕快而尖的「flee…flee…」叫聲。

Colourful and eye-catching. Larger than Grey-throated Minivet. Black, bill and legs. Male has black head, mantle and wings, with a red "inverted tick plus a spot" mark on each wing. Red rump and underparts. Tail is black above and red below. Female has grey head and mantle, yellow rump. Forehead, cheeks and throat to underparts yellow, and with yellow mark on wings. Juvenile male has yellowish plumage with some red. Often flocks with Grey-throated Minivets in mature woodland. In flight, gives a 4-note high-pitched "flee…flee…" call.

1 male 雄鳥：Chinese University of HK 香港中文大學：Oct-08, 08 年 10 月：Ng Lin Yau 吳璉宥
2 female 雌鳥：Tai Po Kau 大埔滘：Jul-08, 08 年 7 月：Michelle and Peter Wong 江敏兒、黃理沛
3 female 雌鳥：Tai Po Kau 大埔滘：Apr-04, 04 年 4 月：Henry Lui 呂德恒
4 male 雄鳥：Tai Po Kau 大埔滘：Apr-04, 04 年 4 月：Henry Lui 呂德恒
5 male 雄鳥：Shing Mun 城門水塘：Jul-07, 07 年 7 月：Wallace Tse 謝鑑超
6 female 雌鳥：Tai Po Kau 大埔滘：Jan-09, 09 年 1 月：Michelle and Peter Wong 江敏兒、黃理沛
7 male 雄鳥：Tai Po Kau 大埔滘：Jul-08, 08 年 7 月：Michelle and Peter Wong 江敏兒、黃理沛

春季過境遷徙鳥 Spring Passage Migrant	夏候鳥 Summer Visitor	秋季過境遷徙鳥 Autumn Passage Migrant	冬候鳥 Winter Visitor

常見月份 1 2 3 4 5 6 7 8 9 10 11 12

留鳥 Resident	迷鳥 Vagrant	偶見鳥 Occasional Visitor

3
4
5
6

7

灰山椒鳥

㊟ huī shān jiāo niǎo

體長 length：18cm

Ashy Minivet | *Pericrocotus divaricatus*

頭部黑色，前額、喉至面頰白色，嘴短黑色，上體深灰色，翼黑色，下體白色沾灰，腳黑色。幼鳥整體偏淡灰褐色。

Black head. Forehead, throat and cheeks white. Short dark bill. Upperparts dark grey with black wings. White underparts with some grey. Black legs. Juvenile is pale greyish brown.

1 male 雄鳥：Po Toi 蒲台：Apr-07, 07 年 4 月：Allen Chan 陳志雄
2 male 雄鳥：Po Toi 蒲台：Apr-07, 07 年 4 月：Michelle and Peter Wong 江敏兒、黃理沛
3 male 雄鳥：Po Toi 蒲台：Apr-07, 07 年 4 月：Michelle and Peter Wong 江敏兒、黃理沛
4 female 雌鳥：Po Toi 蒲台：Apr-08, 08 年 4 月：Michelle and Peter Wong 江敏兒、黃理沛
5 male 雄鳥：Po Toi 蒲台：Apr-08, 08 年 4 月：Danny Ho 何國海
6 female 雌鳥：Po Toi 蒲台：Apr-08, 08 年 4 月：Cherry Wong 黃卓研
7 female 雌鳥：Po Toi 蒲台：Oct-06, 06 年 10 月：Cherry Wong 黃卓研

春季過境遷徙鳥 Spring Passage Migrant	夏候鳥 Summer Visitor	秋季過境遷徙鳥 Autumn Passage Migrant	冬候鳥 Winter Visitor

常見月份	1	2	3	4	5	6	7	8	9	10	11	12

留鳥 Resident	迷鳥 Vagrant	偶見鳥 Occasional Visitor

粉紅山椒鳥

普 fěn hóng shān jiāo niǎo

體長 length：18-20cm

Rosy Minivet | *Pericrocotus roseus*

頭頂至背部灰色，有灰色貫眼紋，下體灰白色，翼邊及尾上黑色，嘴和腳黑色。雄鳥翼斑和尾底淡橙紅色，眉、胸、脇、腰部沾淡橙紅色；雌鳥翼斑和尾底淡黃色，胸、脇、臀、腰部沾淡黃色。

Grey from head to back with darker grey eye-stripes. Greyish-white underparts. Black wing edges and upper tail. Black bill and legs. Male has pale reddish-orange wing patches and under tail. Supercilia, breast, flanks and rump tinged with pale reddish-orange. Female has pale yellow wing patches and under tail. Breast, flanks, vent and rump tinged with pale yellow.

1 adult male 雄成鳥：Po Toi 蒲台：Apr-14, 14 年 4 月：Oldcar Lee 李啟康
2 adult male 雄成鳥：Po Toi 蒲台：Apr-14, 14 年 4 月：Oldcar Lee 李啟康
3 adult male 雄成鳥：Po Toi 蒲台：Apr-14, 14 年 4 月：Oldcar Lee 李啟康

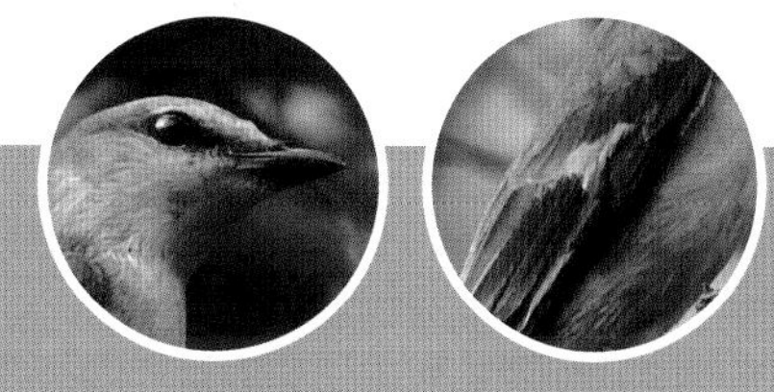

春季過境遷徙鳥 Spring Passage Migrant	夏候鳥 Summer Visitor	秋季過境遷徙鳥 Autumn Passage Migrant	冬候鳥 Winter Visitor

1 2 3 4 5 6 7 8 9 10 11 12 常見月份

留鳥 Resident	迷鳥 Vagrant	偶見鳥 Occasional Visitor

小灰山椒鳥

[普] xiǎo huī shān jiāo niǎo

體長 length：18-19cm

Swinhoe's Minivet | *Pericrocotus cantonensis*

1

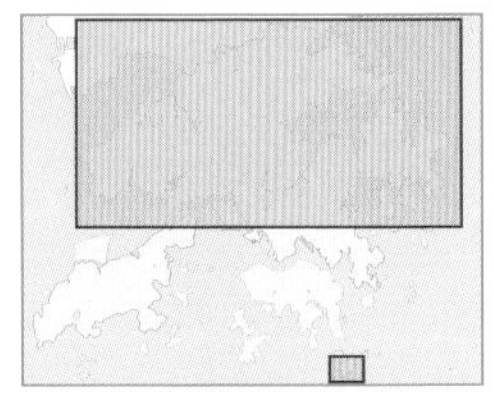

嘴短而黑色，腳黑色。上體深灰色，腰部淡黃褐色。喉部白色，胸部淡灰，腋下帶淡黃褐色。雄鳥前額至冠的中間白色，與黑色過眼線和耳羽成強烈對比。雌鳥前額和冠灰色，有淡白色、短和不明顯的眼眉。

Short black bill, black legs. Upperparts are dark grey with pale buff rump. White throat, pale grey wash over the breast and some buff tinged on lower flanks. Male has white forehead to mid-crown, which contrasts with dark eye-stripe and ear coverts. Female has greyish forehead and crown, with short and indistinct whitish eyebrows.

1 male 雄鳥；Po Toi 蒲台；Apr-07, 07 年 4 月；Allen Chan 陳志雄
2 male 雄鳥；Po Toi 蒲台；Apr-07, 07 年 4 月；Allen Chan 陳志雄
3 female 雌鳥；Po Toi 蒲台；Oct-08, 08 年 10 月；Andy Kwok 郭匯昌
4 male 雄鳥；Po Toi 蒲台；6-Apr-07, 07 年 4 月 6 日；Michelle and Peter Wong 江敏兒、黃理沛
5 female 雌鳥；Po Toi 蒲台；Mar-06, 06 年 3 月；Cherry Wong 黃卓研
6 male 雄鳥；Mai Po 米埔；9-Apr-04, 04 年 4 月 9 日；Michelle and Peter Wong 江敏兒、黃理沛
7 female 雌鳥；Po Toi 蒲台；Apr-06, 06 年 4 月；Matthew and TH Kwan 關朗曦、關子凱

春季過境遷徙鳥 Spring Passage Migrant	夏候鳥 Summer Visitor	秋季過境遷徙鳥 Autumn Passage Migrant	冬候鳥 Winter Visitor

常見月份 1 2 3 4 5 6 7 8 9 10 11 12

留鳥 Resident	迷鳥 Vagrant	偶見鳥 Occasional Visitor

2
3
4
5
6

7

暗灰鵑鵙

[普]àn huī juān juē
㊟鵙：音缺

體長length：19.5-24cm

Black-winged Cuckooshrike | *Lalage melaschistos*

全身灰色，上體較深，下體較淡。兩翼及尾部黑色，尾下羽毛末端有白斑。飛行時翼底有白斑。雌鳥有不完整白眼圈，下體有深色橫紋。

Grey overall, upperparts darker than underparts. Black wings and tail, white tips on underside of tail feathers. In flight, it shows a white patch on the underwing. Female has broken white eye-rings and variable dark bars on underparts.

1 adult male 雄成鳥；Shek Kong 石崗；Feb-09, 09 年 2 月；Sammy Sam and Winnie Wong 森美與雲妮
2 adult male 雄成鳥；Kowloon Park 九龍公園；Mar-04, 04 年 3 月；Cherry Wong 黃卓研
3 adult female 雌成鳥；Po Toi 蒲台；Oct-08, 08 年 10 月；Michelle and Peter Wong 江敏兒、黃理沛
4 adult female 雌成鳥；Po Toi 蒲台；Oct-08, 08 年 10 月；Michelle and Peter Wong 江敏兒、黃理沛
5 adult female 雌成鳥；Po Toi 蒲台；Oct-08, 08 年 10 月；Cherry Wong 黃卓研
6 adult male 雄成鳥；Shek Kong 石崗；Feb-09, 09 年 2 月；Sammy Sam and Winnie Wong 森美與雲妮
7 juvenile 幼鳥；Po Toi 蒲台；Sep-07, 07 年 9 月；Allen Chan 陳志雄
8 adult male 雄成鳥；Po Toi 蒲台；Sep-07, 07 年 9 月；Sammy Sam and Winnie Wong 森美與雲妮

春季過境遷徙鳥 Spring Passage Migrant	夏候鳥 Summer Visitor	秋季過境遷徙鳥 Autumn Passage Migrant	冬候鳥 Winter Visitor

常見月份：1 2 3 4 5 6 7 8 9 10 11 12

留鳥 Resident	迷鳥 Vagrant	偶見鳥 Occasional Visitor

虎紋伯勞

㊟hǔ wén bó láo

體長length：17-18.5cm

Tiger Shrike | *Lanius tigrinus*

其他名稱Other names：Thick-billed Shrike

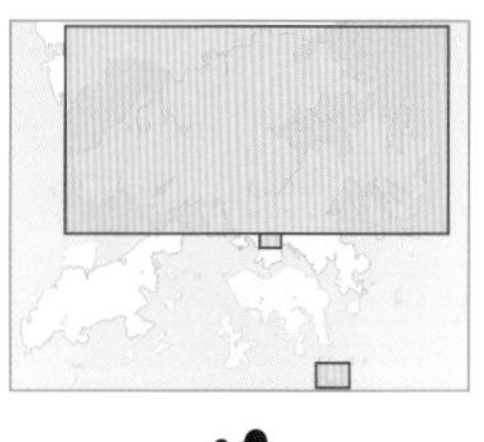

成鳥頭頂灰色，嘴黑色，有黑色貫眼紋。上體紅褐色，翼和上體背部滿佈黑色幼紋，下體白色，雌鳥脇部有黑色幼紋。幼鳥頭頂紅褐色，滿佈黑色幼紋。

Adult with grey crown, black bill and black eye-stripes. Reddish brown upperparts and wings scattered with narrow black stripes. White underparts. Female with narrow black stripes on flanks. Juvenile with reddish brown crown scattered with narrow black stripes.

1 adult female 雌成鳥：Ho Man Tin 何文田：Sep-17, 17 年 9 月：Matthew Kwan 關朗曦
2 adult female 雌成鳥：Ho Man Tin 何文田：Sep-18, 18 年 9 月：Roman Lo 羅文凱
3 adult female 雌成鳥：Ho Man Tin 何文田：Sep-18, 18 年 9 月：Roman Lo 羅文凱
4 adult female 雌成鳥：Ho Man Tin 何文田：Sep-18, 18 年 9 月：Roman Lo 羅文凱
5 juvenile 幼鳥：Po Toi 蒲台：14-Sep-06, 06 年 9 月 14 日：Geoff Welch
6 juvenile 幼鳥：Po Toi 蒲台：Sep-06, 06 年 9 月：Michelle and Peter Wong 江敏兒、黃理沛
7 juvenile 幼鳥：Mai Po 米埔：Sep-06, 06 年 9 月：Cherry Wong 黃卓研

春季過境遷徙鳥 Spring Passage Migrant	夏候鳥 Summer Visitor	秋季過境遷徙鳥 Autumn Passage Migrant	冬候鳥 Winter Visitor

常見月份 1 2 3 4 **5** 6 7 **8** **9** **10** 11 12

留鳥 Resident	迷鳥 Vagrant	偶見鳥 Occasional Visitor

灰背伯勞

㊟hūi bèi bó láo

體長length：21-23cm

Grey-backed Shrike | *Lanius tephronotus*

額至背部灰色，有明顯黑眼罩，嘴和腳黑色。下體白色，翼邊黑色，尾上黑褐色，腰部褐色，脇部沾褐色。和常見的棕背伯勞相似，但背部全灰色，翼邊缺乏白點，尾部較短及褐色較重。

Grey from forehead to back. Prominent black eye-stripes extend to ear-coverts. Black bill and legs. White underparts. Black wing edges. Dark brown upper tail. Brown rump. Flanks tinged with brown. Distinguishes from common Long-tailed Shrike by full grey back and lack of white spot on the edge of wing. Shorter and darker brown tail.

1 adult 成鳥：Lam Tsuen 林村：Sep-15, 15 年 9 月：Allen Chan 陳志雄
2 adult 成鳥：Lam Tsuen 林村：Sep-15, 15 年 9 月：Allen Chan 陳志雄

春季過境遷徙鳥 Spring Passage Migrant			夏候鳥 Summer Visitor			秋季過境遷徙鳥 Autumn Passage Migrant			冬候鳥 Winter Visitor		
1	2	3	4	5	6	7	8	9	10	11	12

常見月份

留鳥 Resident	迷鳥 Vagrant	偶見鳥 Occasional Visitor

牛頭伯勞

[普]niú tóu bó láo

體長 length：19-20cm

Bull-headed Shrike | *Lanius bucephalus*

全身偏褐。頭大，嘴和腳黑色。雄鳥頭頂及後枕紅褐色，有黑色貫眼紋伸延至耳羽，眼眉淡而不明顯，上背及腰灰褐色，尾及翼黑色。喉至下體淡褐色，脇沾褐色。雌鳥及幼鳥上體偏褐，貫眼紋不明顯，幼鳥下體有鱗狀斑紋。

Overall plumage brown. Big head, black bill and legs. Male has rufous head and nape. Black eye-stripes extend to ear coverts. Pale supercilium. Greyish brown mantle and rump. Black tail and wings. Throat to underparts buffy-yellow, flanks brown. Female and juvenile have brownish upperparts, eye-stripes indistinct. Juvenile has scaly underparts.

1 adult female 雌成鳥；Tai Po Kau 大埔滘；Dec-06, 06 年 12 月；Michelle and Peter Wong 江敏兒、黃理沛
2 1st winter 第一年冬天；Sai Kung 西貢；Dec-08, 08 年 12 月；Ken Fung 馮漢城
3 1st winter 第一年冬天；Sai Kung 西貢；Jan-09, 09 年 1 月；Herman Ip 葉紀江
4 adult female 雌成鳥；Tai Po Kau 大埔滘；Dec-06, 06 年 12 月；Michelle and Peter Wong 江敏兒、黃理沛
5 adult male 雄成鳥；Tai Po Kau 大埔滘；Mar-07, 07 年 3 月；Martin Hale 夏敖天
6 adult female 雌成鳥；Tai Po Kau 大埔滘；Dec-06, 06 年 12 月；Lee Kai Hong 李啟康
7 1st winter 第一年冬天；Sai Kung 西貢；Dec-08, 08 年 12 月；Danny Ho 何國海
8 adult female 雌成鳥；Mai Po 米埔；Oct-06, 06 年 10 月；Michelle and Peter Wong 江敏兒、黃理沛

	春季過境遷徙鳥 Spring Passage Migrant			夏候鳥 Summer Visitor			秋季過境遷徙鳥 Autumn Passage Migrant			冬候鳥 Winter Visitor		
常見月份	1	2	3	4	5	6	7	8	9	10	11	12

留鳥 Resident | 迷鳥 Vagrant | 偶見鳥 Occasional Visitor

紅尾伯勞 ㊟hóng wěi bó láo

體長 length：17-20cm

Brown Shrike | *Lanius cristatus*

1

2

全身偏褐，黑色貫眼紋伸延至耳羽，眼眉白色，嘴短黑色，腰及尾紅褐色，腳黑色。常見的*lucionensis*亞種上體灰褐色，頭部淡灰色，下體淡褐色。*superciliosus*亞種上體褐色均勻，有白色粗眼紋。幼鳥的胸及脇有黑色鱗狀斑，上體也有褐色的鱗狀斑。*confusus*亞種，頭及上體灰褐色，白色眼紋較其他亞種小。

Mainly brown. Black eye-stripes extend to ear coverts. Prominent white supercilium, short black bill, rufous rump and tail, black legs. Common subspecies *lucionensis* has greish brown upperparts with contrasting pale greyish forecrown and creamy-buff underparts. Race *superciliosus* has uniform rufous upperparts and a broad white supercilium. Juvenile has dark scaly markings on sides of breast and flanks, with upperparts also scaly. Race *confusus* has greyish head and upperparts with a narrow supercilium.

1 adult male, race *lucionensis* 雄成鳥, *lucionensis* 亞種；Po Toi 蒲台；May-08, 08 年 5 月；Eling Lee 李佩玲
2 adult female, race *confusus* 雌成鳥, *confusus* 亞種；Tung Ping Chau 東平洲；21-Sep-08, 08 年 9 月 21 日；Sung Yik Hei 宋亦希
3 adult male, race *lucionensis* 雄成鳥, *lucionensis* 亞種；Po Toi 蒲台；May-08, 08 年 5 月；Chan Kai Wai 陳佳瑋
4 1st summer male, race *lucionensis* 第一年夏天雄鳥, *lucionensis* 亞種；Po Toi 蒲台；Apr-07, 07 年 4 月；Cherry Wong 黃卓研
5 adult male, race *lucionensis* 雄成鳥, *lucionensis* 亞種；Pui O 貝澳；Apr-04, 04 年 4 月；Henry Lui 呂德恒
6 juvenile 幼鳥；Lai Chi Kok 荔枝角；Jan-09, 09 年 1 月；Michelle and Peter Wong 江敏兒、黃理沛
7 juvenile 幼鳥；Long Valley 塱原；Nov-07, 07 年 11 月；Andy Kwok 郭匯昌
8 juvenile 幼鳥；Long Valley 塱原；Nov-07, 07 年 11 月；Danny Ho 何國海

春季過境遷徙鳥 Spring Passage Migrant	夏候鳥 Summer Visitor	秋季過境遷徙鳥 Autumn Passage Migrant	冬候鳥 Winter Visitor

常見月份	1	2	3	4	5	6	7	8	9	10	11	12

留鳥 Resident	迷鳥 Vagrant	偶見鳥 Occasional Visitor

3
4
5
6
7

8

紅背伯勞

㊟hóng bèi bó láo

體長 length：17cm

Red-backed Shrike | *Lanius collurio*

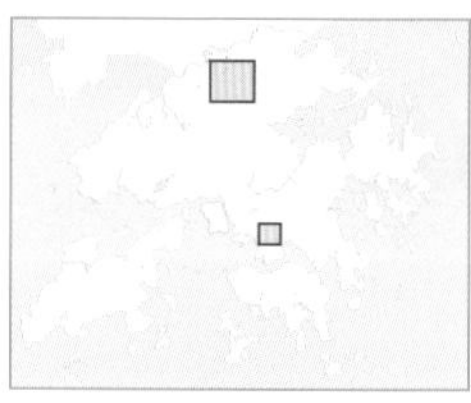

頭頂灰色，有粗黑貫眼紋，上體褐色，飛羽和尾羽黑色。喉部至下體白色，脇部沾淡褐色，雌鳥脇部有深色紋。

Grey crown and black thick eye-stripes. Brown upperparts with black flying feathers and tail. Throat to underparts white. Flanks tinted pale brown. Flanks of female darkly striped.

1 juvenile 幼鳥：Ho Man Tin 何文田：Owen Chiang 深藍
2 juvenile 幼鳥：Ho Man Tin 何文田：Oct-08, 08 年 10 月：Cheng Nok Ming 鄭諾銘
3 juvenile 幼鳥：Ho Man Tin 何文田：Oct-08, 08 年 10 月：Andy Kwok 郭匯昌
4 juvenile 幼鳥：Ho Man Tin 何文田：Oct-08, 08 年 10 月：Michelle and Peter Wong 江敏兒、黃理沛
5 juvenile 幼鳥：Ho Man Tin 何文田：Oct-08, 08 年 10 月：Ann To 陶偉意
6 juvenile 幼鳥：Ho Man Tin 何文田：Oct-08, 08 年 10 月：James Lam 林文華
7 juvenile 幼鳥：Ho Man Tin 何文田：Oct-08, 08 年 10 月：Michelle and Peter Wong 江敏兒、黃理沛

	春季過境遷徙鳥 Spring Passage Migrant			夏候鳥 Summer Visitor			秋季過境遷徙鳥 Autumn Passage Migrant			冬候鳥 Winter Visitor		
常見月份	1	2	3	4	5	6	7	8	9	10	11	12
	留鳥 Resident				迷鳥 Vagrant				偶見鳥 Occasional Visitor			

2
3
4
5
6

7

棕背伯勞

[普]zōng bèi bó láo

體長 length：20-25cm

Long-tailed Shrike | *Lanius schach*

其他名稱 Other names：Rufous-backed Shrike

體型較大而尾長，羽色為褐、黑、灰、白四色配搭。明顯的黑色眼罩伸延至耳羽，嘴和腳黑色。頭部灰色，背部褐色，對比鮮明；翼和尾黑色。幼鳥顏色偏淡，隱約有斑紋。深色型的「黑伯勞」全身深灰，面、翼和尾部全黑。

Large-sized shrike with long tail. Plumage a combination of brown, black, grey and white. Prominent dark eye-stripes extend to ear coverts. Black bill and legs. Grey head contrasts with rufous upperparts. Wings and tail black. Juvenile is paler in colour with faint bars. A fairly common dark morph "Dusky Shrike" is completely dark grey with black face, wings and tail.

1 adult 成鳥；Kam Tin 錦田；Jan-09, 09 年 1 月；Aka Ho
2 adult intermediate morph 中間型成鳥；Fung Lok Wai 豐樂圍；Nov-04, 04 年 11 月；Jemi and John Holmes 孔思義、黃亞萍
3 adult 成鳥；Long Valley 塱原；Apr-08, 08 年 4 月；Isaac Chan 陳家強
4 adult 成鳥；Long Valley 塱原；Feb-07, 07 年 2 月；James Lam 林文華
5 adult 成鳥；Mai Po 米埔；Dec-06, 06 年 12 月；Cherry Wong 黃卓研
6 adult 成鳥；Fung Lok Wai 豐樂圍；Apr-06, 06 年 4 月；Jemi and John Holmes 孔思義、黃亞萍
7 juvenile 幼鳥；Francis Chu 崔汝棠
8 juvenile 幼鳥；Long Valley 塱原；May-07, 07 年 5 月；Eling Lee 李佩玲
9 adult 成鳥；Chinese University of HK 香港中文大學；Dec-08, 08 年 12 月；Christine and Samuel Ma 馬志榮、蔡美蓮

	春季過境遷徙鳥 Spring Passage Migrant			夏候鳥 Summer Visitor			秋季過境遷徙鳥 Autumn Passage Migrant			冬候鳥 Winter Visitor		
常見月份	1	2	3	4	5	6	7	8	9	10	11	12
	留鳥 Resident				迷鳥 Vagrant				偶見鳥 Occasional Visitor			

白腹鳳鶥

[普]bāi fú fèng méi
[粵]鶥：音眉

體長 length：11-13cm

White-bellied Erpornis | *Erpornis zantholeuca*

1

外形頗似鶯類，但可憑短冠羽和黃色尾下覆羽區別。頭頂、背部至尾部綠色，喉至腹部灰白色。常混在其他鳥群中出沒，叫聲為「tsee-tsee-tsee-tsee」。喜在森林中層活動。

Resembles a warbler, but distinguished by short crest and yellow vent. Crown, mantle to tail are green. Throat to belly greyish-white. Often in mixed flocks. Calls are "tsee-tsee-tsee-tsee". Prefers mid-storey in forest.

1 adult 成鳥：Tai Po Kau 大埔滘：Apr-04, 04 年 4 月：Michelle and Peter Wong 江敏兒、黃理沛
2 adult 成鳥：Tai Po Kau 大埔滘：Feb-07, 07 年 2 月：Henry Lui 呂德恒
3 adult 成鳥：Kowloon Hills Catchment 九龍水塘引水道：Feb-04, 04 年 2 月：Henry Lui 呂德恒
4 juvenile 幼鳥：Tai Po Kau 大埔滘：Sep-08, 08 年 9 月：Owen Chiang 深藍
5 adult 成鳥：Tai Po Kau 大埔滘：Apr-04, 04 年 4 月：Michelle and Peter Wong 江敏兒、黃理沛
6 adult 成鳥：Kowloon Hills Catchment 九龍水塘引水道：Jan-04, 04 年 1 月：Henry Lui 呂德恒

	春季過境遷徙鳥 Spring Passage Migrant			夏候鳥 Summer Visitor			秋季過境遷徙鳥 Autumn Passage Migrant			冬候鳥 Winter Visitor		
常見月份	1	2	3	4	5	6	7	8	9	10	11	12
	留鳥 Resident				迷鳥 Vagrant				偶見鳥 Occasional Visitor			

2
3
4
5

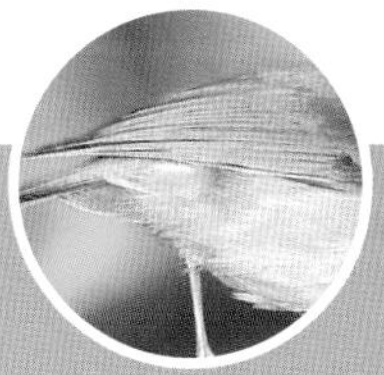

6

黑枕黃鸝

[普]hēi zhěn huáng lí
[粵]鸝：音梨

體長length：23-28cm

Black-naped Oriole | *Oriolus chinensis*

1

雄性成鳥全身金黃色，有粗黑貫眼紋，與後枕的粗黑紋相連，飛羽和尾羽黑色。未成年或雌鳥接近黃綠色，眼部粗紋不明顯，胸部有深色縱紋。

Adult male golden yellow. Eyestripe thick and black, connecting on the nape. Flight feathers and tail black. Immature or female is more greenish yellow, eyestripe indistinct, with dark streaks on the breast.

1 adult male 雄成鳥；Po Toi 蒲台；Oct-08, 08 年 10 月；Michelle and Peter Wong 江敏兒、黃理沛
2 immature 未成年鳥；Mai Po 米埔；30-Sep-04, 04 年 9 月 30 日；Michelle and Peter Wong 江敏兒、黃理沛
3 immature 未成年鳥；Mai Po 米埔；Oct-06, 06 年 10 月；Jemi and John Holmes 孔思義、黃亞萍
4 immature 未成年鳥；Mai Po 米埔；Sep-04, 04 年 9 月；Henry Lui 呂德恆
5 immature 未成年鳥；Po Toi 蒲台；Sep-07, 07 年 9 月；Sammy Sam and Winnie Wong 森美與雲妮
6 immature 未成年鳥；Mai Po 米埔；Sep-04, 04 年 9 月；Henry Lui 呂德恆
7 immature 未成年鳥；Mai Po 米埔；Oct-06, 06 年 10 月；Jemi and John Holmes 孔思義、黃亞萍

春季過境遷徙鳥 Spring Passage Migrant	夏候鳥 Summer Visitor	秋季過境遷徙鳥 Autumn Passage Migrant	冬候鳥 Winter Visitor

常見月份 1 2 3 4 5 6 7 8 9 10 11 12

留鳥 Resident	迷鳥 Vagrant	偶見鳥 Occasional Visitor

2
3
4
5
6

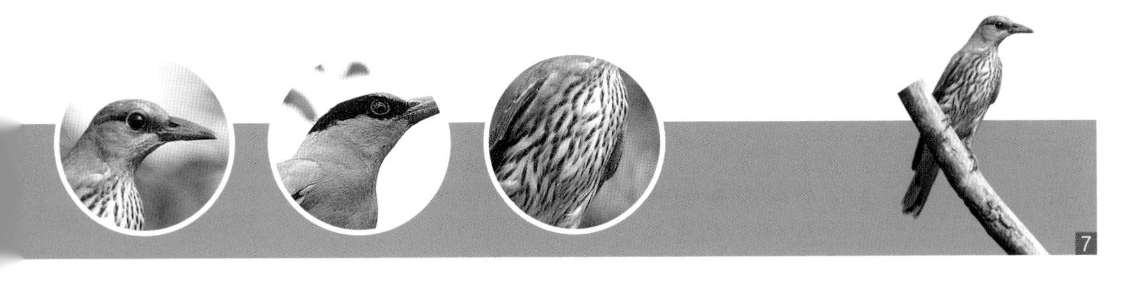

7

黑卷尾

㊟hēi juǎn wěi

體長 length：30-31cm

Black Drongo | *Dicrurus macrocercus*

1

2

全身黑色，上體羽毛有光澤。虹膜紅色，尾開叉呈V形。喜開闊地方，常立於顯眼處如電橙柱、電線、或樹枝上。有很強的領域保護行為，經常驅趕其他鳥類如麻鷹和喜鵲。

Overall plumage black with glossy upperparts. Red iris, tail is distinctively "V" shaped. Favours open areas, perches prominently on electric wires, poles and branches. Strong territorial behavior, often chases off larger birds including Black Kite and Common Magpie.

1 adult 成鳥；Kam Tin 錦田；Apr-05, 05 年 4 月；Martin Hale 夏敖天
2 1st winter 第一年冬天；Lamma Island 南丫島；Sep-07, 07 年 9 月；Harry Li 李炳偉
3 adult 成鳥；Po Toi 蒲台；Sep-08, 08 年 9 月；Cherry Wong 黃卓研
4 1st winter 第一年冬天；Fung Lok Wai 豐樂圍；Oct-03, 03 年 10 月；Jemi and John Holmes 孔思義、黃亞萍
5 adult 成鳥；Nam Sang Wai 南生圍；Jan-07, 07 年 1 月；Hoon Kwok Wai 洪國偉
6 1st winter 第一年冬天；Po Toi 蒲台；Oct-06, 06 年 10 月；Allen Chan 陳志雄
7 adult 成鳥；Long Valley 塱原；Oct-04, 04 年 10 月；Henry Lui 呂德恒

	春季過境遷徙鳥 Spring Passage Migrant			夏候鳥 Summer Visitor			秋季過境遷徙鳥 Autumn Passage Migrant			冬候鳥 Winter Visitor		
常見月份	1	2	3	4	5	6	7	8	9	10	11	12
	留鳥 Resident				迷鳥 Vagrant				偶見鳥 Occasional Visitor			

3
4
5
6

7

灰卷尾 ㊟huī juǎn wěi

體長 length：23-26cm

Ashy Drongo | *Dicrurus leucophaeus*

1

外形似黑卷尾，但全身灰色。上體羽色深灰色，下體較淡。本地有 *salangensis* 和 *leucogenis* 兩個亞種，後者眼部周圍和面頰白色。

Resembles Black Drongo but grey in colour, darker above and paler below. Two sub-species, *salangensis* and *leucogenis*, occur locally. The latter has a white patch around the eye and cheeks.

1 race *salangensis salangensis* 亞種：Po Toi 蒲台：Oct-06, 06 年 10 月：Cherry Wong 黃卓研
2 race *salangensis salangensis* 亞種：Tai Po Kau 大埔滘：Dec-04, 04 年 12 月：Jemi and John Holmes 孔思義、黃亞萍
3 race *hopwoodi hopwoodi* 亞種：Shing Mun 城門水塘：Jan-08, 08 年 1 月：Ken Fung 馮漢城
4 race *leucogensis leucogensis* 亞種：Kowloon Park 九龍公園：Sep-06, 06 年 9 月：Law Kam Man 羅錦文
5 race *hopwoodi hopwoodi* 亞種：Kowloon Hills Catchment 九龍水塘引水道：Mar-05, 05 年 3 月：Henry Lui 呂德恆
6 race *leucogensis leucogensis* 亞種：Po Toi 蒲台：13-Oct-06, 06 年 10 月 13 日：Lo Kar Man 盧嘉孟

春季過境遷徙鳥 Spring Passage Migrant	夏候鳥 Summer Visitor	秋季過境遷徙鳥 Autumn Passage Migrant	冬候鳥 Winter Visitor

常見月份 1 2 3 4 5 6 7 8 9 10 11 12

留鳥 Resident	迷鳥 Vagrant	偶見鳥 Occasional Visitor

2
3
4
5

6

鴉嘴卷尾

[普] yā zuǐ juǎn wěi

體長 length：27-29cm

Crow-billed Drongo | *Dicrurus annectans*

全身羽毛黑色帶金屬藍綠色，嘴和腳黑色。和常見的黑卷尾相似，但身軀較壯，尾較粗短，尾端分叉較淺並稍向上卷曲，嘴部較長及嘴基較粗。

Black body with metallic greenish-blue sheen. Black bill and legs. Distinguishes from common Black Drongo by chunkier body and shorter, thicker tail with shallower fork curling upwards. Relatively longer bill with thicker base.

1 adult 成鳥：Po Toi 蒲台：Sep-14, 14 年 9 月：Allen Chan 陳志雄
2 adult 成鳥：Po Toi 蒲台：Sep-14, 14 年 9 月：Allen Chan 陳志雄
3 adult 成鳥：Po Toi 蒲台：Sep-14, 14 年 9 月：Yam Wing Yiu 任永耀
4 adult 成鳥：Po Toi 蒲台：Sep-14, 14 年 9 月：Kinni Ho 何建業
5 adult 成鳥：Po Toi 蒲台：Sep-14, 14 年 9 月：Yam Wing Yiu 任永耀

	春季過境遷徙鳥 Spring Passage Migrant			夏候鳥 Summer Visitor			秋季過境遷徙鳥 Autumn Passage Migrant			冬候鳥 Winter Visitor		
常見月份	1	2	3	4	5	6	7	8	9	10	11	12
	留鳥 Resident			迷鳥 Vagrant				偶見鳥 Occasional Visitor				

2
3
4
5

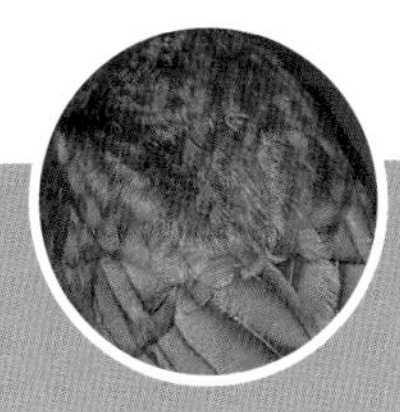

髮冠卷尾

㊟fà guān juǎn wěi

體長 length：28cm

Hair-crested Drongo | *Dicrurus hottentottus*

頭和嘴部大而粗厚，嘴後的細長羽毛伸延至頭後面。全身黑色，在陽光下可見到閃亮的銅輝。尾部闊大，沒有開叉，末端向兩旁翹起。有很強的領域保護行為，經常驅趕其他使用相同生境的鳥類。叫聲沙啞刺耳。

Head and bill look heavy. Elongated crest trails backward from behind the bill. Shiny black overall, with a copper-green sheen under sunlight. Broad tail feathers upturned at the ends but not forked. Strong territorial behavior, often chases off other larger birds. Rasping call.

1 adult 成鳥：Po Toi 蒲台：Oct-08, 08 年 10 月：Michelle and Peter Wong 江敏兒、黃理沛
2 adult 成鳥：Po Toi 蒲台：Oct-08, 08 年 10 月：Michelle and Peter Wong 江敏兒、黃理沛
3 adult 成鳥：Po Toi 蒲台：Apr-07, 07 年 4 月：Sammy Sam and Winnie Wong 森美與雲妮
4 adult 成鳥：Po Toi 蒲台：May-07, 07 年 5 月：Law Kam Man 羅錦文
5 adult 成鳥：Po Toi 蒲台：13-Nov-08, 08 年 11 月 13 日：Geoff Welch
6 adult 成鳥：Tsuen Wan 荃灣：Nov-06, 06 年 11 月：Christina Chan 陳燕明

	春季過境遷徙鳥 Spring Passage Migrant			夏候鳥 Summer Visitor			秋季過境遷徙鳥 Autumn Passage Migrant			冬候鳥 Winter Visitor		
常見月份	1	2	3	4	5	6	7	8	9	10	11	12
	留鳥 Resident				迷鳥 Vagrant				偶見鳥 Occasional Visitor			

2
3
4
5

6

黑枕王鶲

㊟ hēi zhěn wáng wēng
㊟ 鶲：音翁

體長 length：15-17cm

Black-naped Monarch | *Hypothymis azurea*

1

頭部和嘴藍色，眼深色，尾下覆羽白色，站立時身體靠近水平方式橫置。雄鳥全身藍色，枕部有黑色斑塊，前頸有一道幼黑色領紋。雌鳥上體至尾部灰褐色，腹部偏白，帶有藍和褐色。喜愛在樹林下層和林地邊緣活動。

Head and bill bright blue, dark eyes, white undertail coverts. Horizontal posture. Male is overall blue in plumage, with a black patch at nape and a narrow black collar in front neck. Female has greyish brown upperparts to tail, white belly tint with blue and brown. Favours lower forest or forest edge.

1 adult male 雄成鳥：Shek Kong 石崗：Jan-09, 09 年 1 月：Michelle and Peter Wong 江敏兒、黃理沛
2 adult male 雄成鳥：Shek Kong 石崗：Jan-09, 09 年 1 月：Danny Ho 何國海
3 adult male 雄成鳥：Shek Kong 石崗：Jan-09, 09 年 1 月：Michelle and Peter Wong 江敏兒、黃理沛
4 adult male 雄成鳥：Shek Kong 石崗：Jan-09, 09 年 1 月：James Lam 林文華
5 adult male 雄成鳥：Shek Kong 石崗：Jan-09, 09 年 1 月：Michelle and Peter Wong 江敏兒、黃理沛
6 adult male 雄成鳥：Shek Kong 石崗：Jan-09, 09 年 1 月：Andy Kwok 郭匯昌
7 adult male 雄成鳥：Shek Kong 石崗：Jan-09, 09 年 1 月：Andy Kwok 郭匯昌

春季過境遷徙鳥 Spring Passage Migrant	夏候鳥 Summer Visitor	秋季過境遷徙鳥 Autumn Passage Migrant	冬候鳥 Winter Visitor

常見月份：1 2 3 4 5 6 7 8 9 10 11 12

留鳥 Resident	迷鳥 Vagrant	偶見鳥 Occasional Visitor

2
3
4
5
6

7

綬帶

㊟ shòu dài
㊟ 綬：音受

體長 length：20cm

Amur Paradise Flycatcher | *Terpsiphone incei*

其他名稱 Other names：壽帶鳥

頭及喉部灰藍色，眼圈藍色，上體、上翼和尾部深栗色。胸部深灰色，腹部白色。香港可找到較為罕有的白色型，頭部深色有光澤，全身和尾羽白色。雌雄相似，不過冠羽較短和沒有長尾羽。活躍於樹冠間。

Head and throat greyish blue, and distinctive blue eye-ring. It has dark chestnut mantle, upperwings and tail, dark grey breast, and white belly. White morph is rare in Hong Kong, which has dark glossy head with white body and tail. Female has shorter crest and lacks long tail feathers. Active in canopy.

1 1st winter 第一年冬天：Mai Po 米埔：Oct-06, 06 年 10 月：Cherry Wong 黃卓研
2 1st winter 第一年冬天：Shek Kong 石崗：Feb-07, 07 年 2 月：Michelle and Peter Wong 江敏兒、黃理沛
3 1st winter 第一年冬天：Shek Kong 石崗：Feb-07, 07 年 2 月：Pippen Ho 何志剛
4 1st winter 第一年冬天：Shek Kong 石崗：Feb-07, 07 年 2 月：Michelle and Peter Wong 江敏兒、黃理沛
5 1st winter 第一年冬天：Shek Kong 石崗：Feb-07, 07 年 2 月：Michelle and Peter Wong 江敏兒、黃理沛
6 1st winter 第一年冬天：Shek Kong 石崗：Sep-06, 06 年 9 月：Michelle and Peter Wong 江敏兒、黃理沛

春季過境遷徙鳥 Spring Passage Migrant	夏候鳥 Summer Visitor	秋季過境遷徙鳥 Autumn Passage Migrant	冬候鳥 Winter Visitor

常見月份 1 2 3 4 5 6 7 8 9 10 11 12

留鳥 Resident	迷鳥 Vagrant	偶見鳥 Occasional Visitor

2
3
4
5
6

紫綬帶

㊟ zǐ shòu dài
㊞ 綬：音受

體長 length：20cm

Japanese Paradise Flycatcher | *Terpsiphone atrocaudata*

其他名稱 Other names：紫壽帶鳥

體型較綬帶小一點，但看來較深色。頭及喉部黑藍色，有明顯藍眼圈，胸部偏黑，上體深紫紅色。雄鳥繁殖時有特長的尾羽。活躍於樹冠間。

Marginally smaller than Amur Paradise Flycatcher, and overall plumage is darker. Head and throat blackish-blue. Prominent blue eye-rings. Blackish breast, deep maroon-purple upperparts, white belly. In breeding plumage, male bird has elongated central tail feathers. Active in canopy.

1 adult male 雄成鳥；Po Toi 蒲台；Apr-07, 07 年 4 月；Matthew and TH Kwan 關朗曦、關子凱
2 1st winter 第一年冬天；Mai Po 米埔；Apr-06, 06 年 4 月；Henry Lui 呂德恒
3 1st winter 第一年冬天；Mai Po 米埔；Apr-06, 06 年 4 月；Henry Lui 呂德恒
4 1st winter 第一年冬天；16-Oct-05, 05 年 10 月 16 日；Michelle and Peter Wong 江敏兒、黃理沛
5 adult male 雄成鳥；Po Toi 蒲台；Apr-07, 07 年 4 月；Matthew and TH Kwan 關朗曦、關子凱
6 adult female 雌成鳥；Po Toi 蒲台；Oct-06, 06 年 10 月；Allen Chan 陳志雄
7 adult male 雄成鳥；Po Toi 蒲台；Apr-07, 07 年 4 月；Law Kam Man 羅錦文

春季過境遷徙鳥 Spring Passage Migrant	夏候鳥 Summer Visitor	秋季過境遷徙鳥 Autumn Passage Migrant	冬候鳥 Winter Visitor

常見月份 1 2 3 4 5 6 7 8 9 10 11 12

留鳥 Resident	迷鳥 Vagrant	偶見鳥 Occasional Visitor

2
3
4
5
6
7

松鴉 ㊟sōng yā

體長length：32-36cm

Eurasian Jay | *Garrulus glandarius*

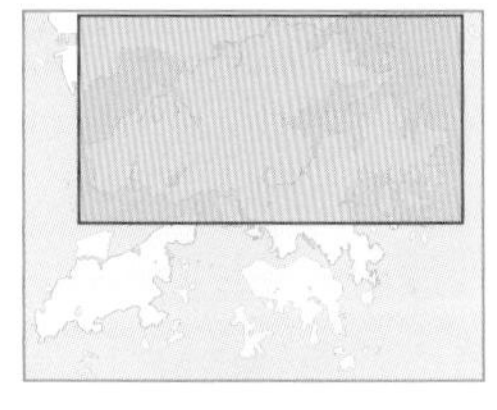

容易辨認，全身淡褐色。虹膜白色，有黑色眼圈。嘴黑色，有粗黑頰紋。飛羽和尾羽黑色。翼上有明顯藍、白、黑三色色斑，腰白色。近年十分稀少。

Unmistakeable. Overall pinkish brown plumage. White iris, black eye-rings. Black bill with thick black submoustachial stripes. Flight feathers and tail black, with distinctive blue, black and white patches on wings. White rump. Very scarce in recent years.

1 race *sinensis sinensis* 亞種：Lok Ma Chau 落馬洲：Jemi and John Holmes 孔思義、黃亞萍

春季過境遷徙鳥 Spring Passage Migrant			夏候鳥 Summer Visitor			秋季過境遷徙鳥 Autumn Passage Migrant			冬候鳥 Winter Visitor		
1	2	3	4	5	6	7	8	9	10	11	12

常見月份

留鳥 Resident	迷鳥 Vagrant	偶見鳥 Occasional Visitor

灰喜鵲

㊕huī xǐ què
㊚鵲：音爵

體長length：36-38cm

Azure-winged Magpie | *Cyanopica cyanus*

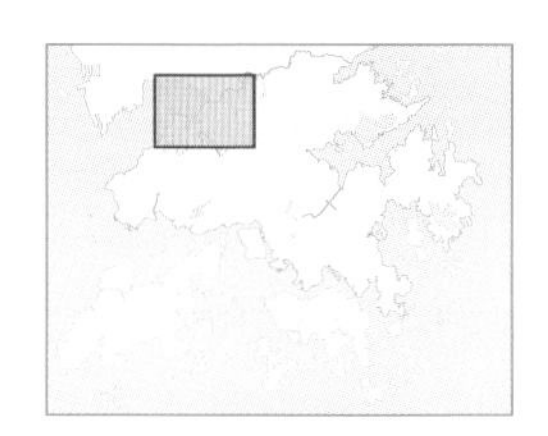

頭、嘴和腳黑色，上體淡褐灰色，飛羽和尾部天藍色，尾羽末端有白點。喉至下體淡白色，沾有褐灰色斑。最近米埔有一引入種群。

Head, bill and legs black. Upperparts pale brownish grey. Flight feathers and tail skyblue, with white at tip of tail. Throat to underparts buffy white, with pale brownish grey. Recently introduced resident at Mai Po.

1 adult 成鳥：Mai Po 米埔：Jul-07, 07 年 7 月：Cherry Wong 黃卓研
2 adult 成鳥：Mai Po 米埔：Oct-06, 06 年 10 月：Bill Man 文權溢
3 adult 成鳥：Mai Po 米埔：Oct-08, 08 年 10 月：Cherry Wong 黃卓研
4 adult 成鳥：Mai Po 米埔：Apr-06, 06 年 4 月：Henry Lui 呂德恒
5 adult 成鳥：Mai Po 米埔：Jul-07, 07 年 7 月：Cherry Wong 黃卓研

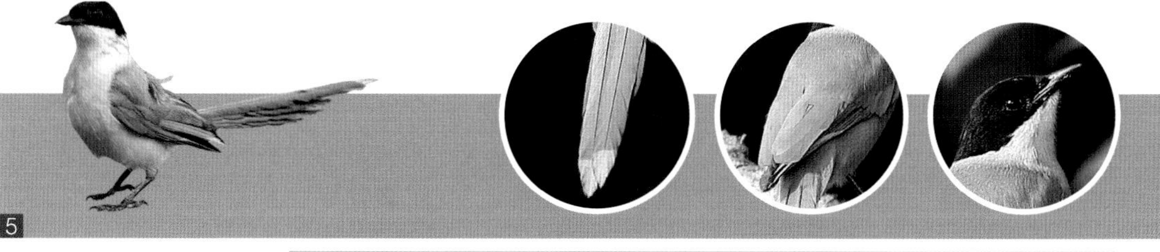

春季過境遷徙鳥 Spring Passage Migrant			夏候鳥 Summer Visitor			秋季過境遷徙鳥 Autumn Passage Migrant			冬候鳥 Winter Visitor		
1	2	3	4	5	6	7	8	9	10	11	12

常見月份

留鳥 Resident	迷鳥 Vagrant	偶見鳥 Occasional Visitor

紅嘴藍鵲

普 hóng zuǐ lán què
粵 鵲：音爵

體長 length：53-64cm

Red-billed Blue Magpie | *Urocissa erythrorhyncha*

引人注目的大型長尾鳥類。頭至胸部黑色，頭頂至後枕白色，嘴、虹膜和腳紅色。上體藍色，下體白色。有很長而帶藍色的尾羽，尾下羽毛黑色，末端有大白斑。常發出高音響亮的叫聲。通常一、兩隻以至一小群活動。

Conspicuous large and long tail. Black from head to breast, Crown to nape white. Bill, iris and legs are red. Blue above and white below. Long blue tail tipped white. Calls loudly with high pitch. Usually singles or in small groups.

1 adult 成鳥：Tsing Yi 青衣：Jan-09, 09 年 1 月：Andy Kwok 郭匯昌
2 adult 成鳥：Kowloon Park 九龍公園：Jan-06, 06 年 1 月：Cherry Wong 黃卓研
3 adult 成鳥：Discovery Bay 愉景灣：Jul-07, 07 年 7 月：Sonia and Kenneth Fung 馮啟文、蕭敏晶
4 juvenile 幼鳥：Chinese University of HK 香港中文大學：May-07, 07 年 5 月：Eling Lee 李佩玲
5 adult 成鳥：Cheung Chau 長洲：Oct-07, 07 年 10 月：Matthew and TH Kwan 關朗曦、關子凱
6 adult 成鳥：Pok Fu Lam 薄扶林：Feb-08, 08 年 2 月：Herman Ip 葉紀江
7 adult 成鳥：Starfish Bay 海星灣：Jun-04, 04 年 6 月：Henry Lui 呂德恒

春季過境遷徙鳥 Spring Passage Migrant			夏候鳥 Summer Visitor			秋季過境遷徙鳥 Autumn Passage Migrant			冬候鳥 Winter Visitor		
常見月份 1	2	3	4	5	6	7	8	9	10	11	12
留鳥 Resident				迷鳥 Vagrant				偶見鳥 Occasional Visitor			

2
3
4
5
6

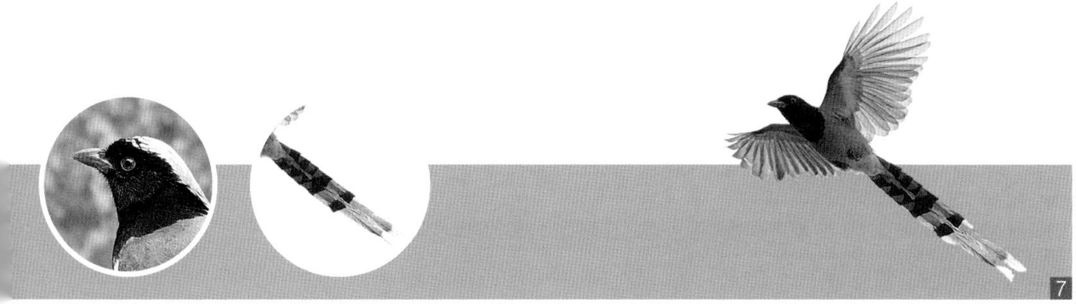
7

灰樹鵲

㊟huī shù què
㊔鵲：音爵

Grey Treepie | *Dendrocitta formosae*

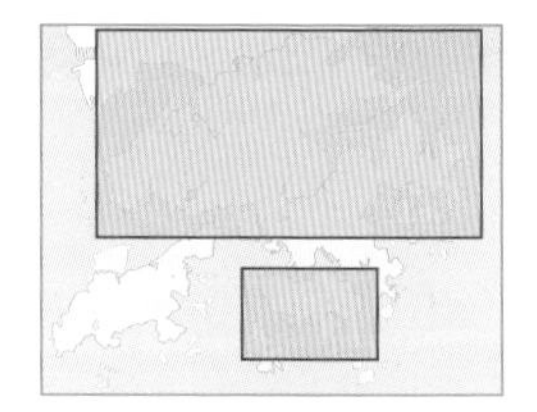

尾長而嘴粗，整體深色而尾部黑色，背部栗色。叫聲嘈吵而帶機器的聲音。通常小群活動。

Long tail and thick bill. Dull-colored body with long black tail and chestnut mantle. Noisy, with a variety of mechanical-sounding calls. Usually in small groups.

1 adult 成鳥；Siu Lek Yuen 小瀝源；May-08, 08 年 5 月；Ken Fung 馮漢城
2 adult moulting from juvenile plumage 幼鳥轉換繁殖羽；Ng Tung Chai 梧桐寨；Mar-07, 07 年 3 月；Henry Lui 呂德恒
3 adult 成鳥；Siu Lek Yuen 小瀝源；May-08, 08 年 5 月；Ken Fung 馮漢城

	春季過境遷徙鳥 Spring Passage Migrant			夏候鳥 Summer Visitor			秋季過境遷徙鳥 Autumn Passage Migrant			冬候鳥 Winter Visitor		
常見月份	1	2	3	4	5	6	7	8	9	10	11	12
	留鳥 Resident				迷鳥 Vagrant				偶見鳥 Occasional Visitor			

達烏里寒鴉

㊟ dá wū lǐ hán yā

體長 length：34-36cm

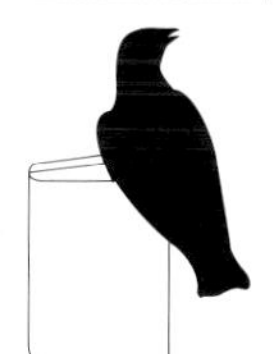

Daurian Jackdaw | *Coloeus dauuricus*

小型烏鴉。全身黑色，嘴尖而短。頭兩側有白色的細紋是最明顯的特徵。頸部至下體毛色淡褐。幼鳥全身偏黑。

Small black corvid. Short and pointed bill. Characterised by prominent whitish streaks on both sides of the head. Collar to underparts pale brown or white. Juvenile is darker overall.

1 juvenile 幼鳥；3-Dec-06, 06 年 12 月 3 日；Michelle and Peter Wong 江敏兒．黃理沛
2 immature 未成年鳥；Penfold Park 彭福公園；31 May-14, 14年5月31日；Beetle Cheng 鄭諾銘

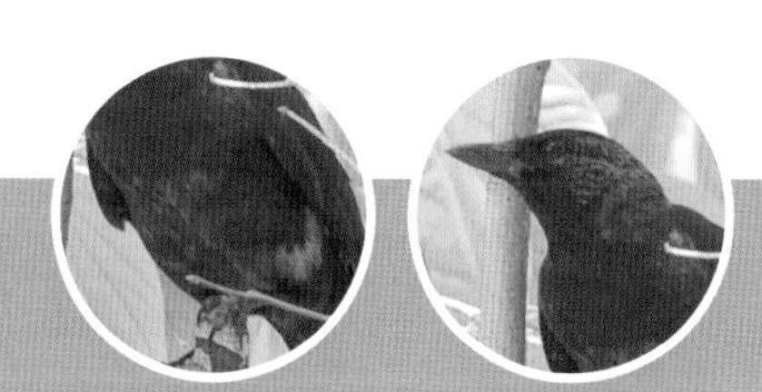

春季過境遷徙鳥 Spring Passage Migrant	夏候鳥 Summer Visitor	秋季過境遷徙鳥 Autumn Passage Migrant	冬候鳥 Winter Visitor

1 2 3 4 5 6 7 8 **9** 10 **11** **12** 常見月份

留鳥 Resident	迷鳥 Vagrant	偶見鳥 Occasional Visitor

喜鵲

㊟ xǐ què
㊟ 鵲：音爵

體長 length：46-50cm

Oriental Magpie | *Pica serica*

其他名稱 Other names：Magpie, Common Magpie

大型鴉科鳥類。全身黑色，在陽光下上體有藍色金屬光澤。肩及腹部白色。胸部、嘴和腳黑色。飛行時可見白色飛羽。叫聲為響亮的「格－格－」聲，喜在開闊地方的樹頂或電纜塔頂營巢。

Large-sized corvid. Black and white overall. Shiny metallic blue on black upperparts under sunlight. White on shoulders and belly. Black on breast, bill and legs. White flight feathers can be seen in flight. Call a loud and continuous "glar-glar-". Nests in open areas on top of structures such as trees and electric pylons.

1 adult 成鳥：Nam Sang Wai 南生圍：Jan-06, 06 年 1 月：Jemi and John Holmes 孔思義、黃亞萍
2 adult 成鳥：Jan-05, 05 年 1 月：Marcus Ho 何萬邦
3 adult 成鳥：Pui O 貝澳：28-Jan-09, 09 年 1 月 28 日：Owen Chiang 深藍
4 adult 成鳥：Pui O 貝澳：28-Jan-09, 09 年 1 月 28 日：Owen Chiang 深藍
5 adult 成鳥：Penfold Park 彭福公園：Dec-04, 04 年 12 月：Henry Lui 呂德恒
6 immature 未成年鳥：Long Valley 塱原：Oct-04, 04 年 10 月：Cherry Wong 黃卓研
7 adult 成鳥：Pui O 貝澳：Jul-08, 08 年 7 月：Chan Kai Wai 陳佳瑋

	春季過境遷徙鳥 Spring Passage Migrant			夏候鳥 Summer Visitor			秋季過境遷徙鳥 Autumn Passage Migrant			冬候鳥 Winter Visitor		
常見月份	1	2	3	4	5	6	7	8	9	10	11	12
	留鳥 Resident				迷鳥 Vagrant				偶見鳥 Occasional Visitor			

2
3
4
5
6

7

家鴉 ㊟ jiā yā

體長 length：40-43cm

House Crow | *Corvus splendens*

1

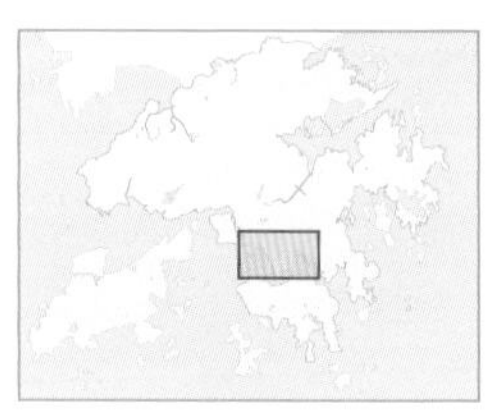

大型鴉類。全身黑色，嘴黑色粗厚，頸部灰褐色，與灰褐色胸帶連接。為外來引入鳥種，在市區內局部地區常見。

Large corvid. Black overall, bill black and thick, greyish-brown collar extends to form breast band. Introduced, locally common in some urban areas.

1 adult 成鳥：Tsuen Wan 荃灣：Jan-05, 05 年 1 月：Cherry Wong 黃卓研
2 adult 成鳥：Kowloon Tsai Park 九龍仔公園：Aug-03, 03 年 8 月：Henry Lui 呂德恒
3 juvenile 幼鳥：Kowloon Tsai Park 九龍仔公園：Aug-03, 03 年 8 月：Henry Lui 呂德恒
4 adult 成鳥：Kowloon Tsai Park 九龍仔公園：Aug-03, 03 年 8 月：Henry Lui 呂德恒
5 adult 成鳥：Tsuen Wan 荃灣：Jan-05, 05 年 1 月：Cherry Wong 黃卓研
6 adult 成鳥：Kowloon Tsai Park 九龍仔公園：Aug-03, 03 年 8 月：Henry Lui 呂德恒

	春季過境遷徙鳥 Spring Passage Migrant			夏候鳥 Summer Visitor			秋季過境遷徙鳥 Autumn Passage Migrant			冬候鳥 Winter Visitor		
常見月份	1	2	3	4	5	6	7	8	9	10	11	12
	留鳥 Resident				迷鳥 Vagrant				偶見鳥 Occasional Visitor			

2
3
4
5

6

禿鼻烏鴉 [普]tū bí wū yā

體長 length：44-47cm

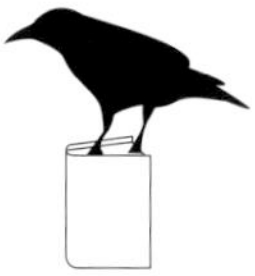

Rook | *Corvus frugilegus*

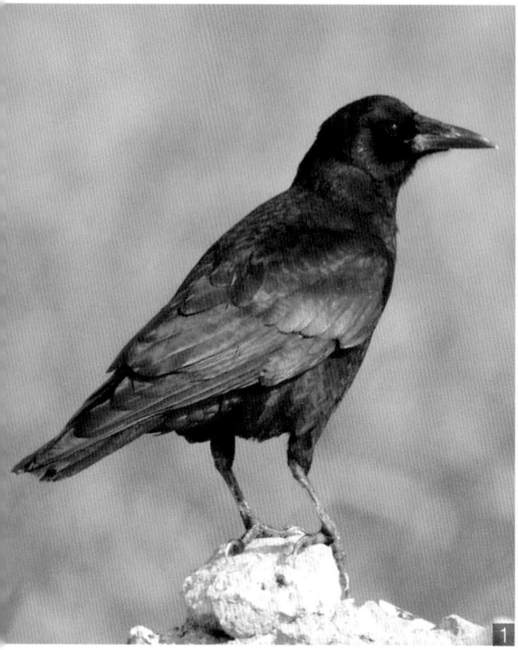
1

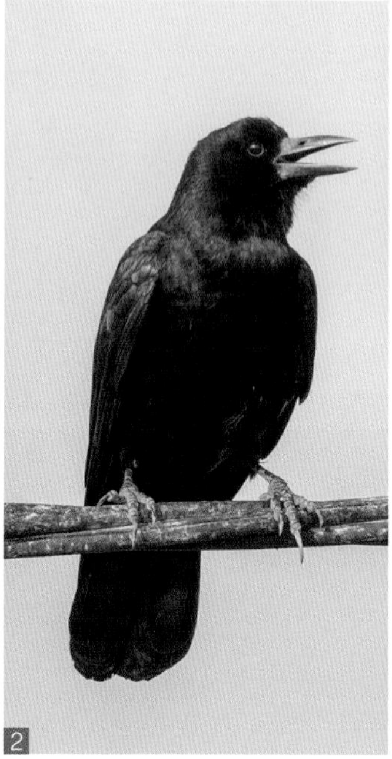
2

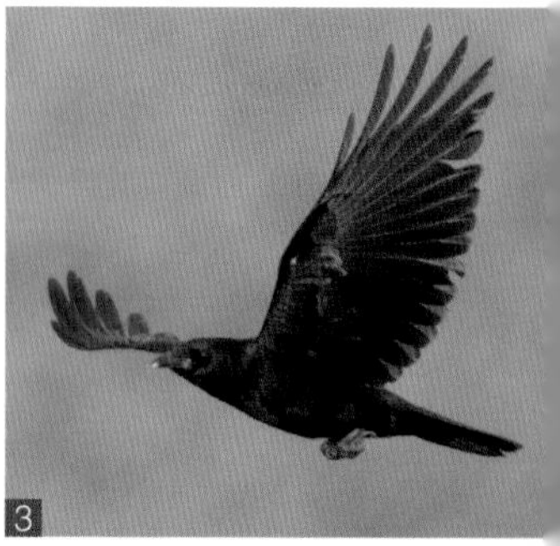
3

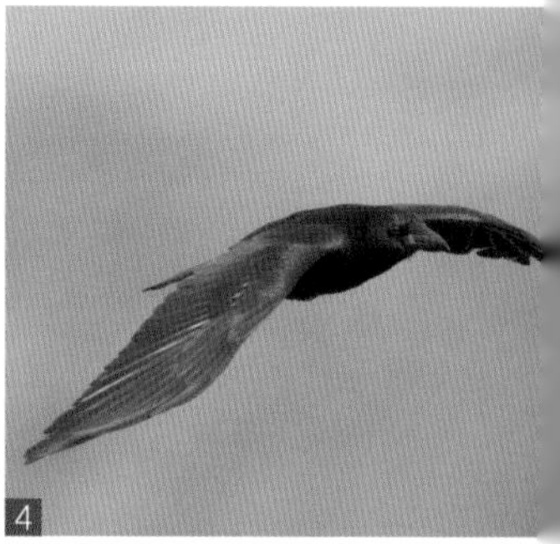
4

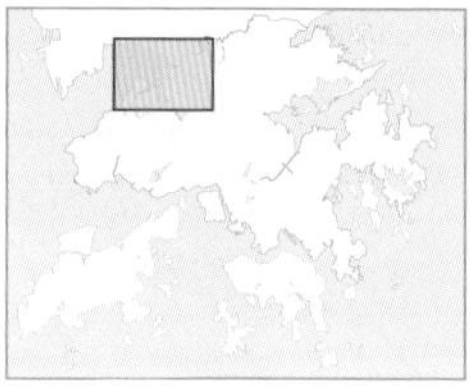

🌳

全身羽毛黑色，黑色的嘴直且尖，嘴基較淡色，腳黑色。成鳥嘴基旁有裸露灰白皮膚，嘴上無毛，前額較高。

Entirely black body. Straight and pointed black bill paler at base. Black legs. Adult has white bare skin at base of bill. Lack of feather at the base of upper bill. High forehead.

[1] adult 成鳥：Lut Chau 甩洲：Nov-17, 17 年 11 月：Kinni Ho 何建業
[2] adult 成鳥：Long Valley 塱原：Nov-18, 18 年 11 月：Matthew Kwan 關朗曦
[3] adult 成鳥：Lut Chau 甩洲：Nov-17, 17 年 11 月：Sam Chan 陳巨輝
[4] adult 成鳥：Lut Chau 甩洲：Nov-17, 17 年 11 月：Sam Chan 陳巨輝
[5] adult 成鳥：Lut Chau 甩洲：Nov-17, 17 年 11 月：Sam Chan 陳巨輝

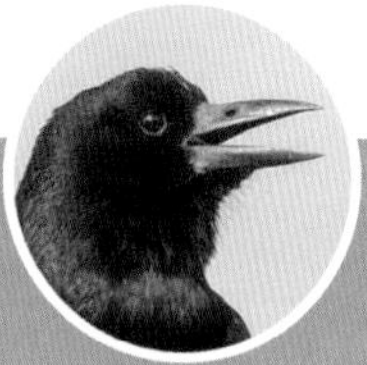

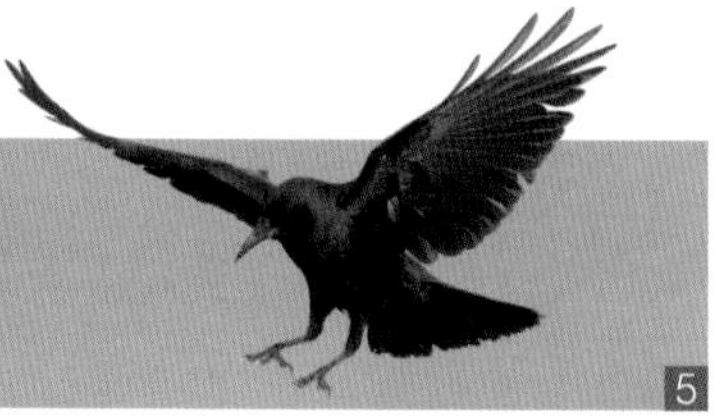
5

春季過境遷徙鳥 Spring Passage Migrant	夏候鳥 Summer Visitor	秋季過境遷徙鳥 Autumn Passage Migrant	冬候鳥 Winter Visitor

常見月份 1 2 3 4 5 6 7 8 9 10 11 12

留鳥 Resident	**迷鳥 Vagrant**	偶見鳥 Occasional Visitor

小嘴烏鴉 [普]xiǎo zuǐ wū yā

體長 length：48-53cm

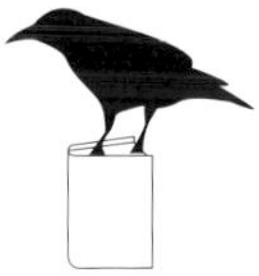

Carrion Crow | *Corvus corone*

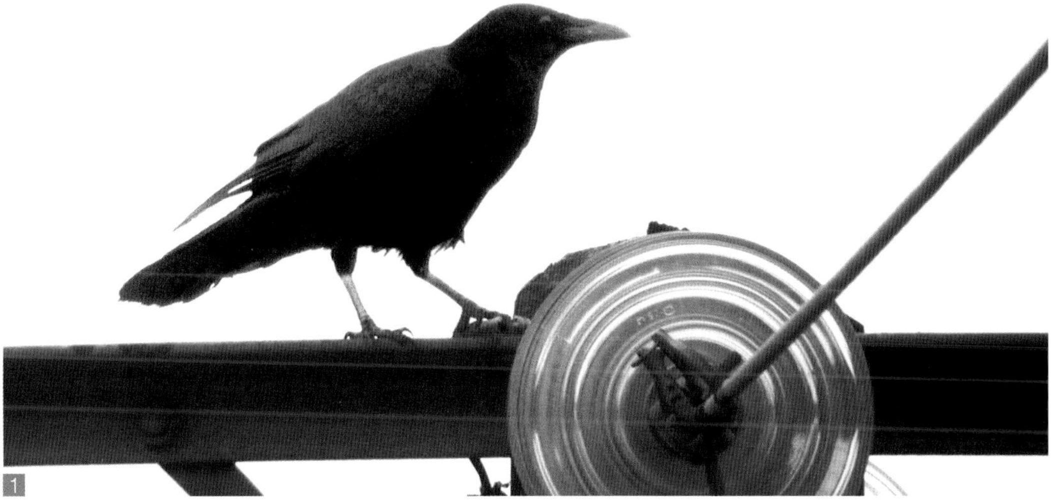

1

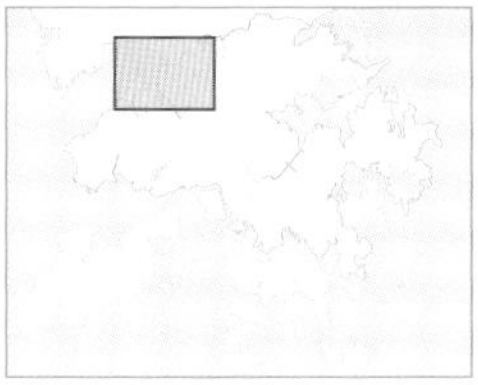

🌊

大型黑色烏鴉。全身羽毛黑色，嘴部較扁平，與前額大致銜接成直線。

Large corvid, black overall. Forehead more sloped than Large-billed Crow, and bill thinner, more pointed.

[1] Mai Po 米埔：Cheung Ho Fai 張浩輝

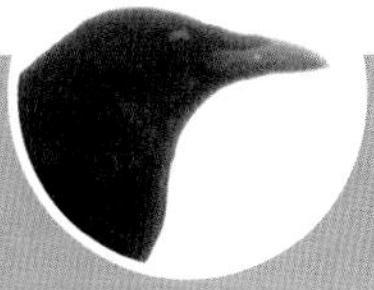

春季過境遷徙鳥 Spring Passage Migrant	夏候鳥 Summer Visitor	秋季過境遷徙鳥 Autumn Passage Migrant	冬候鳥 Winter Visitor

1 2 3 4 5 6 7 8 9 10 11 12 常見月份

留鳥 Resident	迷鳥 Vagrant	偶見鳥 Occasional Visitor

白頸鴉

[普] bái jǐng yā

體長 length：50-55cm

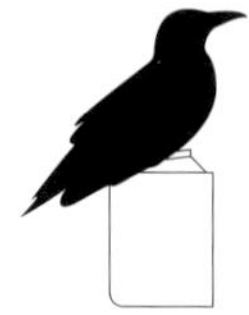

Collared Crow | *Corvus torquatus*

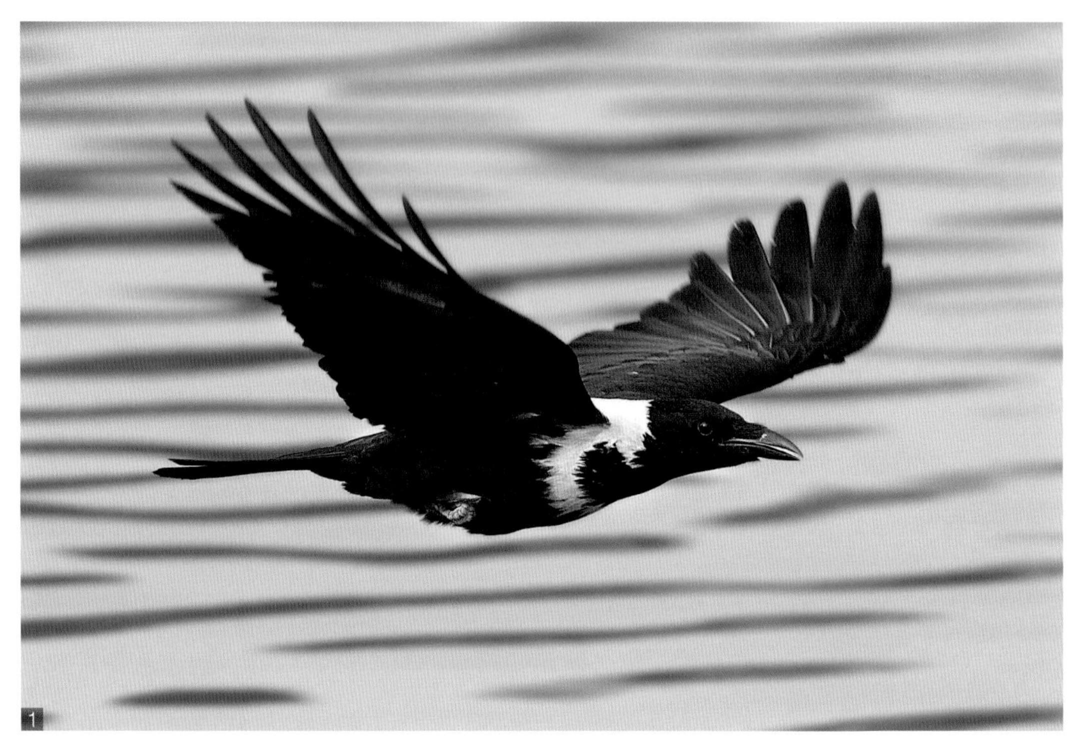

1

大型鴉類。全身黑色，頸部白色，與白色胸帶連接。嘴黑色而粗厚。通常一對或小群在海岸線附近或沿海魚塘活動。

Large corvid. Mainly black, with white collar connected to white breast band. Thick black bill. Usually near the sea coast at fishponds or shoreline in pairs or small parties.

1 adult 成鳥：Nam Sang Wai 南生圍：Dec-07, 07 年 12 月：Andy Kwok 郭匯昌
2 adult 成鳥：Nam Sang Wai 南生圍：Dec-07, 07 年 12 月：Andy Kwok 郭匯昌
3 adult 成鳥：Mai Po 米埔：3-Dec-06, 06 年 12 月 3 日：Michelle and Peter Wong 江敏兒、黃理沛
4 adult 成鳥：Fung Lok Wai 豐樂圍：Dec-04, 04 年 12 月：Jemi and John Holmes 孔思義、黃亞萍
5 adult 成鳥：Mai Po 米埔：Nov-08, 08 年 11 月：Cheng Nok Ming 鄭諾銘
6 adult 成鳥：Mai Po 米埔：3-Dec-06, 06 年 12 月 3 日：Michelle and Peter Wong 江敏兒、黃理沛
7 adult 成鳥：Mai Po 米埔：6-Oct-06, 06 年 10 月 6 日：Owen Chiang 深藍

	春季過境遷徙鳥 Spring Passage Migrant			夏候鳥 Summer Visitor			秋季過境遷徙鳥 Autumn Passage Migrant			冬候鳥 Winter Visitor		
常見月份	1	2	3	4	5	6	7	8	9	10	11	12
	留鳥 Resident				迷鳥 Vagrant				偶見鳥 Occasional Visitor			

2
3
4
5
6

7

大嘴烏鴉 [普] dà zuǐ wū yā

體長 length：46-59cm

Large-billed Crow | *Corvus macrorhynchos*

其他名稱 Other names：Jungle Crow

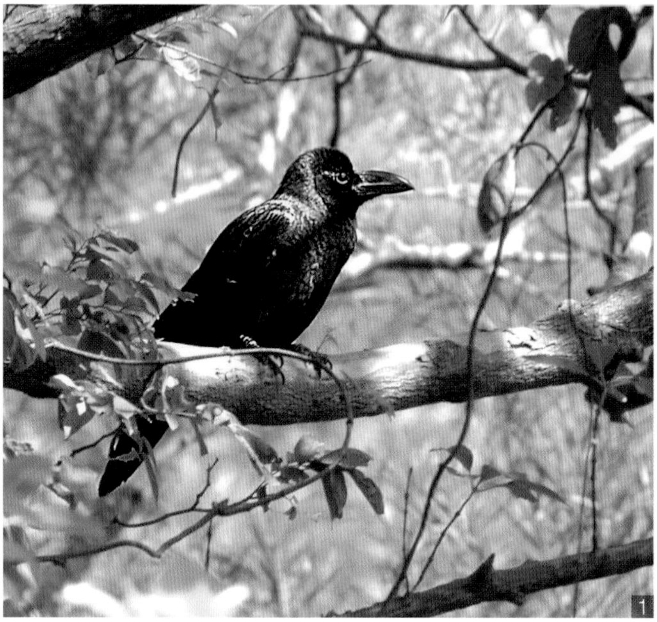

1

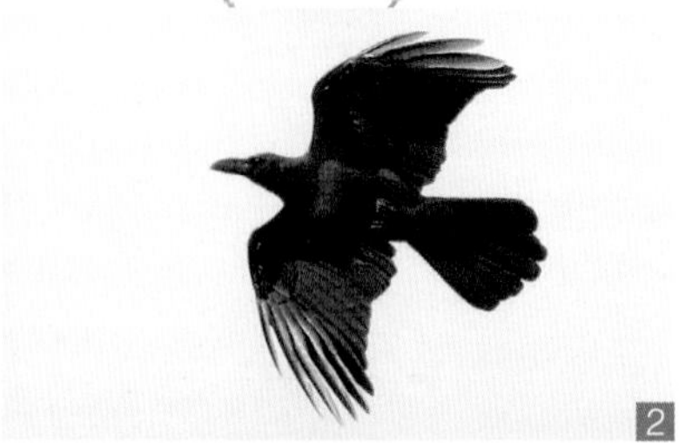

2

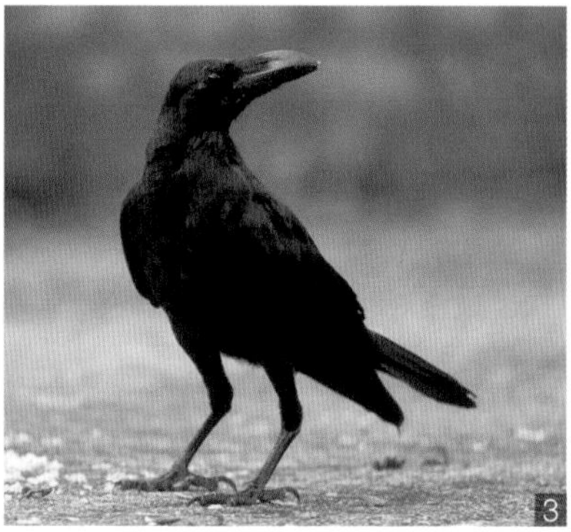

3

大型黑色烏鴉。全身黑色，嘴黑色而粗厚，與前額銜接成台階狀。叫聲為響亮的「鴉－鴉－」聲。香港最常見的烏鴉。

Large corvid, black overall. Bill black and thick, steep angle with forehead. Call is a loud and clear "ar…ar…". The commonest crow in Hong Kong.

[1] Aberdeen 香港仔：Dec-03, 03 年 12 月：Henry Lui 呂德恒
[2] Cheung Chau 長洲：Nov-04, 04 年 11 月：Henry Lui 呂德恒
[3] Sha Lo Tung 沙羅洞：Feb 04, 04年2月：Jemi and John Holmes 孔思義、黃亞萍
[4] Kowloon Hills Catchwater 九龍水塘引水道：Jan-04, 04 年 1 月：Henry Lui 呂德恒

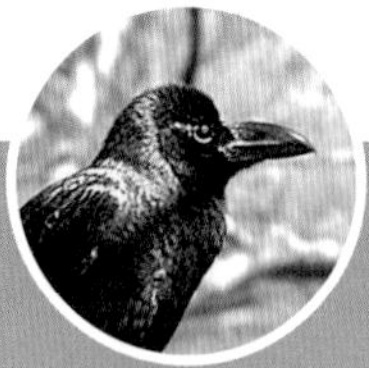

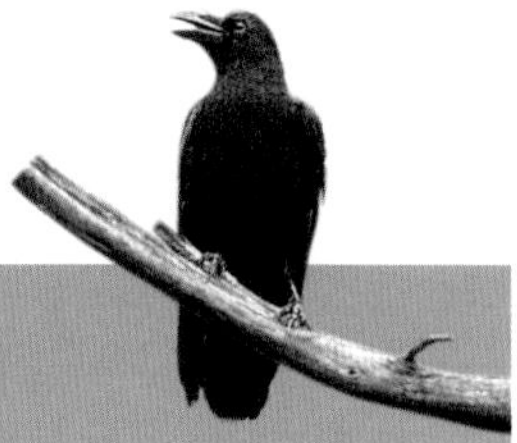

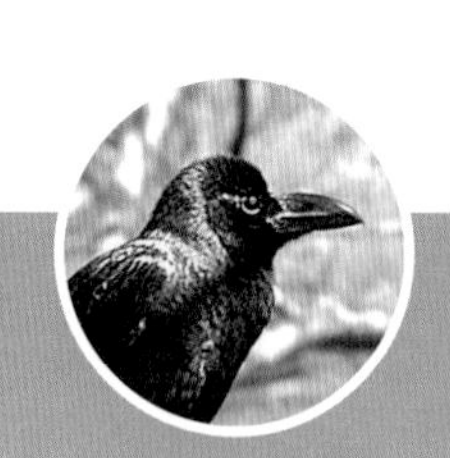

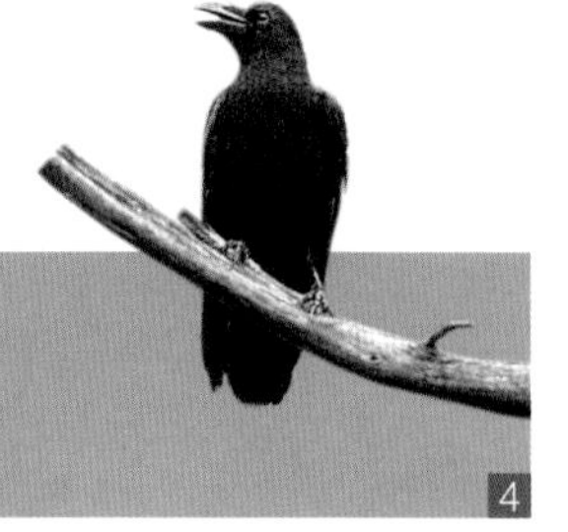

4

	春季過境遷徙鳥 Spring Passage Migrant			夏候鳥 Summer Visitor			秋季過境遷徙鳥 Autumn Passage Migrant			冬候鳥 Winter Visitor		
常見月份	1	2	3	4	5	6	7	8	9	10	11	12
	留鳥 Resident				迷鳥 Vagrant				偶見鳥 Occasional Visitor			

小太平鳥

㊟xiǎo tài píng niǎo

體長 length：15-18cm

Japanese Waxwing | *Bombycilla japonica*

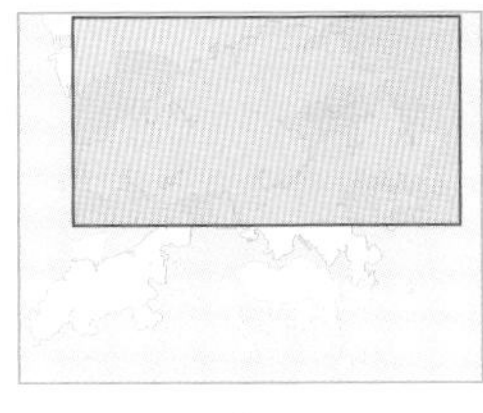

全身桃紅色，嘴和腳黑色。頭部有長冠羽及黑色貫眼紋，由眼先伸延至後枕及冠羽末端。喉部黑色。初級飛羽外側有白點，次級飛羽末端紅色，部分覆羽亦為紅色，飛羽及尾部偏灰。腹部淡黃色，臀及尾端紅色。

Peach-brown. Black bill and legs. Elongated crest trails behind the nape. Black eye-stripes extend from lore through nape up to tip of the crest. Black throat. White on outer tips of primaries, red on the tips and wing coverts of secondaries. Greyish flight feathers and tail. Pale yellowish belly. Red vent and tail tip.

1 Long Valley 塱原；5-Oct-02, 02 年 10 月 5 日；Doris Chu 朱詠兒
2 Long Valley 塱原；Stanley Fok 霍棟豪
3 adult 成鳥；Shek Kong 石崗；Nov-18, 18 年 11 月；Kwok Tsz Ki 郭子祈

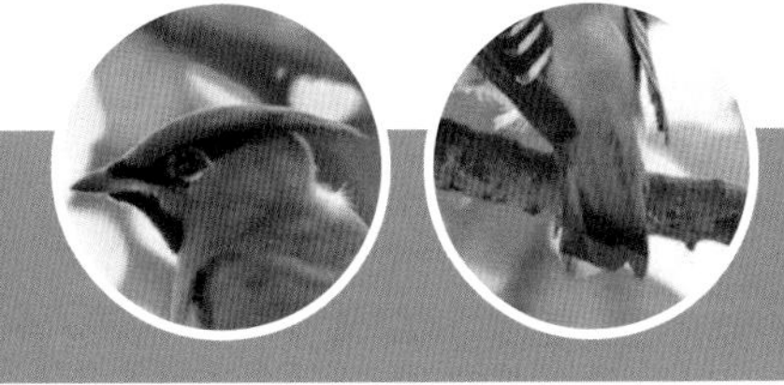

春季過境遷徙鳥 Spring Passage Migrant	夏候鳥 Summer Visitor	秋季過境遷徙鳥 Autumn Passage Migrant	冬候鳥 Winter Visitor

1 2 3 4 5 6 7 8 9 10 11 12 常見月份

留鳥 Resident	迷鳥 Vagrant	偶見鳥 Occasional Visitor

方尾鶲

㊀ fāng wěi wēng
㊁ 鶲：音翁

體長 length：12-13cm

Grey-headed Canary-flycatcher | *Culicicapa ceylonensis*

其他名字 Other names: Grey-headed Flycatcher

頭和胸灰色，上體及翼橄欖綠色，下體黃色。活躍好動如鶯類，但站姿挺直，常混在其他鳥群中。叫聲為獨特的四節「silly billy」。

Grey head and breast. Olive green upperparts and wings, and yellow underparts. Active. Resembles a warbler but identified by its upright posture. Usually in mixed flocks. Distinctive four note call "silly billy".

1 adult 成鳥：Po Toi 蒲台：Oct-08, 08 年 10 月：Lee Kai Hong 李啟康
2 adult 成鳥：Cheung Chau 長洲：Feb-06, 06 年 2 月：Kami Hui 許淑君
3 juvenile 幼鳥：Tai Po Kau 大埔滘：Jan-08, 08 年 1 月：Wallace Tse 謝鑑超
4 adult 成鳥：Po Toi 蒲台：Oct-08, 08 年 10 月：Lee Kai Hong 李啟康
5 Tai Po Kau 大埔滘：Jan-08, 08 年 1 月：Wallace Tse 謝鑑超
6 Shing Mun 城門水塘：Dec-06, 06 年 12 月：Wallace Tse 謝鑑超

	春季過境遷徙鳥 Spring Passage Migrant			夏候鳥 Summer Visitor			秋季過境遷徙鳥 Autumn Passage Migrant			冬候鳥 Winter Visitor		
常見月份	1	2	3	4	5	6	7	8	9	10	11	12
	留鳥 Resident				迷鳥 Vagrant				偶見鳥 Occasional Visitor			

3
4
5

6

黃腹山雀

㊟huáng fù shān què

體長 length：10-11cm

Yellow-bellied Tit | *Pardaliparus venustulus*

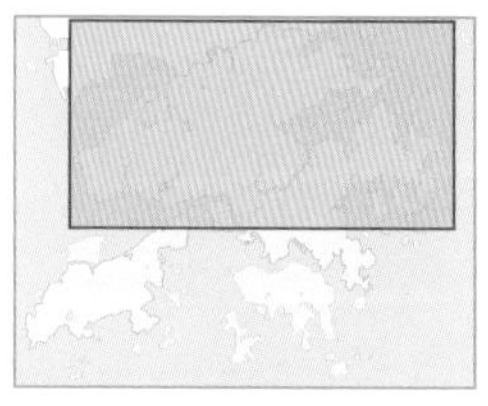

體型細小的山雀。外形與常見的大山雀相似，但身形較小，尾較短。頭黑色，面頰白色，後枕上有一白斑。覆羽及三級飛羽上有白斑，上背灰色。下體鮮黃色，中央沒有黑線。多成群出沒，但也會單獨活動。

Similar to Great Tit but smaller and shorter-tailed. Black head with white cheeks and a white spot on nape. White spots on wing coverts and tertials. Grey mantle. Bright yellow underparts without central black line. Usually in flocks but also as singles.

1 1st winter moulting into breeding male 第一年冬天雄鳥轉換繁殖羽；Lam Tsuen 林村；Jan-06, 06 年 1 月；Martin Hale 夏敖天
2 female / juvenile 雌鳥 / 幼鳥；Lam Tsuen 林村；Jan-06, 06 年 1 月；Martin Hale 夏敖天
3 juvenile male 雄幼鳥；Lam Tsuen 林村；Jan-06, 06 年 1 月；Jemi and John Holmes 孔思義、黃亞萍
4 juvenile 幼鳥；Lam Tsuen 林村；Jan-06, 06 年 1 月；Michelle and Peter Wong 江敏兒、黃理沛
5 juvenile 幼鳥；Lam Tsuen 林村；Jan-06, 06 年 1 月；Michelle and Peter Wong 江敏兒、黃理沛
6 female / juvenile 雌鳥 / 幼鳥；Lam Tsuen 林村；Feb-06, 06 年 2 月；Cherry Wong 黃卓研

春季過境遷徙鳥 Spring Passage Migrant	夏候鳥 Summer Visitor	秋季過境遷徙鳥 Autumn Passage Migrant	冬候鳥 Winter Visitor

常見月份 1 2 3 4 5 6 7 8 9 10 11 12

留鳥 Resident	迷鳥 Vagrant	偶見鳥 Occasional Visitor

2
3
4
5

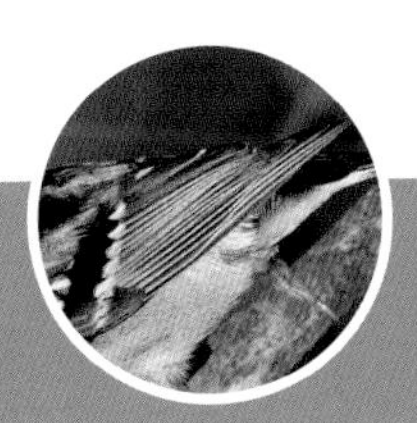

6

雜色山雀 ㊟ zá sè shān què

體長 length：12-14cm

Varied Tit | *Sittiparus varius*

頭頂和喉部的黑色，在臉和胸下的淡棕色襯托下對比鮮明。翅膀和尾部灰色，下體棕褐色，嘴和腳黑色。

Black on crown and throat contrast pale brown face and breast. Grey wing and tail. Brownish-orange underparts. Black bill and legs.

1 adult 成鳥：Tai Tong 大棠：Dec-12, 12 年 12 月：Lee Yat Ming 李逸明
2 adult 成鳥：Tai Tong 大棠：Dec-12, 12 年 12 月：Lee Yat Ming 李逸明
3 adult 成鳥：Tai Tong 大棠：Feb-13, 13 年 2 月：Doris Chu 朱詠兒
4 adult 成鳥：Tai Tong 大棠：Dec-12, 12 年 12 月：Lee Yat Ming 李逸明
5 adult 成鳥：Tai Tong 大棠：Dec-12, 12 年 12 月：Lee Yat Ming 李逸明

春季過境遷徙鳥 Spring Passage Migrant	夏候鳥 Summer Visitor	秋季過境遷徙鳥 Autumn Passage Migrant	冬候鳥 Winter Visitor

常見月份 1 2 3 4 5 6 7 8 9 10 11 12

留鳥 Resident	迷鳥 Vagrant	偶見鳥 Occasional Visitor

遠東山雀

㊥ rì běn shān què

體長 length：12.5-14cm

Japanese Tit | *Parus minor*

頭部黑色而面頰大片白色，嘴和腳黑色。背部墨綠色，翼邊和尾上黑色，飛羽有幼白邊，下體灰白色。和常見的蒼背山雀相似，但背部是墨綠色而非灰色。

Black head with large white patch on each side of face. Black bill and legs. Dark green back. Black wing edge and upper tail. Thin white fringes on flight feathers. Greyish-white underparts. Distinguishes from common Cinereous Tit by dark green instead of grey back.

1 adult male 雄成鳥；Kowloon Park 九龍公園；Sep-18, 18 年 9 月；Allen Chan 陳志雄
2 adult male 雄成鳥；Kowloon Park 九龍公園；Sep-18, 18 年 9 月；Allen Chan 陳志雄
3 adult male 雄成鳥；Kowloon Park 九龍公園；Sep-18, 18 年 9 月；Allen Chan 陳志雄

春季過境遷徙鳥 Spring Passage Migrant	夏候鳥 Summer Visitor	秋季過境遷徙鳥 Autumn Passage Migrant	冬候鳥 Winter Visitor

1 2 3 4 5 6 7 8 9 10 11 12 常見月份

留鳥 Resident	**迷鳥 Vagrant**	偶見鳥 Occasional Visitor

蒼背山雀

㊟ chāng bèi shān què

體長 length：12.5-14cm

Cinereous Tit | *Parus cinereus*

其他名稱 Other names：大山雀，Great Tit

頭黑色，面頰有獨特白斑。上體灰色，翼黑色而邊緣白色。一道黑紋由喉部伸延至腹部中央。叫聲變化多端，如獨特的顫抖「磁－磁－」聲，伴有響亮的「即－即－」歌聲。

Distinctive black head with white cheek patch. Grey upperparts. Black wings with white edges on feathers. Black line extends from throat to centre of belly. Various calls, one a distinctive quivering "tzzz-tzzz-" with loud "chik-chik" notes.

1 adult 成鳥；Ho Man Tin 何文田；Oct-08, 08 年 10 月；Andy Kwok 郭匯昌
2 juvenile 幼鳥；Mai Po 米埔；Jun-07, 07 年 6 月；Cherry Wong 黃卓研
3 adult female 雌成鳥；Tai Po Kau 大埔滘；Dec-06, 06 年 12 月；Wallace Tse 謝鑑超
4 adult male 雄成鳥；Andy Cheung 張玉良
5 adult male 雄成鳥；Andy Cheung 張玉良
6 adult 成鳥；Fung Yuen 鳳園；Dec-08, 08 年 12 月；Ken Fung 馮漢城
7 adult male 雄成鳥；Sai Kung 西貢；Mar-08, 08 年 3 月；Pippen Ho 何志剛
8 adult 成鳥；Long Valley 塱原；Nov-04, 04 年 11 月；Henry Lui 呂德恒

	春季過境遷徙鳥 Spring Passage Migrant			夏候鳥 Summer Visitor			秋季過境遷徙鳥 Autumn Passage Migrant			冬候鳥 Winter Visitor		
常見月份	1	2	3	4	5	6	7	8	9	10	11	12
	留鳥 Resident				迷鳥 Vagrant				偶見鳥 Occasional Visitor			

2
3
4
5
6
7

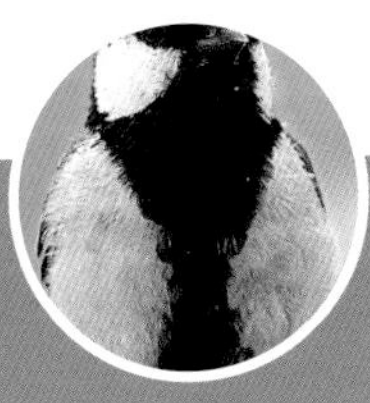

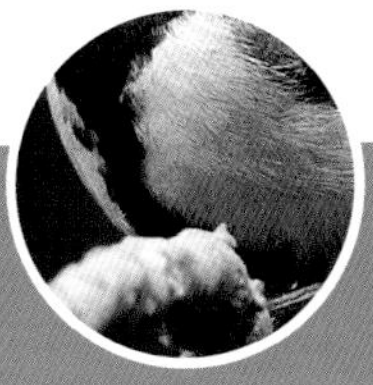

8

黃頰山雀

㊟huáng jiá shān què
㊣頰：音夾

體長length：14-15.5cm

Yellow-cheeked Tit | *Machlolophus spilonotus*

1

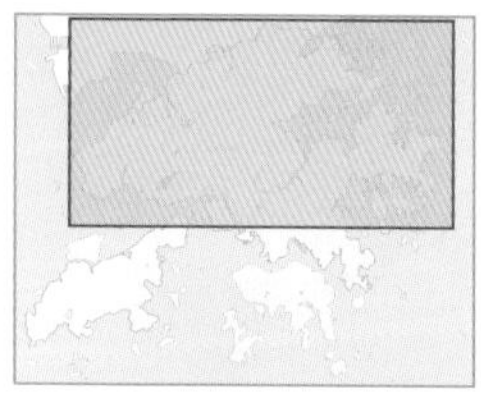

外形獨特的山雀，面頰黃色，頭上有黑色小冠。一道黑紋由喉部伸延至腹部中央。上體偏黑。常混在其他鳥群中，喜在樹林中層或樹冠之間活動。

Distinctive yellow cheeks with short black crest. Black line extends from throat to central belly. Blackish upperparts. Usually in mixed flocks. Favours canopy and mid-storey of forest.

1 adult male, race *rex* 雄成鳥，*rex* 亞種；Tai Po Kau 大埔滘；Feb-08, 08 年 2 月；Andy Kwok 郭匯昌
2 adult male 雄成鳥；Shing Mun 城門水塘；Jan-08, 08 年 1 月；Cherry Wong 黃卓研
3 adult female 雌成鳥；Shing Mun 城門水塘；Dec-08, 08 年 12 月；Ken Fung 馮漢城
4 adult female 雌成鳥；Shing Mun 城門水塘；Mar-08, 08 年 3 月；Jasper Lee 李君哲
5 juvenile female 雌幼鳥；Tai Po Kau 大埔滘；Apr-07, 07 年 4 月；Wallace Tse 謝鑑超
6 adult male 雄成鳥；Tai Po Kau 大埔滘；Mar-07, 07 年 3 月；Martin Hale 夏敖天
7 adult male 雄成鳥；Tai Po Kau 大埔滘；Feb-08, 08 年 2 月；Andy Kwok 郭匯昌
8 adult male 雄成鳥；Tai Po Kau 大埔滘；Aug-04, 04 年 8 月；Michelle and Peter Wong 江敏兒、黃理沛

	春季過境遷徙鳥 Spring Passage Migrant			夏候鳥 Summer Visitor			秋季過境遷徙鳥 Autumn Passage Migrant			冬候鳥 Winter Visitor		
常見月份	1	2	3	4	5	6	7	8	9	10	11	12
	留鳥 Resident				迷鳥 Vagrant				偶見鳥 Occasional Visitor			

2
3
4
5
6
7

8

中華攀雀

[普] zhōng huá pān què

體長 length：10.5cm

Chinese Penduline Tit | *Remiz consobrinus*

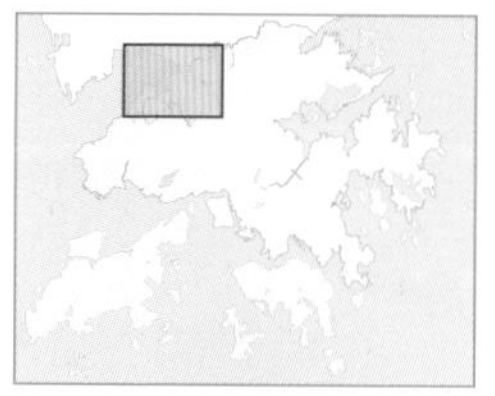

外貌有如一隻細小的伯勞。繁殖時雄鳥頭灰色，有黑色貫眼紋，後頸有偏紅色斑。雌鳥顏色偏褐。叫聲獨特，為音調較高的「tsee-tsee」聲。多成小群，但偶爾也會大群出沒。多在蘆葦叢中活動。

Alike a tiny shrike. Breeding male has grey head, black eye stripe and reddish patch on hindneck. Female is browner. A distinctive high-pitched "tsee-tsee" call. Usually in flocks, sometimes in large ones. Reedbed specialist.

1 adult male 雄成鳥：Mai Po 米埔：Apr-04, 04 年 4 月：Cherry Wong 黃卓研
2 adult female 雌成鳥：Mai Po 米埔：Apr-09, 09 年 4 月：Pippen Ho 何志剛
3 adult male 雄成鳥：Mai Po 米埔：4-Apr-04, 04 年 4 月 4 日：Doris Chu 朱詠兒
4 adult female 雌成鳥：Mai Po 米埔：Apr-09, 09 年 4 月：Pippen Ho 何志剛
5 adult female 雌成鳥：Mai Po 米埔：4-Apr-04, 04 年 4 月 4 日：Lo Kar Man 盧嘉孟
6 adult female 雌成鳥：Mai Po 米埔：Apr-04, 04 年 4 月：Cherry Wong 黃卓研
7 adult female 雌成鳥：Mai Po 米埔：4-Apr-04, 04 年 4 月 4 日：Lo Kar Man 盧嘉孟

	春季過境遷徙鳥 Spring Passage Migrant			夏候鳥 Summer Visitor			秋季過境遷徙鳥 Autumn Passage Migrant			冬候鳥 Winter Visitor		
常見月份	1	2	3	4	5	6	7	8	9	10	11	12
	留鳥 Resident				迷鳥 Vagrant				偶見鳥 Occasional Visitor			

2
3
4
5
6

7

雲雀 ㊟ yún què

體長 length：16-19cm

Eurasian Skylark | *Alauda arvensis*

其他名稱 Other names：Northern Skylark

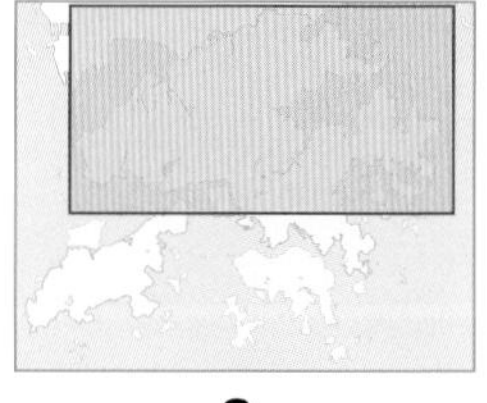

大型雲雀。上身主要灰褐色，有深色斑點，眉紋偏黃色，面頰有一大片明顯的栗色。與小雲雀比較，靜立時黑色初級飛羽伸延超過灰褐色次級飛羽。除初級飛羽外，其他飛羽有淡色邊緣。

Large skylark. Upperparts greyish brown with dark stripes, prominent yellowish eyebrow and chestnut cheeks. Unlike Oriental Skylark, the black primaries extend beyond greyish brown secondaries. Pale fringes on all flight feathers except the primaries.

1 Long Valley 塱原：Oct-05, 05 年 10 月：Michelle and Peter Wong 江敏兒、黃理沛
2 Long Valley 塱原：Nov-07, 07 年 11 月：Owen Chiang 深藍
3 Long Valley 塱原：Nov-07, 07 年 11 月：Isaac Chan 陳家強
4 Long Valley 塱原：Nov-07, 07 年 11 月：Fung Siu Ping 馮少萍
5 Long Valley 塱原：Oct-05, 05 年 10 月：Michelle and Peter Wong 江敏兒、黃理沛

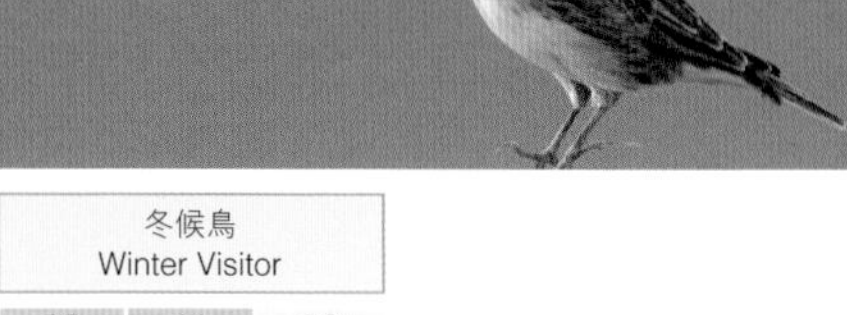

春季過境遷徙鳥 Spring Passage Migrant	夏候鳥 Summer Visitor	秋季過境遷徙鳥 Autumn Passage Migrant	冬候鳥 Winter Visitor

常見月份 1 2 3 4 5 6 7 8 9 10 11 12

留鳥 Resident	迷鳥 Vagrant	偶見鳥 Occasional Visitor

蒙古短趾百靈

[普] méng gǔ duǎn zhǐ bǎi líng

體長 length：14-15cm

Mongolian Short-toed Lark | *Calandrella dukhunensis*

1

2

3

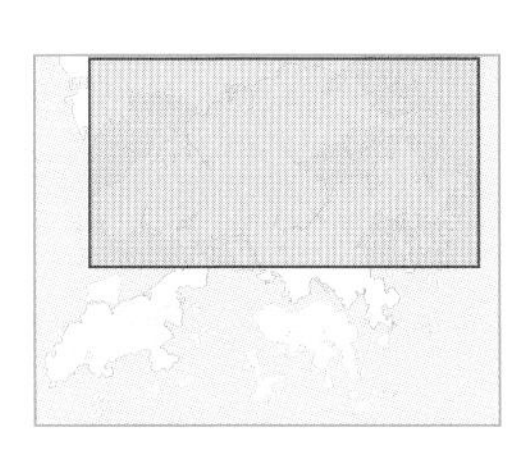

身軀較矮胖的百靈。上體深灰褐色雜有淡褐色，下體白色，胸口有一灰褐色橫線，有深灰色貫眼紋。淡黃褐色的嘴粗短，腳黃褐色。

Chunky-looking lark. Dark greyish-brown upperparts tinged with pale brown. White underparts. Thin greyish brown breast band. Dark grey eye-stripes. Thick and short pale yellowish-brown bill. Yellowish-brown legs.

[1] adult 成鳥：San Tin 新田：Apr-17, 17 年 4 月：Kwok Tsz Ki 郭子祈
[2] adult 成鳥：San Tin 新田：Apr-17, 17 年 4 月：Kwok Tsz Ki 郭子祈
[3] adult 成鳥：San Tin 新田：Apr-17, 17 年 4 月：Kwok Tsz Ki 郭子祈

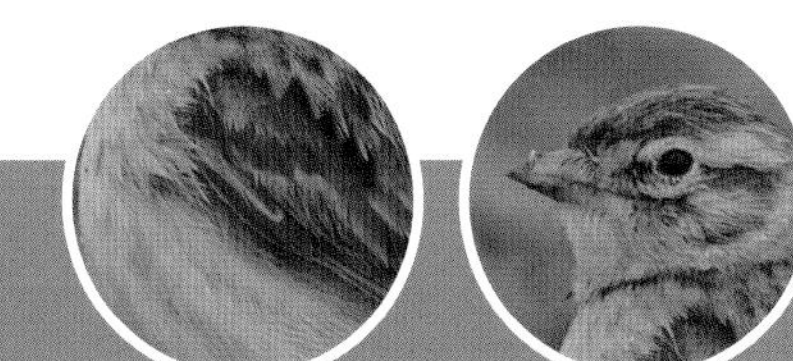

春季過境遷徙鳥 Spring Passage Migrant	夏候鳥 Summer Visitor	秋季過境遷徙鳥 Autumn Passage Migrant	冬候鳥 Winter Visitor

1	2	3	4	5	6	7	8	9	10	11	12	常見月份

留鳥 Resident	**迷鳥 Vagrant**	偶見鳥 Occasional Visitor

歌百靈

[普] gē bǎi líng

體長 length：13.5-15cm

Horsfield's Bush Lark | *Mirafra javanica*

其他名稱 Other names：Singing Bushlark

上體主要灰褐色。頭部有短小冠羽，眉灰褐色，眼睛深色而大，面頰有明顯的深褐色，嘴粗厚呈鈎狀。初級飛羽有紅褐色邊，尾羽短，飛行時可見闊大的翼和顯眼的白色外側尾羽。所有記錄被判斷為逸鳥。

Upperparts greyish brown. It has a small crest, greyish brown supercilium, big and dark eyes, brownish cheeks, and a thick and slightly hooked bill. Primaries fringed reddish brown, tail is short. In flight, it shows broad wings and prominent white outer tail feathers. All records are considered as birds escaped from captivity.

1 Lok Ma Chau 落馬洲：Mar 04, 04 年 3 月：Martin Hale 夏敖天

春季過境遷徙鳥 Spring Passage Migrant	夏候鳥 Summer Visitor	秋季過境遷徙鳥 Autumn Passage Migrant	冬候鳥 Winter Visitor

常見月份 1 2 3 4 5 6 7 8 9 10 11 12

留鳥 Resident	迷鳥 Vagrant	偶見鳥 Occasional Visitor

蒙古百靈

㊟ měng gǔ bǎi líng

體長 length：18-22cm

Mongolian Lark | *Melanocorypha mongolica*

體型大，全身銹褐色，冠紋黃褐色，外面栗色外圈包圍。白色眉紋直延至後頸，上胸具黑色橫紋胸帶，黑色的初級飛羽和白色的次級飛羽與覆羽成對比。所有記錄被判斷為逸鳥。

Large-sized lark. Rufous brown. Yellowish brown crown strip, surrounded by rufous band at the outer edge. White supercilium extending to back of neck. Black secondaries and white primaries contrast with colour of wing coverts. All records are considered as birds escaped from captivity.

1 Chung Mei 涌尾：Nov-04, 04 年 11 月：Jemi and John Holmes 孔思義、黃亞萍
2 Chung Mei 涌尾：Nov-04, 04 年 11 月：Jemi and John Holmes 孔思義、黃亞萍

春季過境遷徙鳥 Spring Passage Migrant	夏候鳥 Summer Visitor	秋季過境遷徙鳥 Autumn Passage Migrant	冬候鳥 Winter Visitor

1 2 3 4 5 6 7 8 9 10 11 12 常見月份

留鳥 Resident	迷鳥 Vagrant	偶見鳥 Occasional Visitor

紅耳鵯

普 hóng ěr bēi
粵 鵯：音卑

體長 length：18-20.5cm

Red-whiskered Bulbul | *Pycnonotus jocosus*

其他名稱 Other names：高髻冠, Crested Bulbul

全身偏褐，頭、嘴和腳黑色，有獨特的直立冠羽，耳羽紅色，面頰及喉白色，有明顯黑色頰紋。上體至尾部褐色，尾羽末端有白點。下體淡褐色，臀部橙紅色。幼鳥無紅色耳羽，臀部紅色較淡。常發出清脆的「bulbit…bulbit…」聲。

Dark brown upperparts. Black head, bill and legs, prominent erect crown feather. Red coverts, white cheeks and throat, with black moustachial stripe. Brown upperparts to tail tipped with white. Underparts pale brown with orange red vent. Juvenile lacks red cheeks and has paler red vent. Call a cheerful "bulbit…bulbit…".

1 adult 成鳥：Siu Lek Yuen 小瀝源：Dec-08, 08 年 12 月：Ken Fung 馮漢城
2 adult 成鳥：Hang Tau 坑頭：Mar-07, 07 年 3 月：Jemi and John Holmes 孔思義、黃亞萍
3 adult 成鳥：Kowloon Park 九龍公園：Jul-08, 08 年 7 月：Lo Chun Fai 勞浚暉
4 adult 成鳥：Cheung Chau 長洲：Dec-04, 04 年 12 月：Henry Lui 呂德恒
5 adult 成鳥：Hong Kong Park 香港公園：8-Sep-07, 07 年 9 月 8 日：Anita Lee 李雅婷
6 juvenile 幼鳥：Mai Po 米埔：Aug-07, 07 年 8 月：Sammy Sam and Winnie Wong 森美與雲妮
7 adult 成鳥：Hong Kong Park 香港公園：Sep-07, 07 年 9 月：Bill Man 文權溢
8 juvenile 幼鳥：Mai Po 米埔：Aug-07, 07 年 8 月：Cherry Wong 黃卓研

	春季過境遷徙鳥 Spring Passage Migrant			夏候鳥 Summer Visitor			秋季過境遷徙鳥 Autumn Passage Migrant			冬候鳥 Winter Visitor		
常見月份	1	2	3	4	5	6	7	8	9	10	11	12

留鳥 Resident	迷鳥 Vagrant	偶見鳥 Occasional Visitor

3
4
5
6
7

8

白頭鵯

㊟ bái tóu bēi
粵 鵯：音卑

體長 length：19cm

Chinese Bulbul | *Pycnonotus sinensis*

全身橄欖綠色，頭、嘴和腳黑色，後枕、面頰和喉部白色，下體至尾下覆羽淡色。幼鳥全身橄欖綠色，沒有白斑。叫聲似紅耳鵯，但較沙啞。

Overall olive green. Black head, bill and legs. Conspicuous white nape, cheeks and throat. Buffy breast to vent. Juvenile is completely olive green and lacks white markings. Call similar to Red-whiskered Bulbul but less musical.

1 adult 成鳥；Mai Po 米埔；Dec-07, 07 年 12 月；Cherry Wong 黃卓研
2 adult 成鳥；Cheung Chau 長洲；Nov-04, 04 年 11 月；Henry Lui 呂德恒
3 adult 成鳥；Cheung Chau 長洲；Dec-04, 04 年 12 月；Henry Lui 呂德恒
4 juvenile 幼鳥；Tai Po Kau 大埔滘；May-04, 04 年 5 月；Henry Lui 呂德恒
5 adult 成鳥；Mai Po 米埔；Apr-06, 06 年 4 月；Cherry Wong 黃卓研
6 adult 成鳥；Mai Po 米埔；Jan-04, 04 年 1 月；Henry Lui 呂德恒
7 adult 成鳥；Fo Tan 火炭；Jan-08, 08 年 1 月；Christine and Samuel Ma 馬志榮、蔡美蓮

	春季過境遷徙鳥 Spring Passage Migrant			夏候鳥 Summer Visitor			秋季過境遷徙鳥 Autumn Passage Migrant			冬候鳥 Winter Visitor		
常見月份	1	2	3	4	5	6	7	8	9	10	11	12
	留鳥 Resident				迷鳥 Vagrant				偶見鳥 Occasional Visitor			

2
3
4
5
6

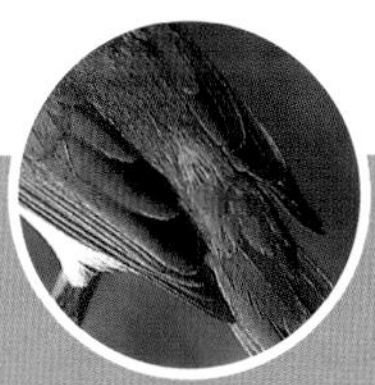

7

白喉紅臀鵯

[普] bái hóu hóng tún bēi
[粵] 鵯：音卑

體長 length：19-21cm

Sooty-headed Bulbul | *Pycnonotus aurigaster*

全身偏褐，頭部黑色，頭頂有時稍為隆起像有冠羽，嘴和腳黑色，面頰有時沾黑。上體至尾部褐色，尾羽末端白色；喉至下體淡褐色，臀部紅色。幼鳥臀部偏黃。叫聲較紅耳鵯和白頭鵯悅耳。

Generally brown. Black crown with a slight crest. Black bill and legs, cheeks sometimes tinted black. Brown upperparts to tail tipped with white. Underparts greish brown with red vent. Juvenile has yellow vent. Call more musical than Red-whiskered or Chinese Bulbul.

1 adult 成鳥：HK Wetland Park 香港濕地公園：Jan-07, 07 年 1 月：Law Kam Man 羅錦文
2 juvenile 幼鳥：Tsing Yi 青衣：Aug-07, 07 年 8 月：Henry Lui 呂德恆
3 adult 成鳥：Long Valley 塱原：Nov-07, 07 年 11 月：Andy Kwok 郭匯昌
4 juvenile 幼鳥：HK Wetland Park 香港濕地公園：Sep-06, 06 年 9 月：Kitty Koo 古愛婉
5 adult 成鳥：Tai Po Kau 大埔滘：Feb-07, 07 年 2 月：Joyce Tang 鄧玉蓮
6 adult 成鳥：Long Valley 塱原：Dec-06, 06 年 12 月：James Lam 林文華
7 adult 成鳥：Cheung Chau 長洲：Apr-03, 03 年 4 月：Henry Lui 呂德恆

	春季過境遷徙鳥 Spring Passage Migrant			夏候鳥 Summer Visitor			秋季過境遷徙鳥 Autumn Passage Migrant			冬候鳥 Winter Visitor		
常見月份	1	2	3	4	5	6	7	8	9	10	11	12
	留鳥 Resident				迷鳥 Vagrant				偶見鳥 Occasional Visitor			

2
3
4
5
6

7

綠翅短腳鵯

[普] lǜ chì duǎn jiǎo bēi
[粵] 鵯：音卑

體長 length：21-24cm

Mountain Bulbul | *Ixos mcclellandii*

1

🌳 ⛰

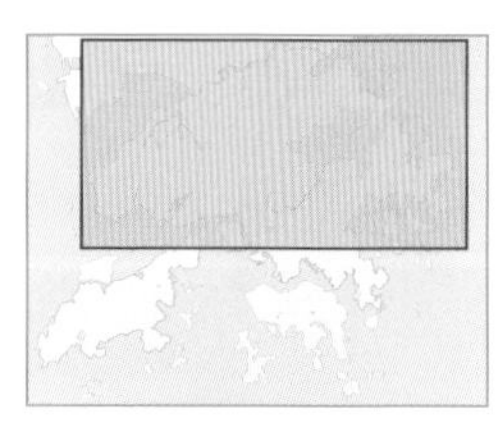

大型鵯類。頭部褐色，深色冠羽短而尖，上體、翼及尾部橄欖色，喉部白色，有黑色縱紋，下體淡褐，胸部有少量縱紋。

Large-sized bulbul. Brownish head, chestnut crest with short and spiky crest. Upperparts, wings and tail olive green. Throat white with brown streaks. Underparts white, lightly streaked on breast.

1 Tai Po Kau 大埔滘：Nov-08, 08 年 11 月：Michelle and Peter Wong 江敏兒、黃理沛
2 Tai Po Kau 大埔滘：Nov-08, 08 年 11 月：Michelle and Peter Wong 江敏兒、黃理沛
3 Tai Po Kau 大埔滘：Nov-08, 08 年 11 月：Michelle and Peter Wong 江敏兒、黃理沛
4 Tai Po Kau 大埔滘：Feb-05, 05 年 2 月：Cherry Wong 黃卓研
5 Tai Po Kau 大埔滘：Dec-05, 05 年 12 月：Jemi and John holmes 孔思義、黃亞萍
6 Tai Po Kau 大埔滘：Feb-06, 06 年 2 月：Henry Lui 呂德恒
7 Tai Po Kau 大埔滘：Apr-07, 07 年 4 月：Allen Chan 陳志雄

	春季過境遷徙鳥 Spring Passage Migrant			夏候鳥 Summer Visitor			秋季過境遷徙鳥 Autumn Passage Migrant			冬候鳥 Winter Visitor		
常見月份	1	2	3	4	5	6	7	8	9	10	11	12
	留鳥 Resident				迷鳥 Vagrant				偶見鳥 Occasional Visitor			

2
3
4
5
6

7

栗背短腳鵯

[普] lì bèi duǎn jiǎo bēi
[粵] 鵯：音卑

體長 length：21.5cm

Chestnut Bulbul | *Hemixos castanonotus*

上體栗褐色，頭頂黑色，有少許冠羽，嘴和腳黑色，翼、腰及尾羽深褐色。喉部及下體白色，與上體對比鮮明，胸和脇有時沾灰。叫聲為響亮並不斷重複的口哨聲，有如問人「去邊處？」。

Bright chestnut upperparts, with black crown and crest. Bill and legs black. Wing, rump and tail dark brown. Throat to underparts white, in good contrast with the upperparts. Breast and flanks sometimes greyish. Loud repetitive whistling calls.

1 adult 成鳥：Tai Po Kau 大埔滘：Oct-03, 03 年 10 月：Michelle and Peter Wong 江敏兒、黃理沛
2 adult 成鳥：Tai Po Kau 大埔滘：May-04, 04 年 5 月：Cherry Wong 黃卓研
3 adult 成鳥：Tai Po Kau 大埔滘：Dec-03, 03 年 12 月：Michelle and Peter Wong 江敏兒、黃理沛
4 adult 成鳥：Tai Po Kau 大埔滘：Dec-04, 04 年 12 月：Jemi and John Holmes 孔思義、黃亞萍
5 adult 成鳥：Tai Po Kau 大埔滘：Jan-09, 09 年 1 月：Cherry Wong 黃卓研
6 adult 成鳥：Tai Po Kau 大埔滘：Jan-07, 07 年 1 月：Andy Kwok 郭匯昌
7 juvenile 幼鳥：Tai Po Kau 大埔滘：Sep-07, 07 年 9 月：Henry Lui 呂德恆

	春季過境遷徙鳥 Spring Passage Migrant			夏候鳥 Summer Visitor			秋季過境遷徙鳥 Autumn Passage Migrant			冬候鳥 Winter Visitor		
常見月份	1	2	3	4	5	6	7	8	9	10	11	12
	留鳥 Resident				迷鳥 Vagrant				偶見鳥 Occasional Visitor			

2
3
4
5
6

7

黑短腳鵯

普 hēi duǎn jiǎo bēi
粵 鵯：音卑

體長 length：23.5-26.5cm

Black Bulbul | *Hypsipetes leucocephalus*

大型鵯類。體色全黑，嘴、腳和眼睛皆赤紅色，對比鮮明。頭頂有時隆起像有冠羽，部分個體頭部全白。未成年鳥有時灰色而非全黑。叫聲拉得很長，帶有鼻音，似小貓沙啞的悲鳴。

Large bulbul. Prominent black plumage contrasts with red bill, feet and eyes. Short crown feathers form a low crest. Head completely white in some individuals. Immature bird has greyish plumage. Call a plaintive mewing.

1 30-Jan-09, 09年1月30日：Lee Yat Ming 李逸明
2 Tai Po Kau大埔滘：Mar-07, 07年3月：Thomas Chan 陳土飛
3 1st winter 第一年冬天：Shek Kong Catchment 石崗引水道：Jan-09, 09 年 1 月：Michelle and Peter Wong 江敏兒、黃理沛
4 adult / 1st winter 成鳥 / 第一年冬天：Tai Po Kau 大埔滘：14-Feb-04, 04 年 2 月 14 日：Lo Kar Man 盧嘉孟
5 1st winter 第一年冬天：Shek Kong Catchment 石崗引水道：Jan-09, 09 年 1 月：Michelle and Peter Wong 江敏兒、黃理沛

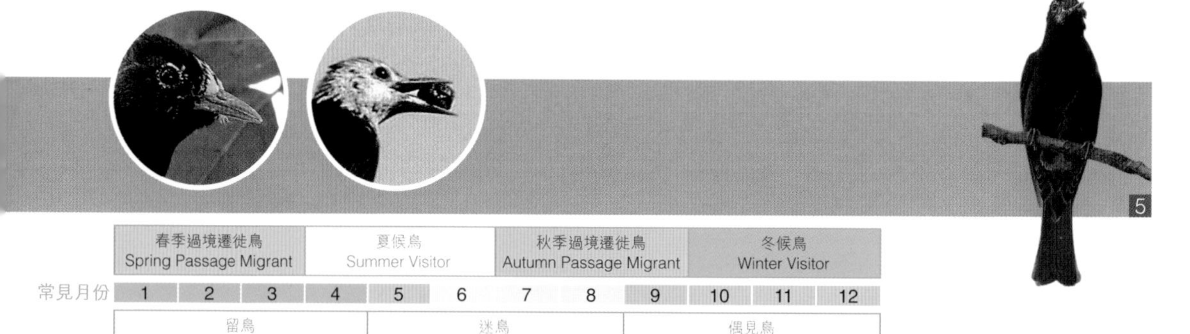

春季過境遷徙鳥 Spring Passage Migrant	夏候鳥 Summer Visitor	秋季過境遷徙鳥 Autumn Passage Migrant	冬候鳥 Winter Visitor

常見月份	1	2	3	4	5	6	7	8	9	10	11	12

留鳥 Resident	迷鳥 Vagrant	偶見鳥 Occasional Visitor

栗耳短腳鵯

㊟ lì ěr duǎn jiǎo bēi
㊟ 鵯：音卑

體長 length：28cm

Brown-eared Bulbul | *Hypsipetes amaurotis*

全身羽毛大致灰黑色，嘴和腳黑色。耳羽有一片黑色，頭部及下體的深灰色羽毛沾灰白色，翅膀羽毛有淡色幼邊。

Mainly greyish-black body. Black bill and legs. Black ear-coverts. Dark grey head and underparts tinged with greyish-white. Thin pale fringes on wing feathers.

1 adult 成鳥；Chai Wan 柴灣；Mar-15, 15 年 3 月；Chung Yan Tang 鄧仲欣

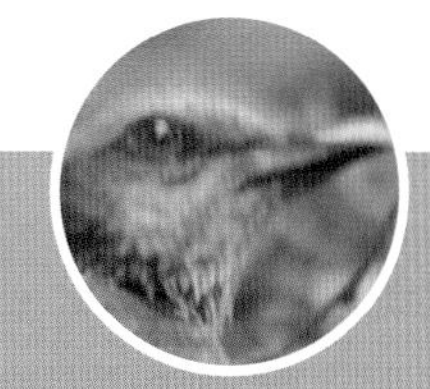

春季過境遷徙鳥 Spring Passage Migrant			夏候鳥 Summer Visitor			秋季過境遷徙鳥 Autumn Passage Migrant			冬候鳥 Winter Visitor			
1	2	3	4	5	6	7	8	9	10	11	12	常見月份
留鳥 Resident				迷鳥 Vagrant				偶見鳥 Occasional Visitor				

崖沙燕

[普] yá shā yàn

體長 length：12-13cm

Sand Martin *Riparia riparia*

其他名稱 Other names：灰沙燕

1

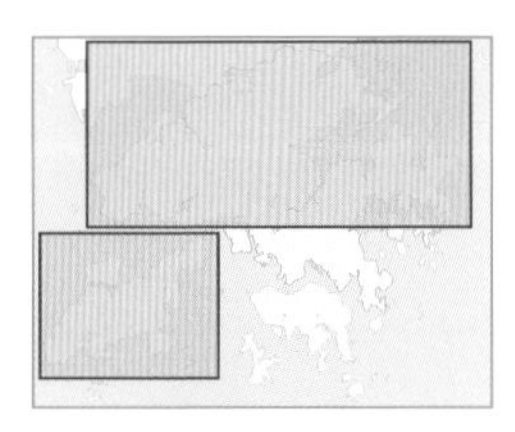

頭部和上體灰褐色，初級飛羽黑色，靜立時長及尾羽。喉部至尾下覆羽白色，胸部沾灰褐色，黑色細嘴短而尖。通常一小群出現，有時和其他雨燕或燕子混在一起。

Greyish brown head and upperparts. Black primaries extends to tip of tail at rest. Throat to undertail coverts white, breast tinted with greyish brown. Black bill is short and pointed. Usually in small flocks, sometimes mixes with other swifts and swallows.

1 Mai Po 米埔：Apr-06, 06 年 4 月：Martin Hale 夏敖天
2 Mai Po 米埔：May-07, 07 年 5 月：James Lam 林文華
3 Mai Po 米埔：Apr-07, 07 年 4 月：Sammy Sam and Winnie Wong 森美與雲妮
4 Mai Po 米埔：17-Apr-06, 06 年 4 月 17 日：Michelle and Peter Wong 江敏兒、黃理沛
5 Mai Po 米埔：May-07, 07 年 5 月：Joyce Tang 鄧玉蓮
6 Mai Po 米埔：Apr-07, 07 年 4 月：Helen Chan 陳燕芳
7 Mai Po 米埔：8-Apr-07, 07 年 4 月 8 日：Michelle and Peter Wong 江敏兒、黃理沛
8 Mai Po 米埔：Apr-06, 06 年 4 月：Henry Lui 呂德恆

春季過境遷徙鳥 Spring Passage Migrant	夏候鳥 Summer Visitor	秋季過境遷徙鳥 Autumn Passage Migrant	冬候鳥 Winter Visitor

常見月份 1 2 3 4 5 6 7 8 9 10 11 12

留鳥 Resident	迷鳥 Vagrant	偶見鳥 Occasional Visitor

2
3
4
5
6
7

8

家燕 ㊣ jiā yàn

體長 length：18cm

Barn Swallow | *Hirundo rustica*

其他名稱 Other names：House Swallow

頭及上體深色而帶藍色光澤，額及喉部紅褐色，胸部有時沾黑，腹至尾下覆羽白色。飛行時有明顯長叉尾，近尾端有白斑。嘴闊黑色，雛鳥嘴明顯黃色。常大群出沒，有時會達一千隻。多營巢於舊屋的簷蓬下。

Head and upperparts dark glossy blue, forehead and throat reddish brown, dark bar on breast. Belly to undertail coverts white. In flight, the long forked tail shows white bars near the end. Broad black bill, yellow on chicks. Appears in large flocks of up to a thousand. Often nests under eaves of old buildings.

1 adult male 雄成鳥：Tsim Bei Tsui 尖鼻咀：Apr-04, 04 年 4 月：Michelle and Peter Wong 江敏兒、黃理沛
2 juvenile moulting into breeding female, 幼羽轉換繁殖羽雌鳥：Tsim Bei Tsui 尖鼻咀：Apr-04, 04 年 4 月：Michelle and Peter Wong 江敏兒、黃理沛
3 juvenile 幼鳥：Siu Lek Yuen 小瀝源：Jul-08, 08 年 7 月：Ken Fung 馮漢城
4 adult famale 雌成鳥：Mai Po 米埔：Mar-05, 05 年 3 月：Pippen Ho 何志剛
5 adult 成鳥：Long Valley 塱原：Feb-08, 08 年 2 月：Cheng Nok Ming 鄭諾銘

春季過境遷徙鳥 Spring Passage Migrant			夏候鳥 Summer Visitor			秋季過境遷徙鳥 Autumn Passage Migrant			冬候鳥 Winter Visitor		
1	2	3	4	5	6	7	8	9	10	11	12

常見月份

留鳥 Resident	迷鳥 Vagrant	偶見鳥 Occasional Visitor

洋斑燕

普 yáng bān yàn

體長 length：13cm

Pacific Swallow | *Hirundo tahitica*

上體近黑色的深藍色，飛羽和尾部黑色。額和喉胸紅褐色，腹部近白色，嘴和腳黑色。和常見的家燕相似，但胸部缺乏藍黑色橫紋，腹部沾淡灰褐色，臀部有黑斑，尾部分叉較淺及沒有尾後飾羽。

Black upperparts with dark blue sheen. Black flight feathers and tail. Reddish-brown forehead, throat and breast. Whitish belly. Black bill and legs. Resembles common Barn Swallow but lacks dark blue breast band. Belly tinged with pale greyish-brown. Black spots on vent. Shallower-forked tail without streamers.

1 adult 成鳥；Tai Sang Wai 大生圍；Nov-16, 16 年 11 月；Wing W.S. Tang 鄧詠詩

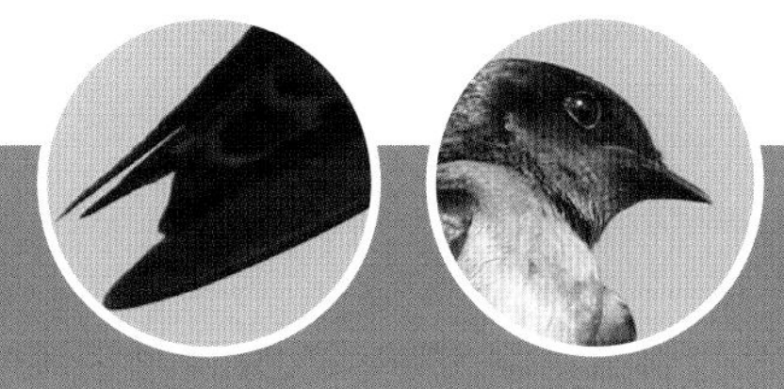

春季過境遷徙鳥 Spring Passage Migrant			夏候鳥 Summer Visitor			秋季過境遷徙鳥 Autumn Passage Migrant			冬候鳥 Winter Visitor		
1	2	3	4	5	6	7	8	9	10	11	12

常見月份

留鳥 Resident	迷鳥 Vagrant	偶見鳥 Occasional Visitor

白腹毛腳燕

㊟bái fù máo jiào yàn

體長 length：13-14cm

Common House Martin | *Delichon urbicum*

1

2

3

4

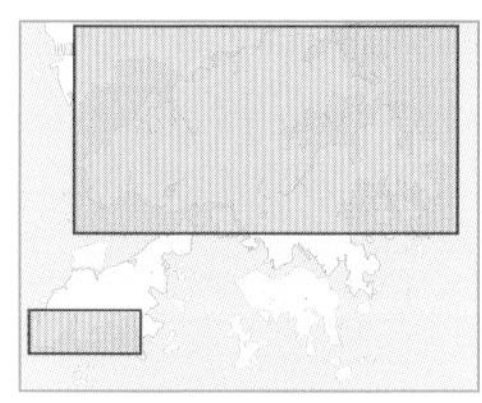

上體近黑色的深藍色，飛羽和尾部黑色，飛行時可見腰部大片白色。下體白色且較其他沙燕白，臀部灰色，尾部分叉較深。

Black upperparts with dark blue sheen. Black flight feathers and tail. White rump visible in flight. White underparts look whiter than other martins. Grey vent. Deep-forked tail.

1 adult 成鳥：Tai Sang Wai 大生圍：Nov-16, 16 年 11 月：Kinni Ho 何建業
2 adult 成鳥：Tam Kon Chau Road 担竿洲路：Dec-15, 15 年 12 月：Kinni Ho 何建業
3 adult 成鳥：Tai Sang Wai 大生圍：Nov-16, 16 年 11 月：Jemi and John Holmes 孔思義、黃亞萍
4 adult 成鳥：Tai Sang Wai 大生圍：Nov-16, 16年 11 月：Jemi and John Holmes 孔思義、黃亞萍
5 adult 成鳥：Mai Po米埔：21 Oct-17, 17 年 10 月 21 日：Beetle Cheng 鄭諾銘

5

春季過境遷徙鳥 Spring Passage Migrant	夏候鳥 Summer Visitor	**秋季過境遷徙鳥 Autumn Passage Migrant**	冬候鳥 Winter Visitor

常見月份	1	2	3	4	5	6	7	8	9	**10**	**11**	**12**

留鳥 Resident	迷鳥 Vagrant	偶見鳥 Occasional Visitor

煙腹毛腳燕

㊥yān fù máo jiào yàn

體長 length：13cm

Asian House Martin | *Delichon dasypus*

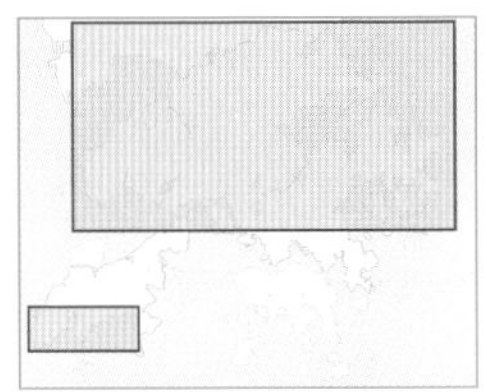

上體深藍色，翼黑色，下體偏白。腳長有偏白色毛，腰白色，尾稍分叉。

Upperparts deep blue with black wings. White underparts. Legs with whitish feathers. White rump. Tail slightly forked.

1 Long Valley 塱原：9-Mar-07, 07 年 3 月 9 日：Owen Chiang 深藍
2 Long Valley 塱原：9-Mar-07, 07 年 3 月 9 日：Owen Chiang 深藍
3 Long Valley 塱原：9-Mar-07, 07 年 3 月 9 日：Owen Chiang 深藍
4 Mar-05, 05 年 3 月：Pippen Ho 何志剛

春季過境遷徙鳥 Spring Passage Migrant			夏候鳥 Summer Visitor			秋季過境遷徙鳥 Autumn Passage Migrant			冬候鳥 Winter Visitor			
1	2	3	4	5	6	7	8	9	10	11	12	常見月份
留鳥 Resident				迷鳥 Vagrant				偶見鳥 Occasional Visitor				

金腰燕

㊕jīn yāo yàn

體長length：16-17cm

Red-rumped Swallow | *Cecropis daurica*

1

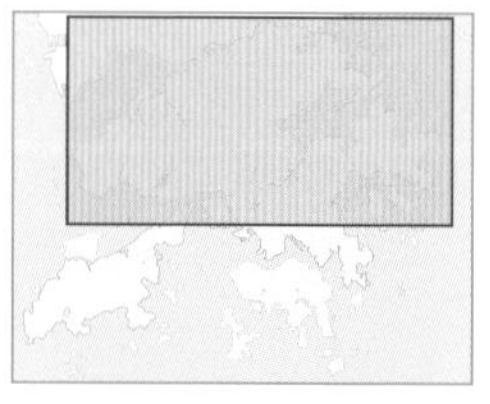

較家燕略大，上體黑色具有金屬藍色光澤，而頰至後枕淡紅褐色。腰部明顯淺色，有時甚至栗紅色，下體淡褐色有黑色縱紋。尾部打開時無明顯白斑。常與大群家燕混在一起，喜在魚塘空中覓食，時而站在電線上。

Slightly larger than Barn Swallow. Upperparts black with metallic blue light reflection. Cheeks to nape pale reddish brown with dark strips, with pale and often reddish rump. Underparts pale brownish variable fine streaking, lacks white spots on tail. Often occurs in mixed flocks with Barn Swallow. Favours flying above fishponds and perching on the electric wires.

1 Mai Po 米埔：Apr-06, 06 年 4 月：Michelle and Peter Wong 江敏兒、黃理沛
2 Mai Po 米埔：Jan-08, 08 年 1 月：Pippen Ho 何志剛
3 Kam Tin 錦田：May-08, 08 年 5 月：Cheng Nok Ming 鄭諾銘
4 Mai Po 米埔：Apr-06, 06 年 4 月：Michelle and Peter Wong 江敏兒、黃理沛
5 Mai Po 米埔：Apr-06, 06 年 4 月：Michelle and Peter Wong 江敏兒、黃理沛
6 Mai Po 米埔：Apr-06, 06 年 4 月：Michelle and Peter Wong 江敏兒、黃理沛

	春季過境遷徙鳥 Spring Passage Migrant			夏候鳥 Summer Visitor			秋季過境遷徙鳥 Autumn Passage Migrant			冬候鳥 Winter Visitor		
常見月份	1	2	3	4	5	6	7	8	9	10	11	12
	留鳥 Resident				迷鳥 Vagrant				偶見鳥 Occasional Visitor			

2
3
4
5

6

小鷦鶥

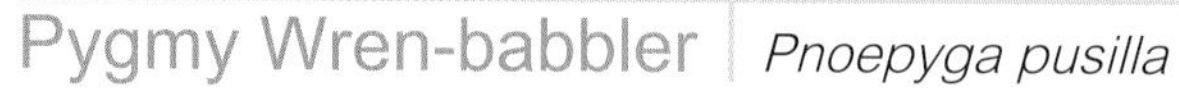

[普] xiǎo jiāo méi
[粵] 鷦鶥：音招眉

體長 length：7.5-9cm

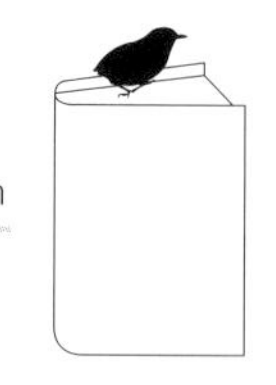

Pygmy Wren-babbler | *Pnoepyga pusilla*

其他名稱 Other names：小鱗胸鷦鶥

🌳

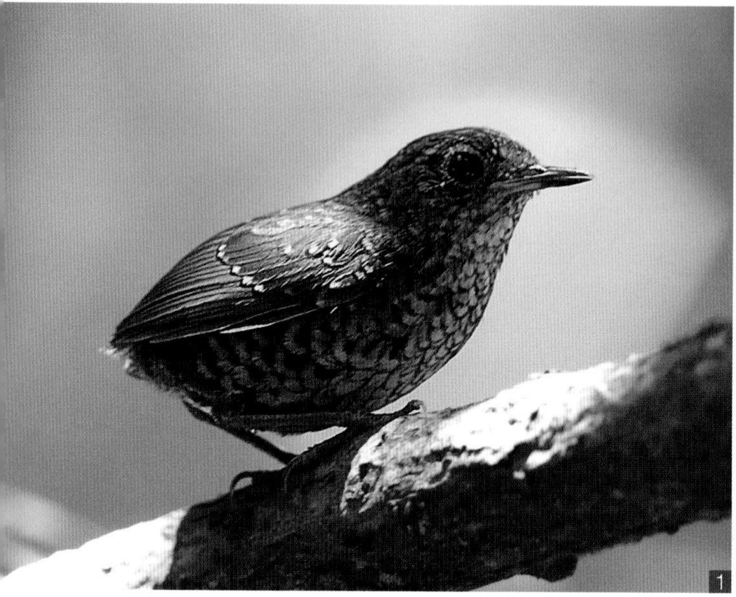

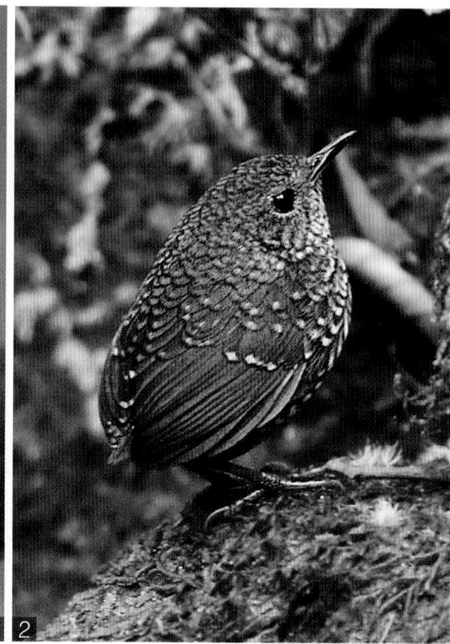

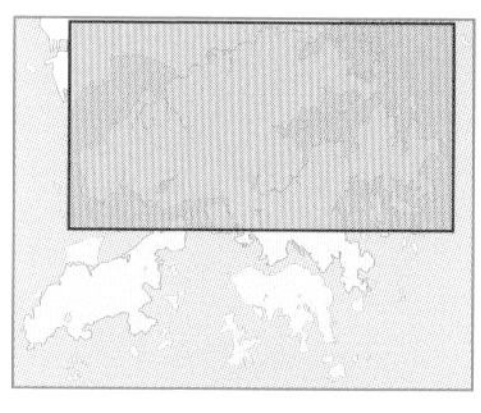

尾部非常短，頭部有淡色眼圈，嘴尖而短小，腳粉紅色。上體深褐色，上背至飛羽具有淡色羽緣。下身深褐，羽緣較白和粗，看似魚鱗紋。於樹林下層叢林活動，叫聲為明顯長而重複的口哨聲「嘶—梳」。

Very short tail, pale eye-ring, short and pointed bill, pink legs. Upperparts dark brown, with light feather fringes from mantle to flight feathers. Underparts dark brown, with thick whitish feather fringes which make it look like patterns of fish scale. Active on forest floor. Call is a long and repeating whistle "see-saw".

[1] adult pale morph 淺色型成鳥；Tai Po Kau 大埔滘；Jul-07, 07 年 7 月；Danny Ho 何國海
[2] adult pale morph 淺色型成鳥；Tai Po Kau 大埔滘；Feb-08, 08 年 2 月；Allen Chan 陳志雄
[3] adult pale morph 淺色型成鳥；Tai Po Kau 大埔滘；Jul-07, 07 年 7 月；Danny Ho 何國海
[4] adult pale morph 淺色型成鳥；Tai Po Kau 大埔滘；27-Sep-04, 04 年 9 月 27 日；Michelle and Peter Wong 江敏兒、黃理沛
[5] adult pale morph 淺色型成鳥；Tai Po Kau 大埔滘；Jul-07, 07 年 7 月；Danny Ho 何國海

春季過境遷徙鳥 Spring Passage Migrant | 夏候鳥 Summer Visitor | 秋季過境遷徙鳥 Autumn Passage Migrant | 冬候鳥 Winter Visitor

常見月份 1 2 3 4 5 6 7 8 9 10 11 12

留鳥 Resident | 迷鳥 Vagrant | 偶見鳥 Occasional Visitor

3
4
5

棕臉鶲鶯

[普] zōng liǎn wēng yīng
[粵] 鶲：音翁

體長 length：8cm

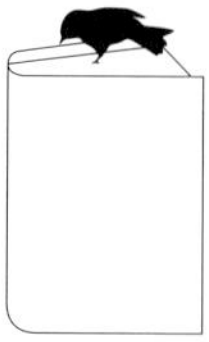

Rufous-faced Warbler | *Abroscopus albogularis*

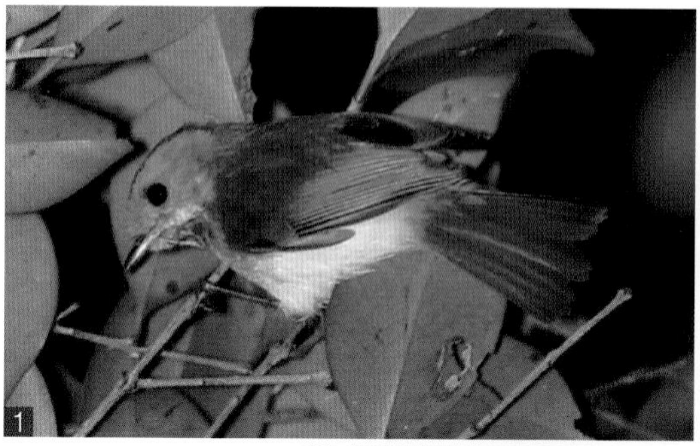
1

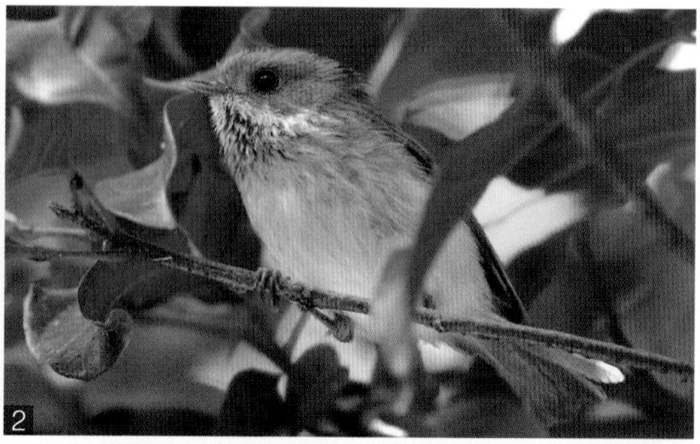
2

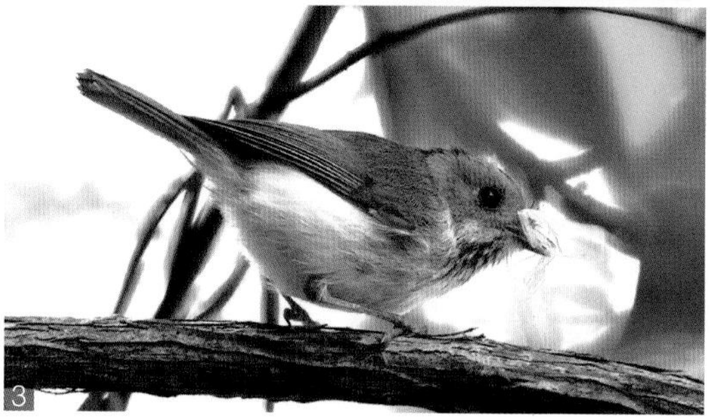
3

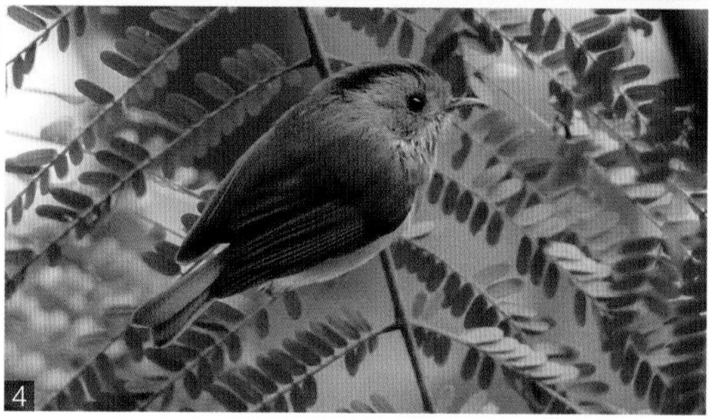
4

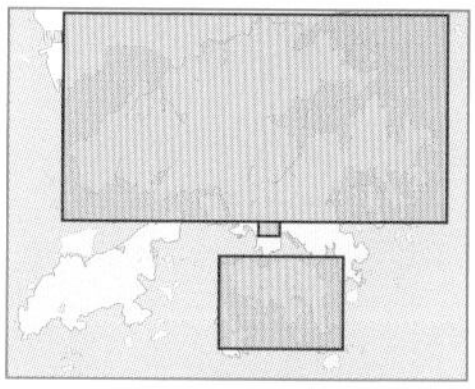

🌳

小型鶯類，有黑色長側冠紋，眼眉、眼先、耳羽和面頰鮮明黃栗色，眼睛大而深色。上體橄欖綠色，腰部黃色。喉部白色，有黑色幼縱紋。胸至尾下覆羽白色，胸部沾有黃斑。上嘴黑色，下嘴淡粉紅色。

Small-sized warbler. Long black lateral crown stripes. Prominent yellowish chestnut supercilium, lores, ear coverts and chin. Big and dark eyes. Upperparts olive-green, rump yellow. Throat is white with narrow black stripes. Breast to vent white, breast tinted yellow. Upper bill black and lower bill pale pink.

[1] Tai Po Kau 大埔滘；Jan-04, 04 年 1 月；Michelle & Peter Wong 江敏兒、黃理沛
[2] adult 成鳥；Sai Kung 西貢；Dec-19, 19 年 12 月；Roman Lo 羅文凱
[3] adult 成鳥；Sai Kung 西貢；Dec-19, 19 年 12 月；Roman Lo 羅文凱
[4] adult 成鳥；Bride's pool 新娘潭；Dec-17, 17 年 12 月；Matthew Kwan 關朗曦
[5] Tai Po Kau 大埔滘；Jan-04, 04 年 1 月；Michelle & Peter Wong 江敏兒、黃理沛

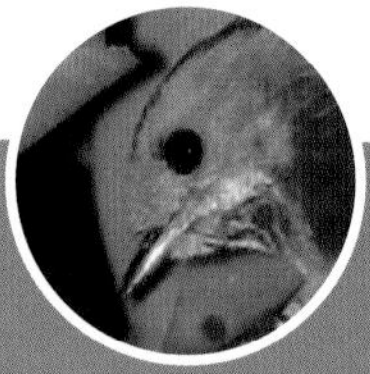

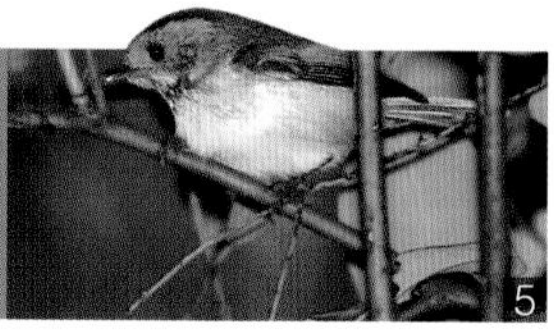
5

春季過境遷徙鳥 Spring Passage Migrant	夏候鳥 Summer Visitor	秋季過境遷徙鳥 Autumn Passage Migrant	冬候鳥 Winter Visitor

常見月份 1 2 3 4 5 6 7 8 9 10 11 12

留鳥 Resident	迷鳥 Vagrant	偶見鳥 Occasional Visitor

遠東樹鶯

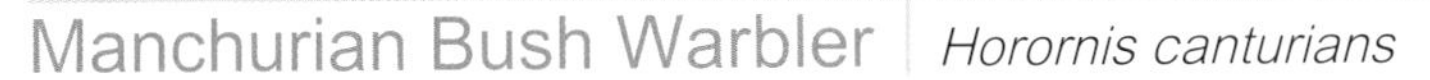

[普] yuǎn dōng shù yīng

體長 length：15-18cm

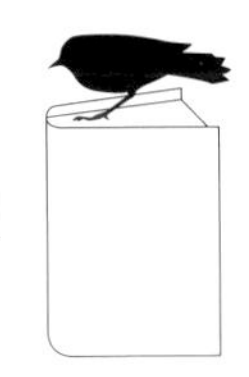

Manchurian Bush Warbler | *Horornis canturians*

其他名稱 Other names：短翅樹鶯

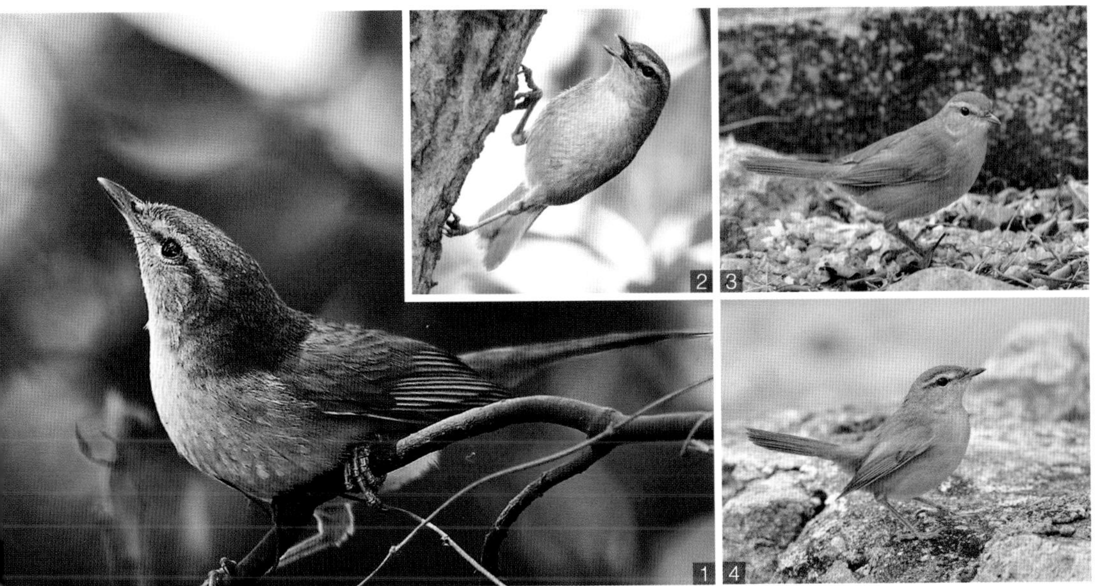

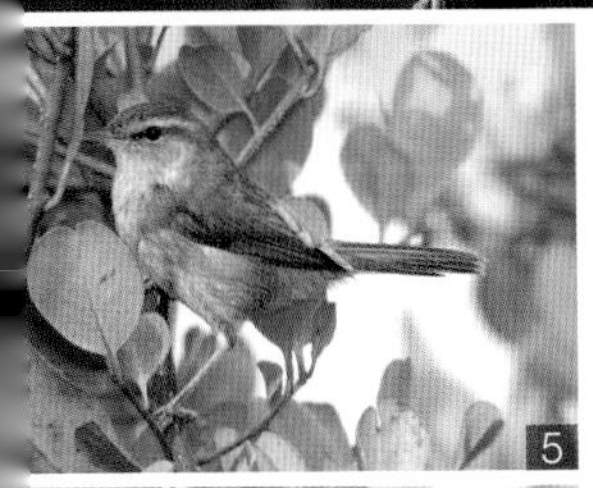

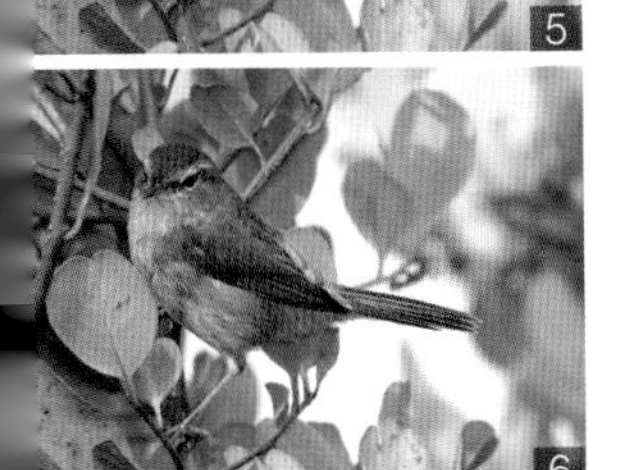

頭頂淡紅褐色，喉部、腹部至尾下覆羽帶淡褐色。有淡黃色眉紋。尾長及向上翹起，腳長呈淺粉紅色。雌雄體型不同，雄鳥明顯較大。叫聲變化多樣，由簡單的「chik-chik」至春天時輕快的「kolo-kolo-lo-witchit-chit」。一般只在樹林或開闊田野的濃密下層叢林活動。

Crown reddish brown. Throat, belly to undertail covert buffish yellow. Supercilium pale yellow. Long tail which is slightly pointed upward. Long and pinkish legs. Sexually dimorphic and male is clearly bigger than female. Call varies from a simple "chik-chik" to a sweetly "kolo-kolo-lo-witchit-chit" during spring. Prefers dense shrubs at open area or forest floor.

1 adult probably male 成鳥, 很可能是雄鳥；Mai Po 米埔；Mar-08, 08 年 3 月；Thomas Chan 陳土飛
2 adult probably female 成鳥, 很可能是雌鳥；Mai Po 米埔；Mar-08, 08 年 3 月；Eling Lee 李佩玲
3 adult 成鳥；Po Toi 蒲台；Nov-18, 18 年 11 月；Wong Wong Yung 王煌容
4 adult 成鳥；Po Toi 蒲台；Nov-18, 18 年 11 月；Wong Wong Yung 王煌容
5 adult probably female 成鳥, 很可能是雌鳥；26-Dec-05, 05 年 12 月 26 日；Michelle and Peter Wong 江敏兒、黃理沛
6 adult probably male 成鳥, 很可能是雄鳥；26-Dec-05, 05 年 12 月 26 日；Michelle and Peter Wong 江敏兒、黃理沛

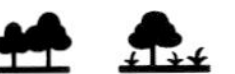

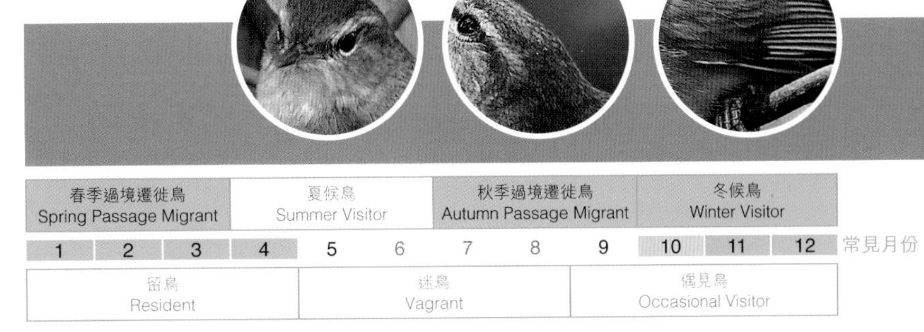

春季過境遷徙鳥 Spring Passage Migrant			夏候鳥 Summer Visitor			秋季過境遷徙鳥 Autumn Passage Migrant			冬候鳥 Winter Visitor		
1	2	3	4	5	6	7	8	9	10	11	12
留鳥 Resident				迷鳥 Vagrant				偶見鳥 Occasional Visitor			

常見月份

金頭縫葉鶯 ㊟ jīng tóu féng yè yīng

Mountain Tailorbird | *Phyllergates cucullatus*

體長length：10-12cm

小型鶯，尾長嘴長。頭頂鮮橙紅色，眼眉黃色，耳羽面頰和後枕深灰色。背部至尾部橄欖綠色，腰部黃色。喉至胸部灰色，脇部、腹部至尾下覆羽黃色。這鳥種近年在冬季定期錄得，分佈漸漸廣泛。

A small-sized warbler with long bill and tail. Crown is bright orange red, supercilium yellow. Ear coverts, chin to nape is dark grey. Olive-green mantle to tail, but the rump is yellow. Throat to breast is grey. Flanks, belly to undertail covert is yellow. This species has been found regularly wintering in Hong Kong, and is becoming widespread.

1 adult 成鳥 18-Sep-04, 04 年 9 月 18 日；Michelle and Peter Wong 江敏兒、黃理沛
2 juvenile 幼鳥；Tai Po Kau 大埔滘；May-07, 07 年 5 月；Wallace Tse 謝鑑超
3 adult 成鳥；Tai Po Kau 大埔滘；May-07, 07 年 5 月；Wallace Tse 謝鑑超
4 adult 成鳥；Tai Po Kau 大埔滘；May-07, 07 年 5 月；Wallace Tse 謝鑑超

春季過境遷徙鳥 Spring Passage Migrant			夏候鳥 Summer Visitor			秋季過境遷徙鳥 Autumn Passage Migrant			冬候鳥 Winter Visitor		
1	2	3	4	5	6	7	8	9	10	11	12
留鳥 Resident				迷鳥 Vagrant				偶見鳥 Occasional Visitor			

常見月份

強腳樹鶯

㊟ qiáng jiǎo shù yīng

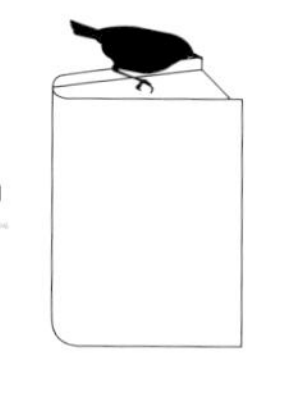

體長 length：11-12.5cm

Brown-flanked Bush Warbler | *Horornis fortipes*

其他名稱 Other names：山樹鶯, Mountain Bush Warbler

1

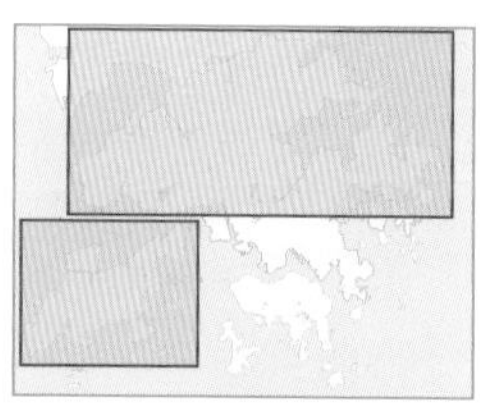

上體和脇部為均勻深褐色。眉紋皮黃色但不明顯，虹膜褐色，上嘴深褐，下嘴淺色。腹部淡色，腳粉褐色。叫聲為持續而上升而急促的「do-me-sol」和「me-sol-do」。於濃密灌叢中活動。

Uniform dark brown upperparts and flanks. Buffish yellow supercilium which is not obvious. Brown iris. Upperbill dark brown, lowerbill pale colour. Pale belly. Brownish pink legs. Calls an increasing and tight musical note "do-me-sol" and "me-sol-do". Hide in dense shrubs.

1 adult 成鳥：Long Valley 塱原：Jan-08, 08 年 1 月：Andy Cheung 張玉良
2 adult 成鳥：Po Toi 蒲台：Dec-08, 08 年 12 月：Michelle and Peter Wong 江敏兒、黃理沛
3 adult 成鳥：Long Valley 塱原：Jan-08, 08 年 1 月：Owen Chiang 深藍
4 adult 成鳥：Long Valley 塱原：Jan-08, 08 年 1 月：Owen Chiang 深藍
5 adult 成鳥：Long Valley 塱原：Jan-08, 08 年 1 月：Andy Cheung 張玉良
6 adult 成鳥：Long Valley 塱原：Jan-08, 08 年 1 月：Andy Cheung 張玉良
7 juvenile 幼鳥：Jul-04, 04 年 7 月：Michelle and Peter Wong 江敏兒、黃理沛

	春季過境遷徙鳥 Spring Passage Migrant			夏候鳥 Summer Visitor			秋季過境遷徙鳥 Autumn Passage Migrant			冬候鳥 Winter Visitor		
常見月份	1	2	3	4	5	6	7	8	9	10	11	12
	留鳥 Resident				迷鳥 Vagrant				偶見鳥 Occasional Visitor			

2
4
3
5
6

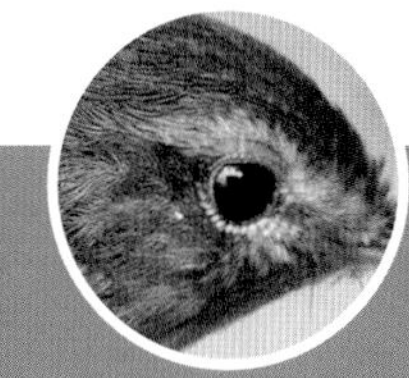

7

鱗頭樹鶯 [普]lín tóu shù yīng

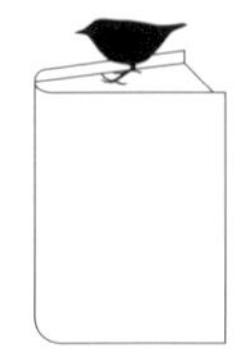

體長 length：9.5-10.5cm

Asian Stubtail | *Urosphena squameiceps*

其他名稱 Other names：Short-tailed Bush Warbler

1

2

3

頭頂、背部至尾部深褐色，喉部至腹部白色，脇及尾下覆羽呈淡黃色。有長而明顯的黃色眉紋，深褐色貫眼紋。尾甚短，腳長呈淺粉紅色。頭頂有鱗狀斑紋，但於野外較難觀察。有類似褐柳鶯的「tchek」叫聲但較短，音調也較高。一般只在樹林的地上出現。

Crown, mantle to tail dark brown. Throat to belly white, with buffish flanks and vent. Long prominent yellowish supercilium with dark brown eyestripe. Very short tail. Long pale pinkish legs. Scaly pattern on head is not easily visible in the field. Call a "tchek", similar to Dusky Warbler but shorter and higher-pitched. Favours forest floor.

[1] adult 成鳥：Wonderland Villas 華景山莊：Dec-08, 08 年 12 月：Matthew Kwan 關朗曦
[2] adult 成鳥：Po Toi 蒲台：Feb-08, 08 年 2 月：Ng Lin Yau 吳璉宥
[3] adult 成鳥：Kowloon Park 九龍公園：Nov-06, 06 年 11 月：Law Kam Man 羅錦文
[4] adult 成鳥：Po Toi 蒲台：Feb-08, 08 年 2 月：Jay Wan 溫柏豪

4

	春季過境遷徙鳥 Spring Passage Migrant			夏候鳥 Summer Visitor			秋季過境遷徙鳥 Autumn Passage Migrant			冬候鳥 Winter Visitor		
常見月份	1	2	3	4	5	6	7	8	9	10	11	12

留鳥 Resident	迷鳥 Vagrant	偶見鳥 Occasional Visitor

淡腳樹鶯

㊟ dàn jiǎo shù yīng

體長 length：11-12.5cm

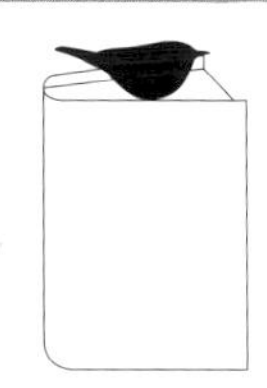

Pale-footed Bush Warbler | *Urosphena pallidipes*

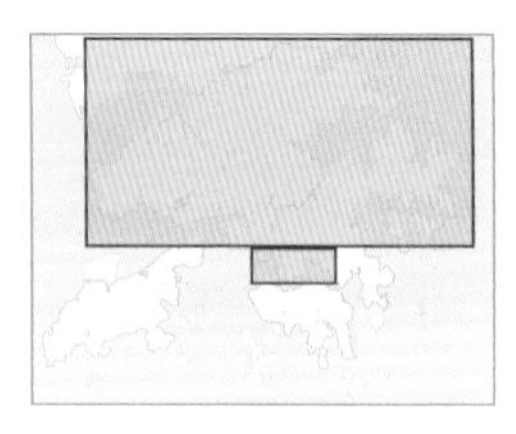

上體深橄欖褐色，頭部有顯眼的皮黃色眉紋和黑色貫眼紋。喉部及下身白色，胸脇部和尾下覆羽皮黃色，腳淺色，尾部呈方形。於密集的草叢間出沒。

Dark olive brown upperparts. Prominent buffish yellow supercilium and dark eye stripes. White from throat to underparts with buffy breast, flanks and vent. Pale pinkish legs, square tail. Prefers dense lower vegetation.

1 adult 成鳥：Lai Chi Kok 荔枝角：Jan-08, 08 年 1 月：Kelvin Yam 任德政
2 adult 成鳥：Lai Chi Kok 荔枝角：Jan-08, 08 年 1 月：Kelvin Yam 任德政
3 adult 成鳥：Lai Chi Kok 荔枝角：Jan-08, 08 年 1 月：Kelvin Yam 任德政

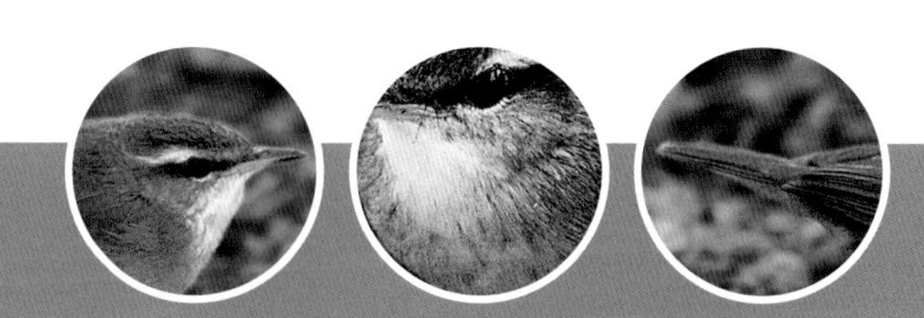

春季過境遷徙鳥 Spring Passage Migrant			夏候鳥 Summer Visitor			秋季過境遷徙鳥 Autumn Passage Migrant			冬候鳥 Winter Visitor			
1	2	3	4	5	6	7	8	9	10	11	12	常見月份
留鳥 Resident				迷鳥 Vagrant				偶見鳥 Occasional Visitor				

紅頭長尾山雀

㊟hóng tóu cháng wěi shān què

體長 length：10.5cm

Black-throated Tit | *Aegithalos concinnus*

其他名稱 Other names：Red-headed Tit

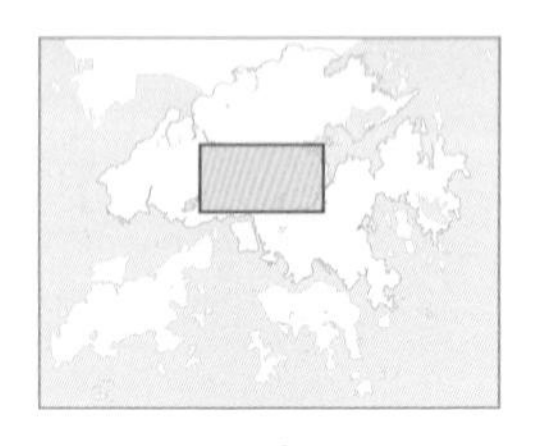

頭頂紅褐色，眼眉至面頰、嘴和喉部黑色，頰下紋白色。上體至尾部灰色，飛羽沾有褐色。脇部紅褐色，伸延至胸部成胸帶。腹部和尾下覆羽白色。幼鳥頭頂和脇部紅褐色，但喉部沒有黑色。

Crown reddish-brown. Eyebrow to cheeks, bill and throat are black. Submoustachial stripe white. Upperparts to tail grey, flight feathers brown. Flanks reddish-brown, extending to form a breast band. Belly and undertail coverts white. Juvenile has distinctive reddish-brown crown and flanks, but no black throat.

1 adult 成鳥；Tai Po Kau 大埔滘；Apr-03, 03 年 4 月；Cherry Wong 黃卓研
2 juvenile 幼鳥；Tai Po Kau 大埔滘；Apr-03, 03 年 4 月；Cherry Wong 黃卓研

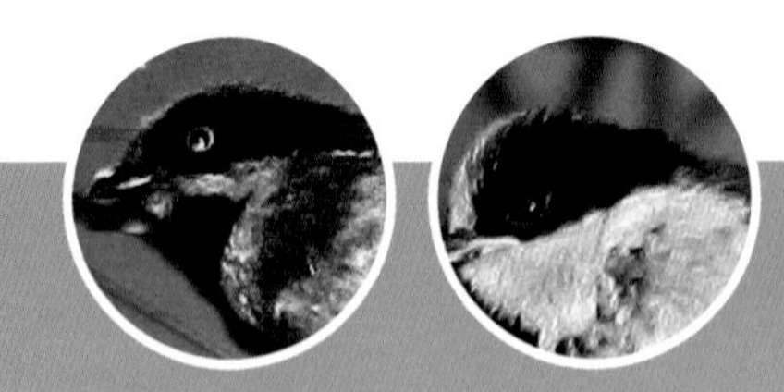

	春季過境遷徙鳥 Spring Passage Migrant			夏候鳥 Summer Visitor			秋季過境遷徙鳥 Autumn Passage Migrant			冬候鳥 Winter Visitor		
常見月份	1	2	3	4	5	6	7	8	9	10	11	12
	留鳥 Resident				迷鳥 Vagrant				偶見鳥 Occasional Visitor			

淡眉柳鶯 [普] dàn méi liǔ yīng

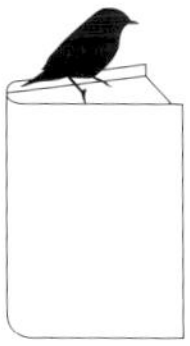

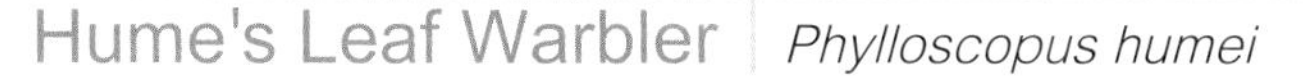

體長 length：10-11cm

Hume's Leaf Warbler | *Phylloscopus humei*

其他名稱 Other names：休氏黃眉柳鶯，休氏柳鶯 Hume's Yellow-browed Warbler

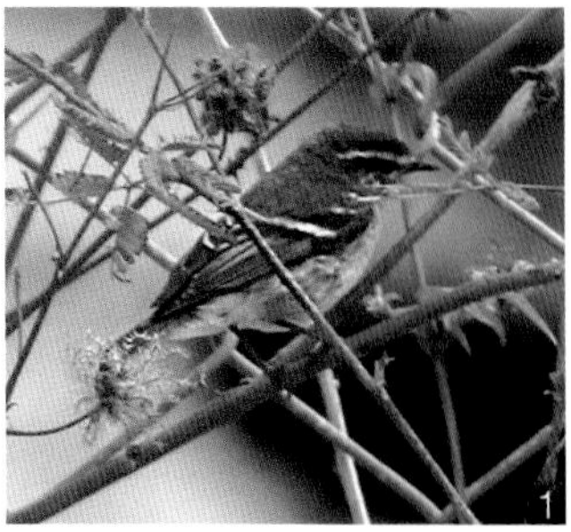
1

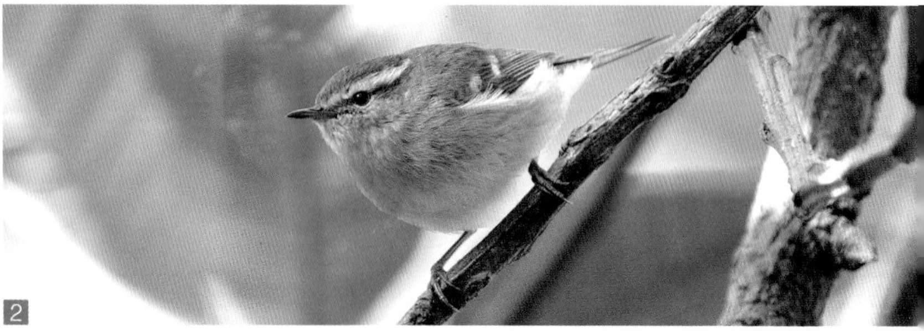
2

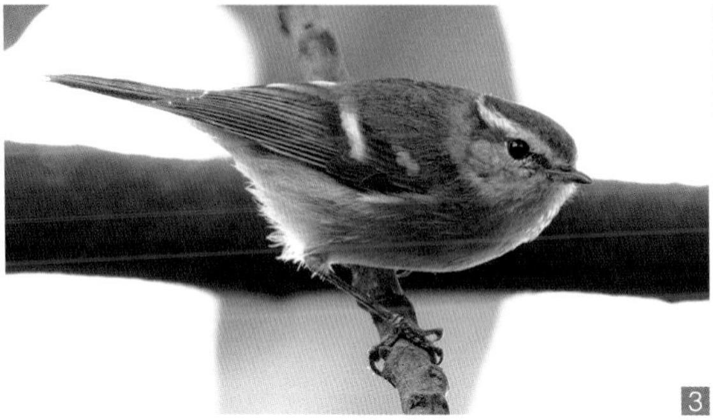

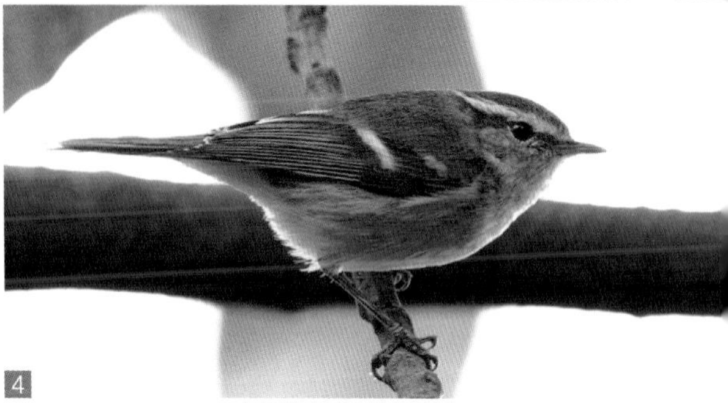
3 4

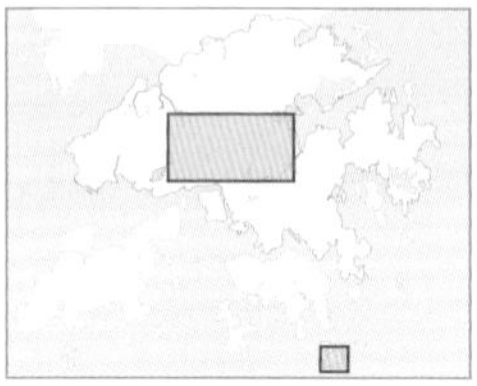

🌳

上體灰色或灰綠色，有淡灰色冠紋，淡黃側冠紋不明顯。眼前方的白色長眉紋轉幼，貫眼紋深色。有兩條白色翼帶，前方翼帶不明顯。下體白色，嘴和腳黑色。

Grey or greyish green upperparts. Pale greyish crown strip, and pale yellowish lateral crown strips which are not prominent. Long white supercilium with thinner front part. Black eye-stripes. Two white wing bars where the upper bar is not obvious. White underparts. Black bill and legs.

[1] adult 成鳥：Kam Tin 錦田：Feb-06, 06 年 2 月：Yu Yat Tung 余日東
[2] adult 成鳥：Tai Lam 大欖：Dec-19, 19 年 12 月：Roman Lo 羅文凱
[3] adult 成鳥：Tai Lam 大欖：Dec-19, 19 年 12 月：Roman Lo 羅文凱
[4] adult 成鳥：Tai Lam 大欖：Dec-19, 19 年 12 月：Roman Lo 羅文凱
[5] adult 成鳥：Tai Lam 大欖：Dec-19, 19 年 12 月：Roman Lo 羅文凱

5

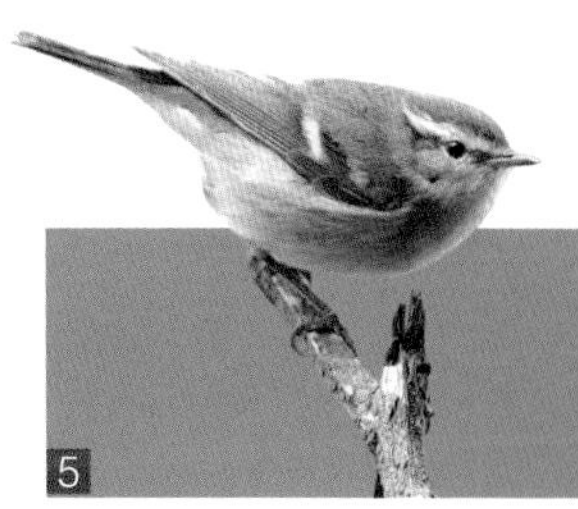

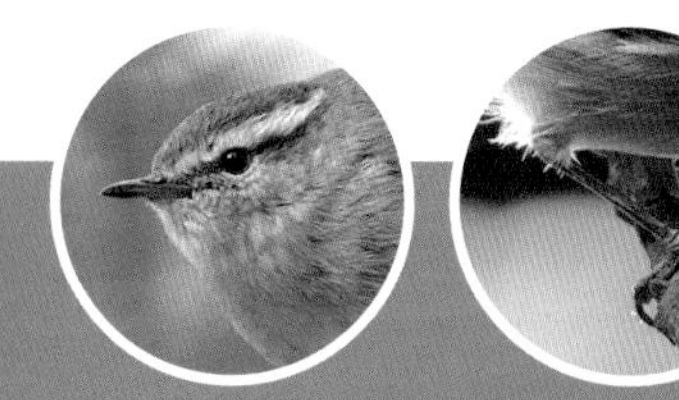

春季過境遷徙鳥 Spring Passage Migrant	夏候鳥 Summer Visitor	秋季過境遷徙鳥 Autumn Passage Migrant	冬候鳥 Winter Visitor

1 2 3 4 5 6 7 8 9 10 11 12 常見月份

留鳥 Resident	迷鳥 Vagrant	偶見鳥 Occasional Visitor

黃眉柳鶯

[普]huáng méi liǔ yīng

體長 length：10-11cm

Yellow-browed Warbler | *Phylloscopus inornatus*

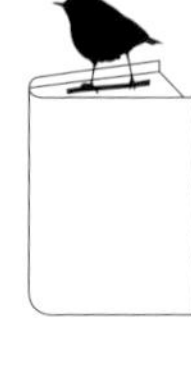

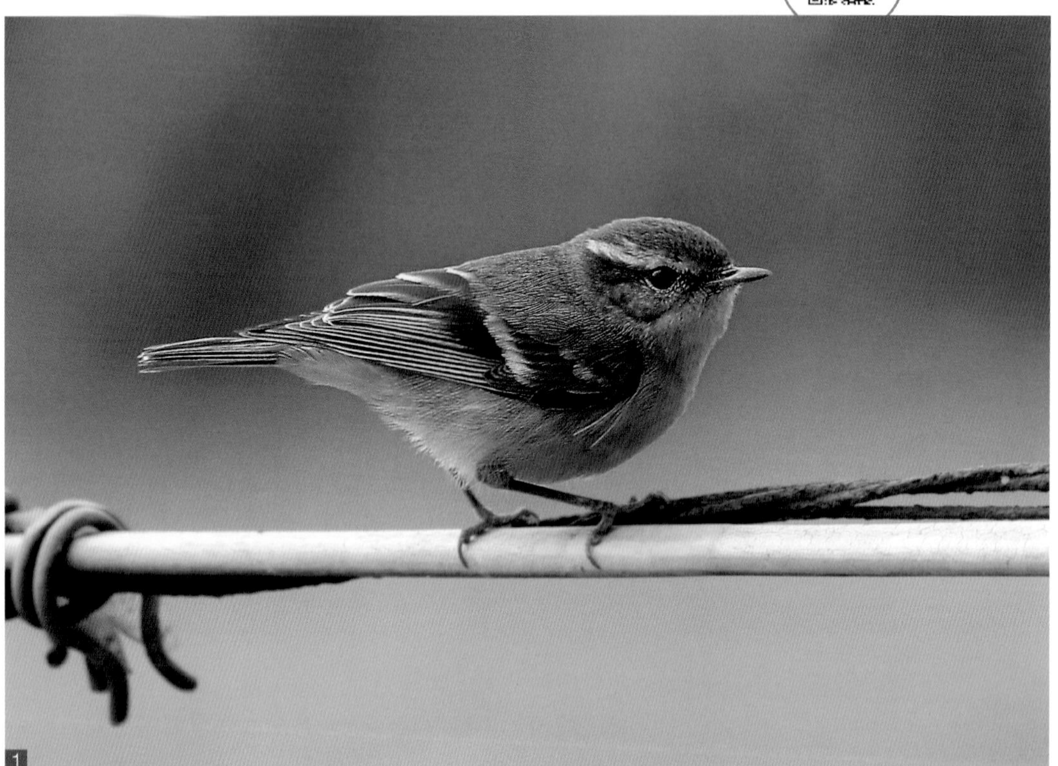

1

香港最常見的柳鶯，眼上有明顯黃色長眉紋。上體淡橄欖綠色，三級飛羽上有白邊，翅膀上有兩道白色翼帶，腰淡黃色，下體污白色，腳粉紅色。活躍好動。叫聲為「tswe-et」，與黃腰柳鶯相近，但沒有明顯輕重音之分。喜於樹林的中上層活動。

Commonest *phylloscopus* warbler in Hong Kong. Prominent yellowish long supercilium. Pale olive-green upperparts. White fringes on tertials. Two white wing bars. Pale yellow rump. Underparts dirty white. Pink legs. Active. Call "tswe-et", similar to Pallas's Leaf Warbler but plaintive. Prefers high and mid storey woodland.

1 adult 成鳥：Long Valley 塱原：6-Feb-08, 08 年 2 月 6 日：Anita Lee 李雅婷
2 adult 成鳥：Cheung Chau 長洲：Jan-06, 06 年 1 月：Henry Lui 呂德恒
3 adult 成鳥：Shek Lei Pui 石籬貝：Nov-08, 08 年 11 月：Michelle and Peter Wong 江敏兒、黃理沛
4 adult 成鳥：Shek Kong 石崗：Feb-07, 07 年 2 月：Martin Hale 夏敖天
5 adult 成鳥：Pui O 貝澳：Feb-08, 08 年 2 月：Cherry Wong 黃卓研
6 adult 成鳥：Kadoorie Farm 嘉道理農場：Apr-07, 07 年 4 月：Michelle and Peter Wong 江敏兒、黃理沛

	春季過境遷徙鳥 Spring Passage Migrant				夏候鳥 Summer Visitor			秋季過境遷徙鳥 Autumn Passage Migrant		冬候鳥 Winter Visitor		
常見月份	1	2	3	4	5	6	7	8	9	10	11	12
	留鳥 Resident				迷鳥 Vagrant				偶見鳥 Occasional Visitor			

2
3
4
5
6

黃腰柳鶯

[普]huáng yāo liǔ yīng

體長length：9-10cm

Pallas's Leaf Warbler | *Phylloscopus proregulus*

其他名稱 Other names：Lemon-rumped Warbler

1

香港體型最小的柳鶯。上體偏綠，腰部顯眼黃色，有淺黃色的冠紋和深綠色的側冠紋，眉長而呈黃色。三級飛羽有白邊，翅膀上有兩道白色翼帶，下體污白色，腳粉紅色。非常活躍好動，叫聲為兩音節的「chu-eet」聲，頭一音節較重。喜於樹林的中上層活動。

Smallest *phylloscopus* warbler in Hong Kong. Greenish upperparts with prominent yellow rump. Distinctive pale yellow crown stripe and two dark green lateral crown stripes. Long yellow supercilium. White fringes on tertials. Two white wing bars. Underparts dirty white, pink legs. Restless. Call a disyllabic "chu-eet", heavy on first note. Favours high and middle storey in wooded areas.

1 adult 成鳥：Shek Kong Catchment 石崗引水道：Jan-09, 09 年 1 月：Isaac Chan 陳家強
2 adult 成鳥：Shek Kong Catchment 石崗引水道：Jan-09, 09 年 1 月：Michelle and Peter Wong 江敏兒、黃理沛
3 adult 成鳥：Shek Kong Catchment 石崗引水道：Jan-09, 09 年 1 月：Michelle and Peter Wong 江敏兒、黃理沛
4 adult 成鳥：Tai Po Kau 大埔滘：Feb-05, 05 年 2 月：Cherry Wong 黃卓研
5 adult 成鳥：Cheung Chau 長洲：Mar-06, 06 年 3 月：Henry Lui 呂德恒
6 adult 成鳥：Po Toi 蒲台：Dec-08, 08 年 12 月：Ng Lin Yau 吳璉宥
7 adult 成鳥：Martin Hale 夏敖天

春季過境遷徙鳥 Spring Passage Migrant	夏候鳥 Summer Visitor	秋季過境遷徙鳥 Autumn Passage Migrant	冬候鳥 Winter Visitor

常見月份 1 2 3 4 5 6 7 8 9 10 11 12

留鳥 Resident	迷鳥 Vagrant	偶見鳥 Occasional Visitor

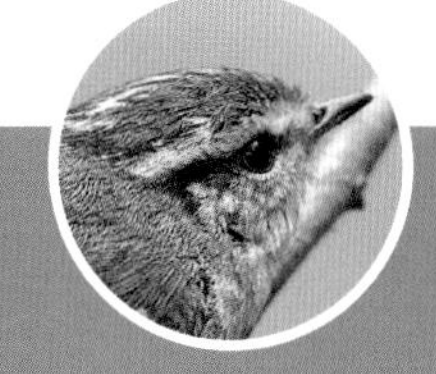

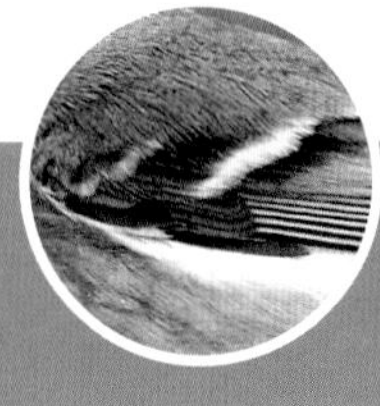

棕眉柳鶯

[普] zōng méi liǔ yīng

體長 length：12cm

Yellow-streaked Warbler | *Phylloscopus armandii*

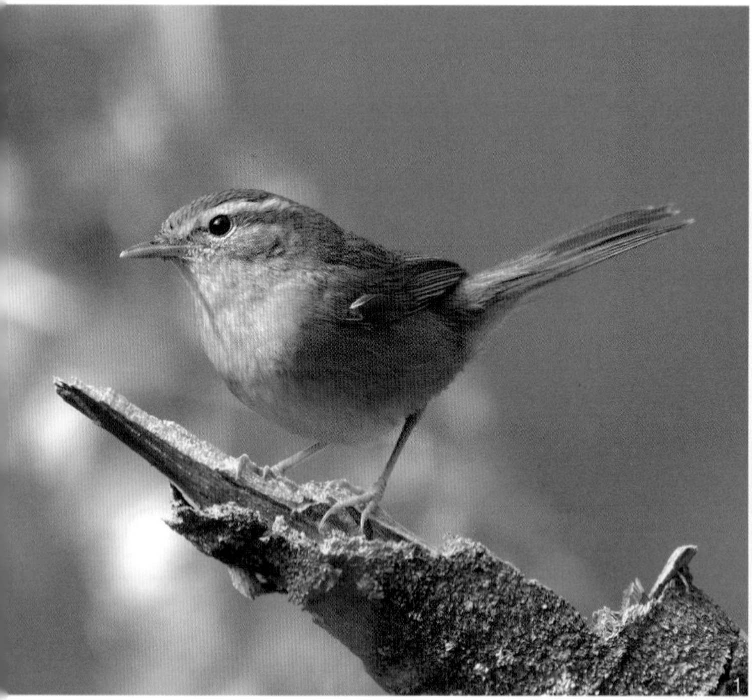

1

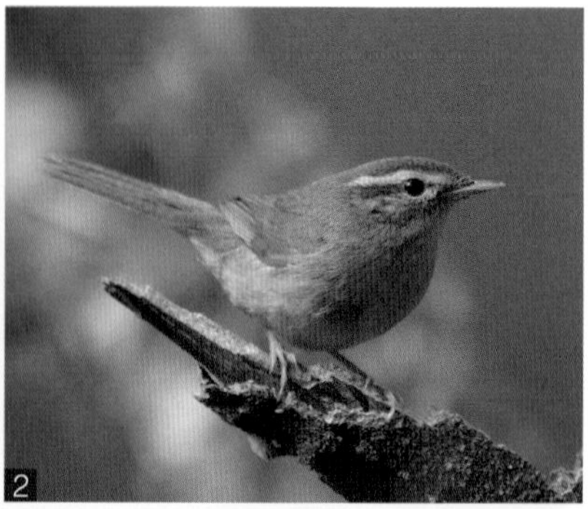

2

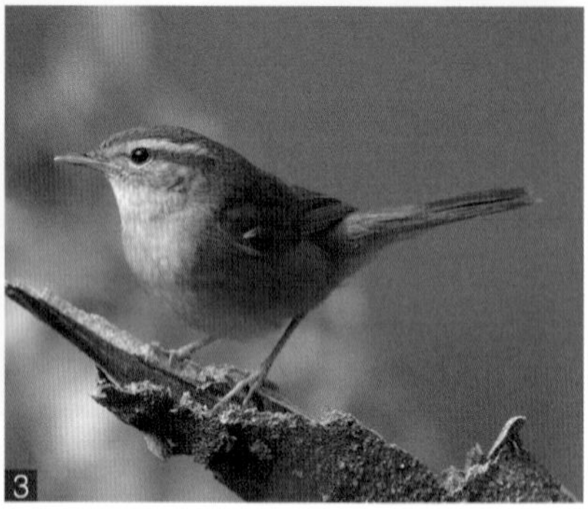

3

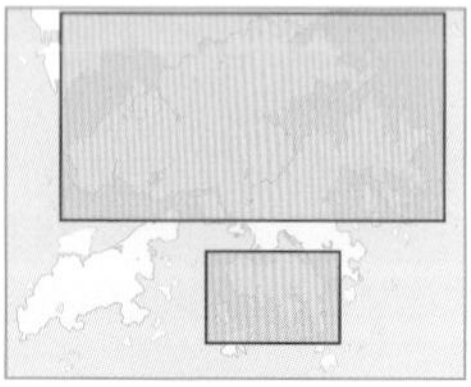

🌳 🌳

上體橄欖褐色，無翼斑，下體黃白色，胸側及脇部沾橄欖色。有偏白色眉紋及灰黑色貫眼紋，臉側有雜斑，上嘴褐色，下嘴較淡，腳黃褐色。

Olive-brown upperparts. No wing bars. Yellowish-white underparts. Sides of breast and flanks tinged with olive green. Whitish supercilia and greyish-black eye-stripes. Streaked face. Brown upper bill with paler lower bill. Yellowish-brown legs.

[1] adult 成鳥：Mount Davis 摩星嶺：Dec-18, 18 年 12 月：Roman Lo 羅文凱
[2] adult 成鳥：Mount Davis 摩星嶺：Dec-18, 18 年 12 月：Roman Lo 羅文凱
[3] adult 成鳥：Mount Davis 摩星嶺：Dec-18, 18 年 12 月：Roman Lo 羅文凱

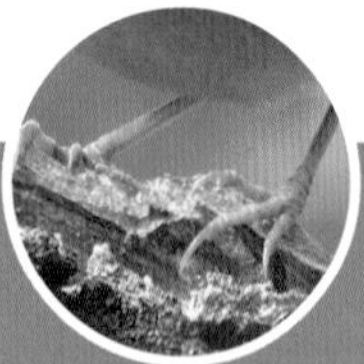

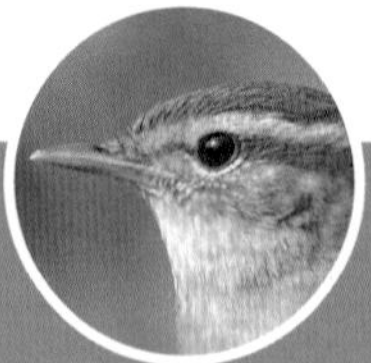

春季過境遷徙鳥 Spring Passage Migrant	夏候鳥 Summer Visitor	秋季過境遷徙鳥 Autumn Passage Migrant	冬候鳥 Winter Visitor

常見月份	1	2	3	4	5	6	7	8	9	10	11	12

留鳥 Resident	迷鳥 Vagrant	偶見鳥 Occasional Visitor

歐柳鶯

㊟ ōu liǔ yīng

體長 length：11-12.5cm

Willow Warbler | *Phylloscopus trochilus*

體型略小無突出的柳鶯，上體綠褐色，無冠紋和翼斑。眉紋淡色，嘴基部淡黃色，有較長的初級飛羽。喉和胸淡黃，腳淡色。

Small, indistinct warbler, with greyish-green upperparts. No crown stripe or wing bar. Pale supercilium. Light yellowish bill base, with long primary projection. Throat and breast pale yellow. Pale legs.

1 adult 成鳥：Long Valley 塱原：Oct-08, 08 年 10 月：Frankie Chu 朱錦滿

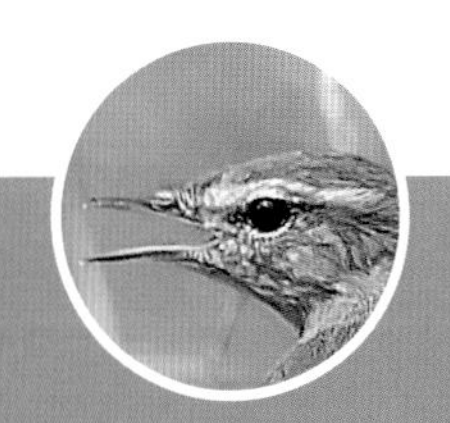

春季過境遷徙鳥 Spring Passage Migrant	夏候鳥 Summer Visitor	秋季過境遷徙鳥 Autumn Passage Migrant	冬候鳥 Winter Visitor

1 2 3 4 5 6 7 8 9 10 11 12 常見月份

留鳥 Resident	迷鳥 Vagrant	偶見鳥 Occasional Visitor

巨嘴柳鶯 ㊟jù zuǐ liǔ yīng

體長 length：12.5-13.5cm

Radde's Warbler | *Phylloscopus schwarzi*

外貌和褐柳鶯十分相似，但體型較大，尾部比例上較褐柳鶯短，嘴較粗壯，但在野外不易分辨。上體橄欖褐色。眉長而闊，淺黃色，在眼前的一段較模糊，一直伸延至後頸。下體沾黃色。叫聲為特別的「quip」聲，喜接近地面活動。

Very similar to Dusky Warbler but is larger, has proportionally shorter tail and stouter bill but not easily distinguishable in the field. Upperparts olive-brown. Long and broad buffish supercilium, ill-defined in front of the eye and extends to nape. Underparts have yellowish wash. Distinctive call "quip". Stays close to ground.

1 adult 成鳥：Shek Kong Catchment 石崗引水道：Jan-09, 09 年 1 月：Michelle and Peter Wong 江敏兒、黃理沛
2 adult 成鳥：Tai Po Kau 大埔滘：1-Nov-03, 03 年 11 月：Michelle and Peter Wong 江敏兒、黃理沛
3 adult 成鳥：Shek Kong Catchment 石崗引水道：Jan-09, 09 年 1 月：Michelle and Peter Wong 江敏兒、黃理沛
4 adult 成鳥：Tai Po Kau 大埔滘：1-Nov-03, 03 年 11 月：Michelle and Peter Wong 江敏兒、黃理沛
5 adult 成鳥：Shek Kong Catchment 石崗引水道：Jan-09, 09 年 1 月：Michelle and Peter Wong 江敏兒、黃理沛
6 adult 成鳥：Po Toi 蒲台：9-Nov-06, 06 年 11 月 9 日：Geoff Welch

春季過境遷徙鳥 Spring Passage Migrant	夏候鳥 Summer Visitor	秋季過境遷徙鳥 Autumn Passage Migrant	冬候鳥 Winter Visitor

常見月份 1 2 3 4 5 6 7 8 9 **10 11 12**

留鳥 Resident	迷鳥 Vagrant	偶見鳥 Occasional Visitor

2
3
4
5
6

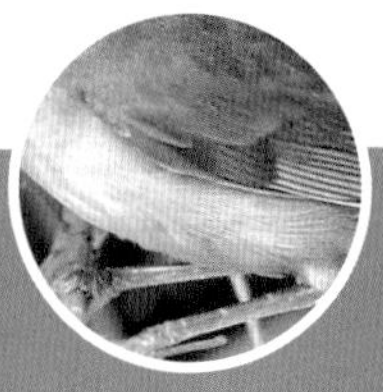

褐柳鶯 ㊟hè liǔ yīng

體長 length：11-12cm

Dusky Warbler | *Phylloscopus fuscatus*

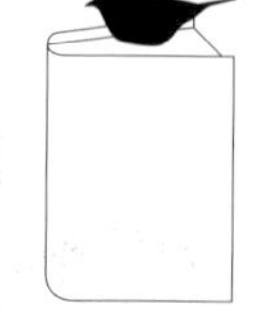

1

全身褐色，有明顯淺黃色眉紋。喉至腹部偏白，脇及尾下覆羽淺黃色。叫聲為重複不斷的「tsak-tsak-tsak」聲，為主要辨認特徵。喜接近地面和水邊活動。

Brown warbler with prominent buff supercilium. Whitish throat to belly, strong buff on flanks and vent. Best identified by its call, which is a distinctive repeating "tsak-tsak-tsak". Stays close to ground and near water.

1 adult 成鳥：Kam Tin 錦田：Jan-07, 07 年 1 月：Winnie Wong and Sammy Sam 森美與雲妮
2 adult 成鳥：HK Wetland Park 香港濕地公園：Oct-06, 06 年 10 月：Law Kam Man 羅錦文
3 adult 成鳥：Mai Po 米埔：Dec-06, 06 年 12 月：Matthew and TH Kwan 關朗曦、關子凱
4 adult 成鳥：Mai Po 米埔：18-Nov-06, 06 年 11 月 18 日：Michelle and Peter Wong 江敏兒、黃理沛
5 adult 成鳥：Long Valley 塱原：Oct-04, 04 年 10 月：Henry Lui 呂德恒
6 adult 成鳥：Fung Lok Wai 豐樂圍：Nov-03, 03 年 11 月：Jemi and John Holmes 孔思義、黃亞萍

	春季過境遷徙鳥 Spring Passage Migrant			夏候鳥 Summer Visitor			秋季過境遷徙鳥 Autumn Passage Migrant			冬候鳥 Winter Visitor		
常見月份	1	2	3	4	5	6	7	8	9	10	11	12
	留鳥 Resident				迷鳥 Vagrant				偶見鳥 Occasional Visitor			

2
3
4
5

6

嘰喳柳鶯

㊉ jī zhā liǔ yīng
㊐ 嘰喳：音機渣

體長 length：11-12cm

Common Chiffchaff | *Phylloscopus collybita*

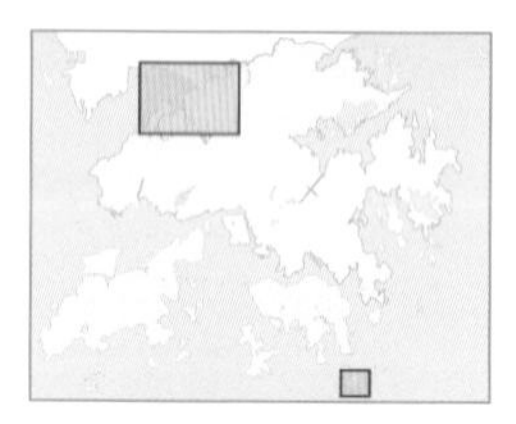

上體綠褐色，無顯著翼斑，下體偏白，脇部帶淡黃褐色。有淡黃褐色眉紋及灰黑色貫眼紋，嘴和腳黑色。

Greenish-brown upperparts. No obvious wing bars. Whitish underparts with pale yellowish-brown flanks. Pale yellowish-brown supercilia and greyish-black eye-stripe. Black bill and legs.

1 adult 成鳥：Long Valley 塱原：Jan-14, 14 年 1 月：Kinni Ho 何建業
2 adult 成鳥：Mai Po 米埔：Dec-17, 17 年 12 月：Chan Jun Siu Brian 陳俊兆
3 adult 成鳥：Long Valley 塱原：Jan-14, 14 年 1 月：Doris Chu 朱詠兒
4 Long Valley 塱原：5 Jan-14, 14 年 1 月 14 日：Beetle Cheng 鄭諾銘

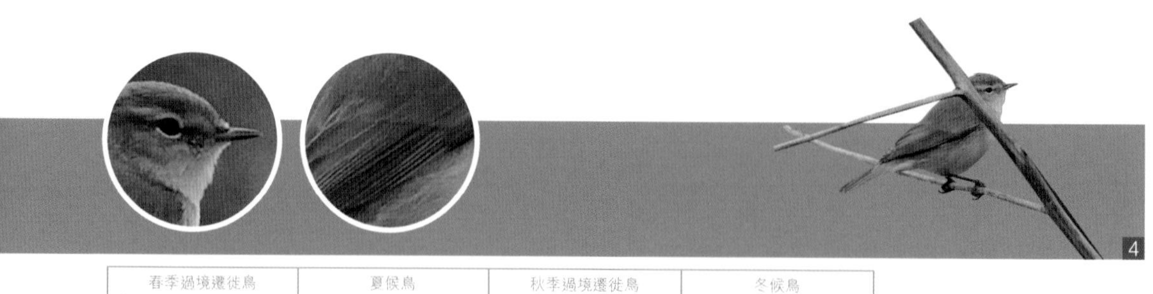

	春季過境遷徙鳥 Spring Passage Migrant			夏候鳥 Summer Visitor			秋季過境遷徙鳥 Autumn Passage Migrant			冬候鳥 Winter Visitor		
常見月份	1	2	3	4	5	6	7	8	9	10	11	12
	留鳥 Resident				迷鳥 Vagrant				偶見鳥 Occasional Visitor			

冕柳鶯

㊟miǎn liǔ yīng
㊟冕：音免

體長length：11-12cm

Eastern Crowned Warbler | *Phylloscopus coronatus*

上體鮮明黃橄欖色，有幼長淡色冠紋和眉紋，眼先和過眼線黑色，飛羽有淡色羽緣，有一條黃色翼帶。下體白色，臀部淡黃色。

Bright yellowish-olive upperparts. Long and narrow pale crown strip and supercilium. Dark lore and eye-stripe. Flight feathers has pale frings with one yellow wing bar. White underparts and yellow vent.

1 adult 成鳥；29-Oct-05, 05 年 10 月 29 日；Michelle and Peter Wong 江敏兒、黃理沛
2 adult 成鳥；Po Toi 蒲台；4-Apr-07, 07 年 4 月 4 日；Geoff Welch
3 adult 成鳥；Tai Po Kau 大埔滘；Aug-08, 08 年 8 月；Allen Chan 陳志雄
4 adult 成鳥；Po Toi 蒲台；13-Oct-06, 06 年 10 月 13 日；Lo Kar Man 盧嘉孟

春季過境遷徙鳥 Spring Passage Migrant	夏候鳥 Summer Visitor	秋季過境遷徙鳥 Autumn Passage Migrant	冬候鳥 Winter Visitor

1 2 3 4 5 6 7 8 9 10 11 12 常見月份

留鳥 Resident	迷鳥 Vagrant	偶見鳥 Occasional Visitor

飯島柳鶯

[普] fàn dǎo liǔ yīng

體長 length：11-12cm

Ijima's Leaf Warbler | *Phylloscopus ijimae*

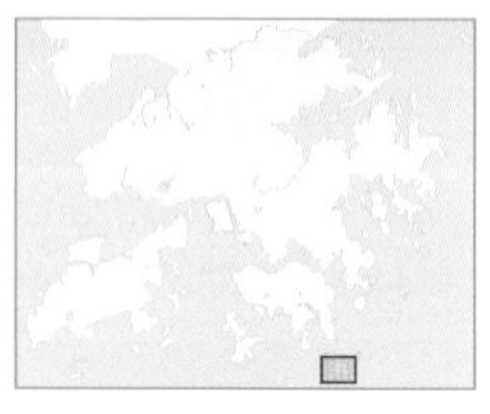

與較常見的冕柳鶯相似但體型較小。上體綠褐色，有一條近白色幼翼斑，下體近白色，尾下覆羽偏黃。頭部偏灰色，無頂冠紋。上嘴黑褐色，下嘴粉紅色，腳肉色。

Resemble more common Eastern Crowned Warbler but smaller-sized. Greenish-brown upperparts. One thin whitish wing bar. Whitish underparts. Yellowish undertail-coverts. Greyish head without median crown-stripe. Black upper bill with pink lower bill. Pinkish-brown legs.

1 adult 成鳥；Po Toi 蒲台；Apr-15, 15 年 4 月；Kinni Ho 何建業
2 adult 成鳥；Po Toi 蒲台；Apr-15, 15 年 4 月；Yam Wing Yiu 任永耀
3 adult 成鳥；Po Toi 蒲台；Apr-15, 15 年 4 月；Yam Wing Yiu 任永耀
4 adult 成鳥；Po Toi 蒲台；Apr-15, 15 年 4 月；Yam Wing Yiu 任永耀

春季過境遷徙鳥 Spring Passage Migrant	夏候鳥 Summer Visitor	秋季過境遷徙鳥 Autumn Passage Migrant	冬候鳥 Winter Visitor

常見月份 1 2 3 4 5 6 7 8 9 10 11 12

留鳥 Resident	迷鳥 Vagrant	偶見鳥 Occasional Visitor

比氏鶲鶯

㊗ bǐ shì wēng yīng

體長 length：11-12cm

Bianchi's Warbler | *Phylloscopus valentini*

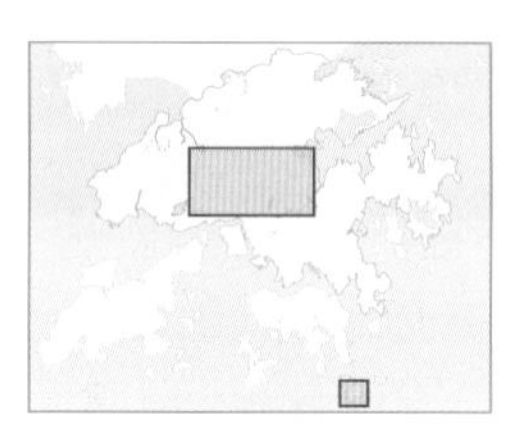

似灰冠鶲鶯，頭頂灰黑相間，有一條不明顯翼斑。頭頂灰色，有顯眼的金黃色眼圈。上體偏綠色，下體檸檬黃色。上嘴黑色，下嘴黃褐色，腳黃褐色。

Similar to Grey-crowned Warbler. Black and grey stripes on crown. One vague wing bar. Grey crown. Conspicuous golden-yellow eye-rings. Greenish upperparts with lemon yellow underparts. Black upper bill with yellowish-brown lower bill. Yellowish-brown legs.

1 adult 成鳥；Tai Po Kau 大埔滘；Feb-20, 20 年 2 月；Roman Lo 羅文凱
2 adult 成鳥；Tai Po Kau 大埔滘；Feb-20, 20 年 2 月；Roman Lo 羅文凱
3 adult 成鳥；Tai Po Kau 大埔滘；Feb-20, 20 年 2 月；Roman Lo 羅文凱

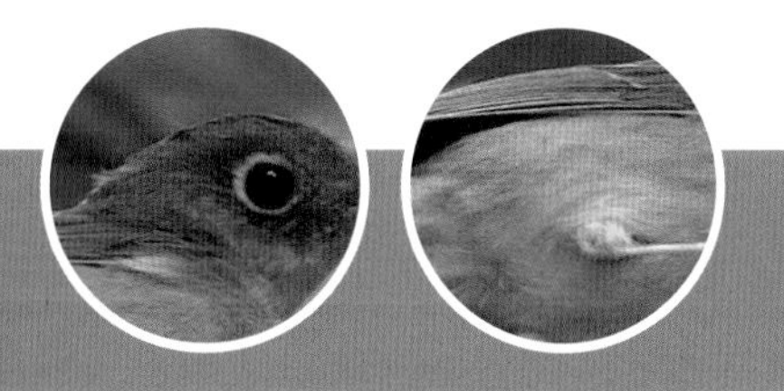

春季過境遷徙鳥 Spring Passage Migrant	夏候鳥 Summer Visitor	秋季過境遷徙鳥 Autumn Passage Migrant	冬候鳥 Winter Visitor

1 2 3 4 5 6 7 8 9 10 11 12 常見月份

留鳥 Resident	迷鳥 Vagrant	偶見鳥 Occasional Visitor

白眶鶲鶯

㊥bái kuàng wēng yīng

體長length：11-12cm

White-spectacled Warbler | *Phylloscopus intermedius*

1

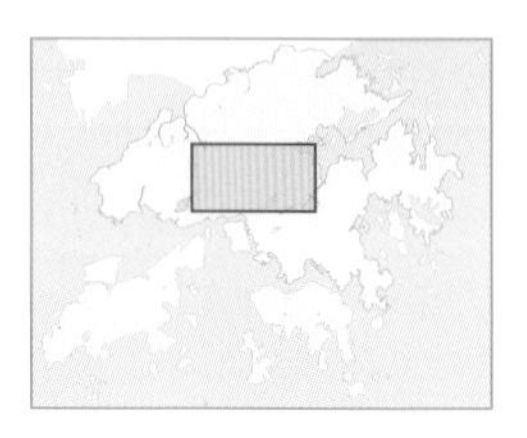

上體橄欖綠色，頭頂偏灰，有黑色長側冠紋，喉至臀部鮮明黃色。明顯上截斷開的黃色眼圈是這鳥種的主要特徵。翼部一般有兩條翼帶，尾下主要偏灰，尾兩側有些白色。上嘴黑色，下嘴黑而有橙色粗邊緣，腳偏粉紅色。偏好於森林的中及下層活動。

Upperparts olive-green, with greyish crown and long black lateral crown stripes. Throat to vent bright yellow. Prominent broken (on top) yellow eye ring is diagnostic. Two wing bars are usually present. Undertail is mainly greyish with some white on sides. Black upper bill and lower bill black with thick orange edge. Pinkish legs. Favours mid and lower storeys in forest.

1 adult 成鳥：Tai Po Kau 大埔滘：16-Jan-04, 04 年 1 月 16 日：Michelle and Peter Wong 江敏兒、黃理沛
2 adult 成鳥：Tai Po Kau 大埔滘：Feb-05, 05 年 2 月：Michelle and Peter Wong 江敏兒、黃理沛
3 adult 成鳥：Tai Po Kau 大埔滘：11-Jan-04, 04 年 1 月 11 日：Michelle and Peter Wong 江敏兒、黃理沛
4 adult 成鳥：Tai Po Kau 大埔滘：Jan-19, 19 年 1 月：Roman Lo 羅文凱
5 adult 成鳥：Tai Po Kau 大埔滘：Jan-20, 20 年 1 月：Roman Lo 羅文凱
6 adult 成鳥：Tai Po Kau 大埔滘：Feb-20, 20 年 2 月：Roman Lo 羅文凱
7 adult 成鳥：Tai Po Kau 大埔滘：Feb-20, 20 年 2 月：Roman Lo 羅文凱
8 adult 成鳥：Tai Po Kau 大埔滘：16-Jan-04, 04 年 1 月 16 日：Michelle and Peter Wong 江敏兒、黃理沛

春季過境遷徙鳥 Spring Passage Migrant			夏候鳥 Summer Visitor			秋季過境遷徙鳥 Autumn Passage Migrant			冬候鳥 Winter Visitor		
常見月份 1	2	3	4	5	6	7	8	9	10	11	12
留鳥 Resident				迷鳥 Vagrant				偶見鳥 Occasional Visitor			

2
3
4
5
6
7

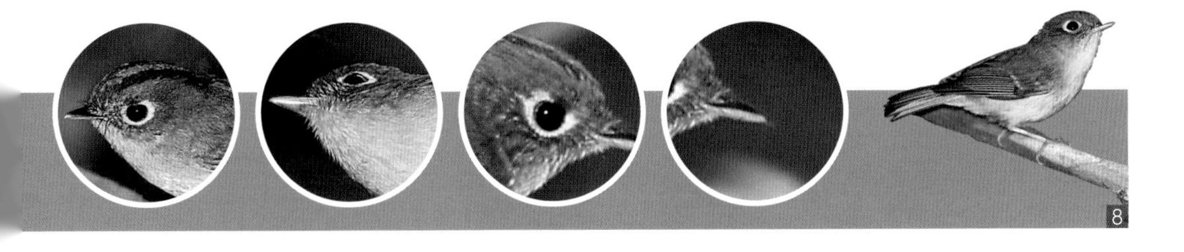

8

純色尾鶲鶯

㊟ chún sè wěi wēng yīng

體長 length：11-12cm

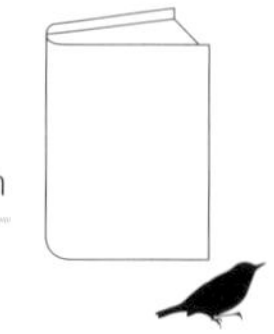

Alstrom's Warbler | *Phylloscopus soror*

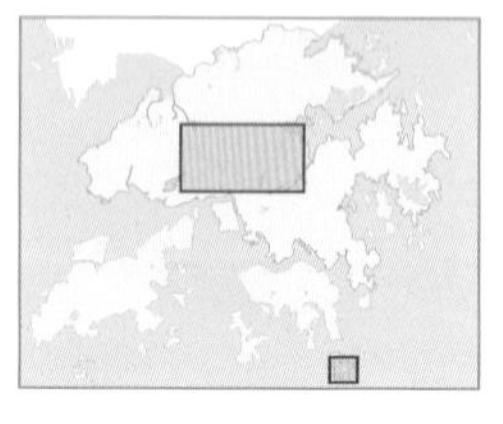

上體灰青色，頭頂灰藍色，有前段模糊的明顯深色冠紋，具金黃色眼圈，外側尾羽白色，下體鮮黃色。上嘴黑色，下嘴黃色，腳黃褐色。

Greyish-green upperparts. Greyish-blue crown. Prominent dark lateral crown-stripes vague near forehead. Golden-yellow eye-rings. White outer tail feathers. Bright yellow underparts. Black upper bill with yellow lower bill. Yellowish-brown legs.

1 adult 成鳥；Po Toi 蒲台；Oct-19, 19 年 10 月；Roman Lo 羅文凱
2 adult 成鳥；Po Toi 蒲台；Oct-19, 19 年 10 月；Roman Lo 羅文凱

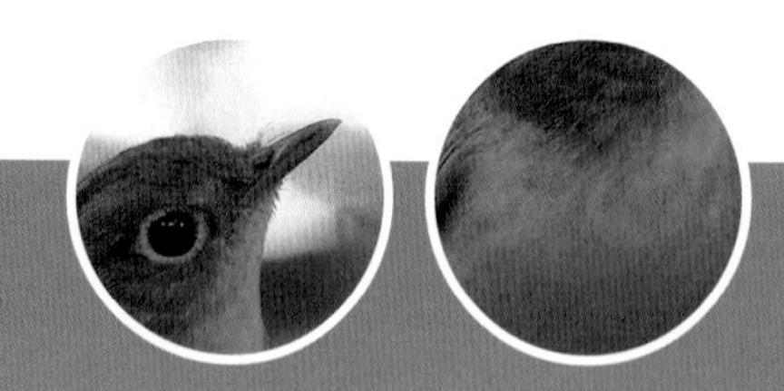

春季過境遷徙鳥 Spring Passage Migrant			夏候鳥 Summer Visitor			秋季過境遷徙鳥 Autumn Passage Migrant			冬候鳥 Winter Visitor		
常見月份 1	2	3	4	5	6	7	8	9	10	11	12
留鳥 Resident				迷鳥 Vagrant				偶見鳥 Occasional Visitor			

雙斑柳鶯

㊟ shuāng bān liǔ yīng

體長 length：11.5-12cm

Two-barred Warbler | *Phylloscopus plumbeitarsus*

上體深綠色，淡色眉紋延至嘴基，下嘴橙粉紅，常見兩條淡色翼帶。下體淡色。與黃眉柳鶯比較，沒有三級飛羽沒有淺色羽緣，翼帶較幼，嘴較長，腳藍灰色。

Green upperparts. The pale supercilium reaches the base of the bill. Lower bill is pinkish orange. Two pale wing bars are usually seen. Pale underparts. Compare with Yellow-browed Warbler, it has no fringes on tertials, thinner wing bars, longer bill and bluish grey legs.

1 adult 成鳥；Fung Yuen 鳳園；Dec-07, 07 年 12 月；Thomas Chan 陳土飛
2 adult 成鳥；Po Toi 蒲台；Oct-08, 08 年 10 月；Isaac Chan 陳家強
3 Shek Kong 石崗；Dec-09, 09 年 12 月；Michelle and Peter Wong 江敏兒、黃理沛

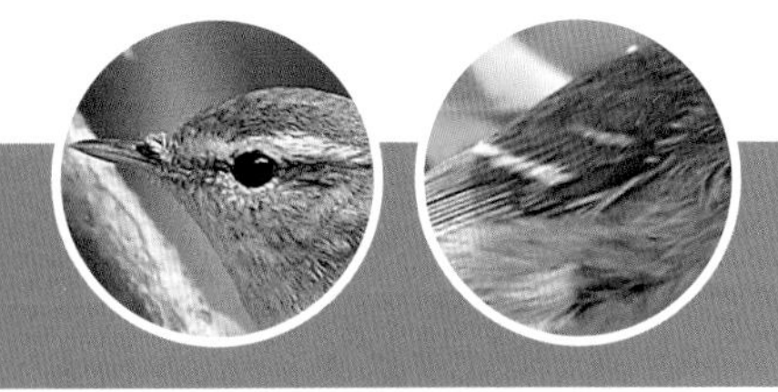

春季過境遷徙鳥 Spring Passage Migrant	夏候鳥 Summer Visitor	秋季過境遷徙鳥 Autumn Passage Migrant	冬候鳥 Winter Visitor

1	2	3	4	5	6	7	8	9	10	11	12	常見月份

留鳥 Resident	迷鳥 Vagrant	偶見鳥 Occasional Visitor

庫頁島柳鶯

[普] kù yè dǎo lǐu yīng

體長 length：11.5cm

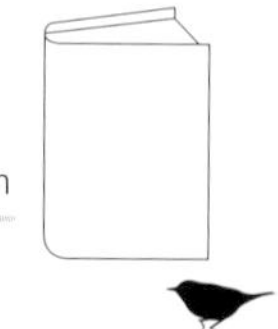

Sakhalin Leaf Warbler | *Phylloscopus borealoides*

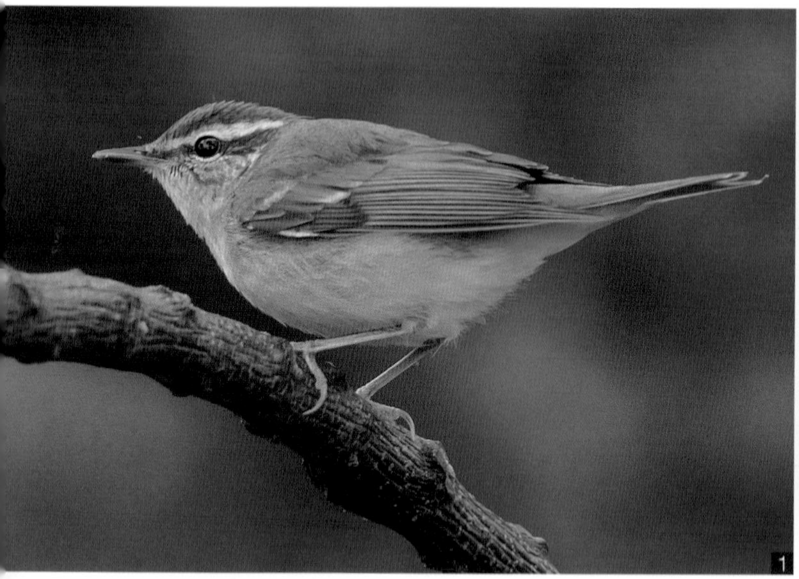

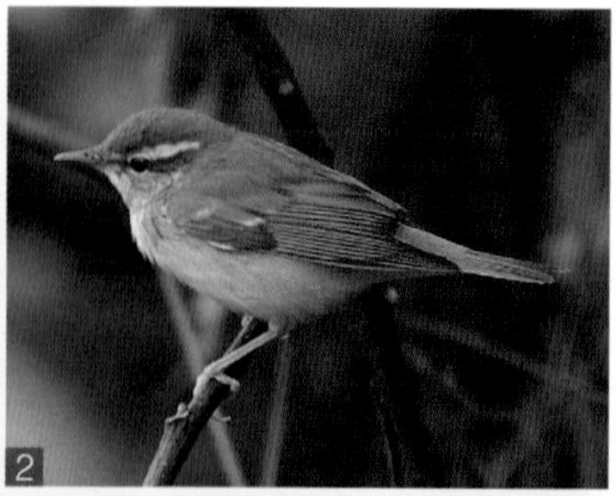

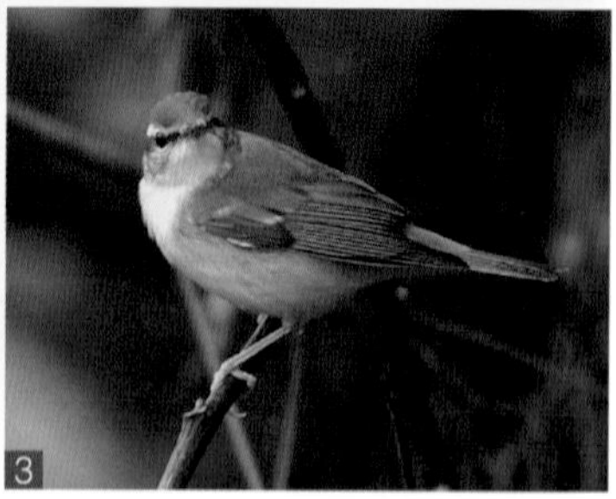

🌳

與淡腳柳鶯相似，上體橄欖褐色，有兩條淡黃色翼斑，脇部沾黃灰色，下體近白色。頭頂灰色，近白色長眉紋前端黃褐色，灰黑色頂冠紋及貫眼紋。上嘴暗褐色，下嘴稍淡色，腳淺粉紅色。

Similar to Pale-legged Leaf Warbler. Olive-brown upperparts. Two thin pale-yellow wing bars. Flanks tinged with yellowish-grey. Whitish underparts. Grey crown. Long whitish supercilia become yellowish-brown near forehead. Greyish-black median crown-stripe and eye-stripes. Dark brown upper bill with paler lower bill. Light pink legs.

[1] adult 成鳥；Ho Man Tin 何文田; Sep-18, 18年9月；Matthew Kwan 關朗曦
[2] adult 成鳥；Mai Po 米埔；Sep-18, 18 年 9 月；Kenneth Lam 林釗
[3] adult 成鳥；Mai Po 米埔；Sep-18, 18 年 9 月；Kenneth Lam 林釗
[4] adult 成鳥；Mai Po 米埔；Sep-18, 18 年 9 月；Kenneth Lam 林釗

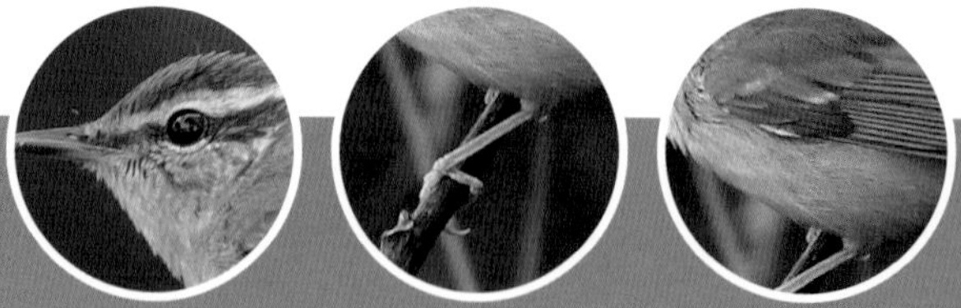

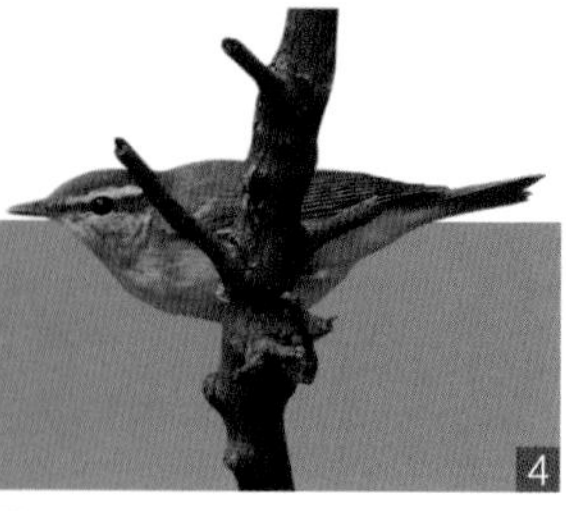

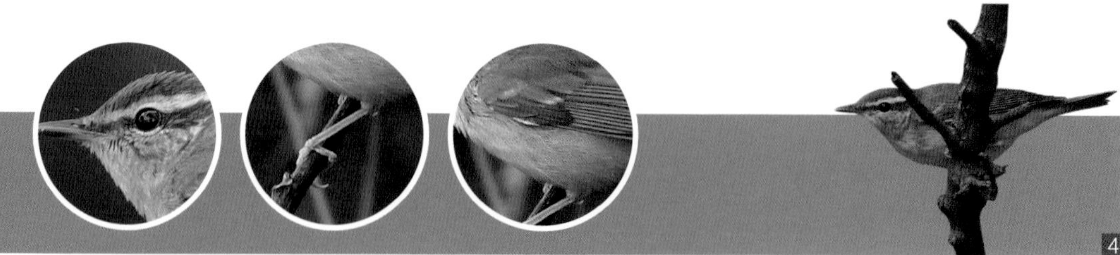

春季過境遷徙鳥 Spring Passage Migrant	夏候鳥 Summer Visitor	秋季過境遷徙鳥 Autumn Passage Migrant	冬候鳥 Winter Visitor

常見月份 1 2 3 4 5 6 7 8 9 10 11 12

留鳥 Resident	迷鳥 Vagrant	偶見鳥 Occasional Visitor

日本柳鶯 [普] rì běn lǐu yīng

體長 length：12-13cm

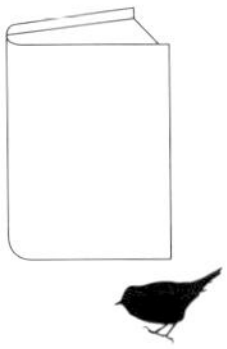

Japanese Leaf Warbler | *Phylloscopus xanthodryas*

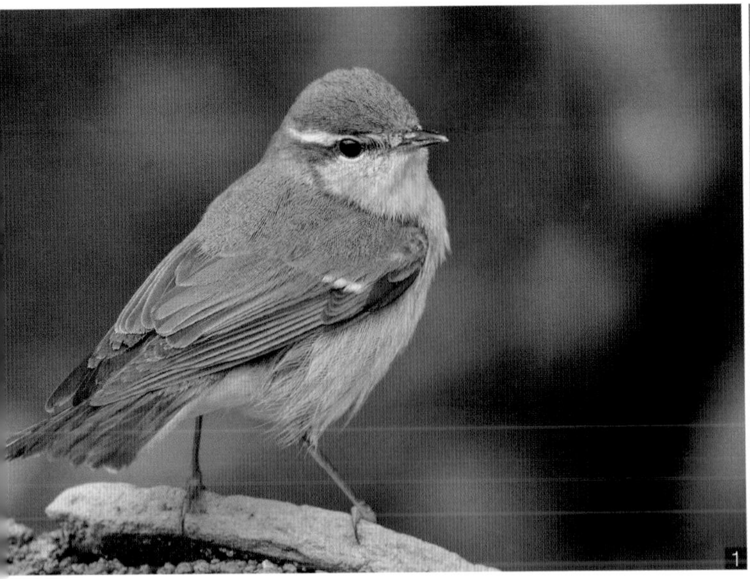
1

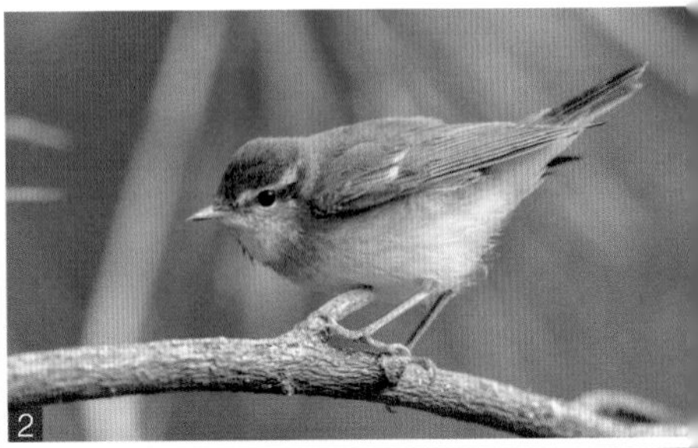
2

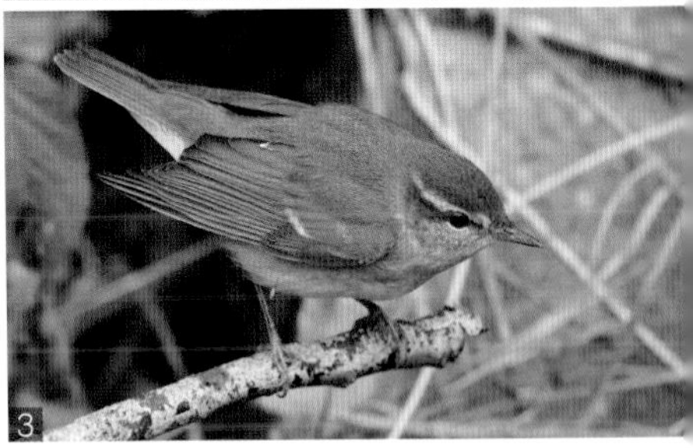
3

🌳

與較常見的極北柳鶯相似，橄欖色的上體稍偏綠，有兩條近白色翼斑，前翼斑模糊，白色的下體偏黃。嘴黑色，腳褐色。

Similar to more common Arctic Warbler but greener olive-coloured upperparts. Two whitish wing bars with the front one being inconspicuous. Yellowish-white underparts. Black bill. Brown legs.

[1] adult 成鳥：Po Toi 蒲台：Nov-18, 18 年 11 月：Wong Wong Yung 王煌容
[2] adult 成鳥：Po Toi 蒲台：Nov-18, 18 年 11 月：Wong Wong Yung 王煌容
[3] adult 成鳥：Po Toi 蒲台：Nov-18, 18 年 11 月：Wong Wong Yung 王煌容

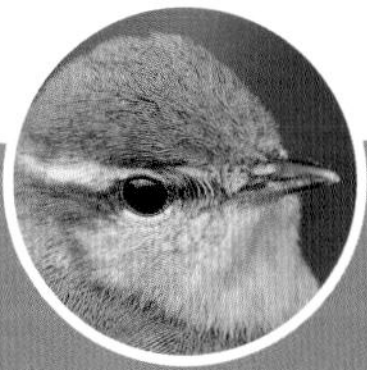

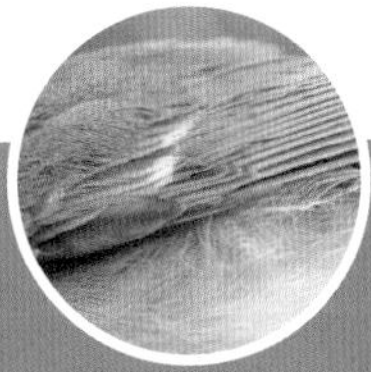

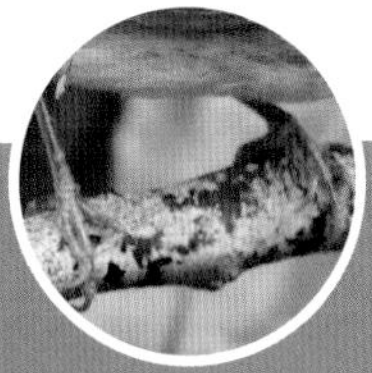

春季過境遷徙鳥 Spring Passage Migrant			夏候鳥 Summer Visitor			秋季過境遷徙鳥 Autumn Passage Migrant			冬候鳥 Winter Visitor			
1	2	3	4	5	6	7	8	9	10	11	12	常見月份
留鳥 Resident				迷鳥 Vagrant				偶見鳥 Occasional Visitor				

淡腳柳鶯 [普] dàn jiǎo liǔ yīng

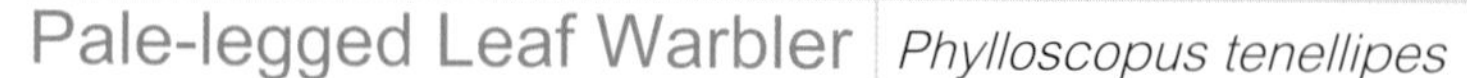

體長 length：10-11cm

Pale-legged Leaf Warbler | *Phylloscopus tenellipes*

其他名稱 Other names：灰腳柳鶯、灰腳樹鶯

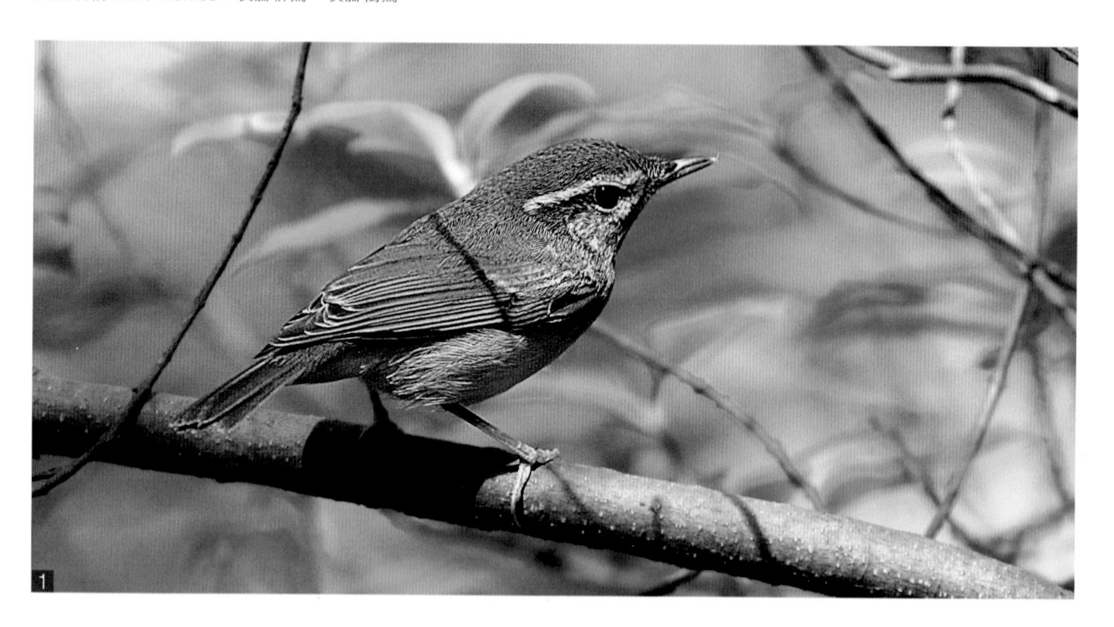

1

淡腳柳鶯和日本淡腳柳鶯外形十分相似，難以在野外分辨。淡腳柳鶯一般體型較小，靜立時初級飛羽突出部分較短。兩者上體皆為橄欖褐或橄欖綠色，眉紋長而闊，淺黃色，後方向上翹起。三級飛羽沒有白邊，只有一道翼帶。上嘴邊緣淺色，令嘴尖看起來似有一淺色斑點。腳呈淡灰粉紅色。叫聲彷似金屬發出的「chink」聲，喜於樹林下層活動。

Pale-legged and Sakhalin Leaf Warbler are very similar. Identification is almost impossible in the field. Pale-legged Leaf Warbler is usually smaller in size and has shorter primary projection. Olive-brown or olive-green upperparts. Long broad pale yellowish supercilium with upward "kink" in the rear. No white fringes on tertials. One wing bar. Pale edge on upper mandible gives the illusion of a pale spot on bill tip. Legs pale grey-pink. Call a metallic "chink". Favours lower storey in wooded area.

1 adult 成鳥：Tai Po Kau 大埔滘：Sep-08, 08 年 9 月：Martin Hale 夏敖天
2 adult 成鳥：Po Toi 蒲台：Sep-07, 07 年 9 月：Michelle and Peter Wong 江敏兒、黃理沛
3 adult 成鳥：Mai Po 米埔：Sep-03, 03 年 9 月：Michelle and Peter Wong 江敏兒、黃理沛
4 Mai Po 米埔：Sep-03, 03 年 9 月：Michelle & Peter Wong 江敏兒、黃理沛
5 adult 成鳥：Po Toi 蒲台：Apr-06, 06 年 4 月：Michelle and Peter Wong 江敏兒、黃理沛
6 adult 成鳥：Tai Po Kau 大埔滘：Sep-03, 03 年 9 月：Michelle and Peter Wong 江敏兒、黃理沛
7 adult 成鳥：Tai Po Kau 大埔滘：Sep-03, 03 年 9 月：Michelle and Peter Wong 江敏兒、黃理沛
8 Mai Po 米埔：Sep-03, 03 年 9 月：Michelle & Peter Wong 江敏兒、黃理沛
9 adult 成鳥：Mai Po 米埔：Sep-07, 07 年 9 月：Michelle and Peter Wong 江敏兒、黃理沛

春季過境遷徙鳥 Spring Passage Migrant	夏候鳥 Summer Visitor	秋季過境遷徙鳥 Autumn Passage Migrant	冬候鳥 Winter Visitor

常見月份	1	2	3	4	5	6	7	8	9	10	11	12

留鳥 Resident	迷鳥 Vagrant	偶見鳥 Occasional Visitor

極北柳鶯 ㊟jí běi liǔ yīng

體長 length：12-13cm

Arctic Warbler | *Phylloscopus borealis*

1

體型比黃眉柳鶯大。身長翼長，但尾相對其他柳鶯短。上體綠色，眉長黃色(差一點才到嘴基)。上嘴深色，下嘴尖端深色而基部淺色。三級飛羽上沒有白邊，大多數只有一道白色翼帶，但個別例子會有兩道短翼帶。腹部灰白。叫聲為「tzick」聲。喜於樹林上層活動。

Larger than Yellow-browed Warbler. A long-bodied, long-winged, apparently short tailed *phylloscopus* warbler. Green upperparts with long yellow supercilium (which does not reach bill base). Dark upper bill, lower bill has dark tip and pale base. No white fringes on tertials. Usually one white wing bar but occasionally a second short one. Underparts greyish-white. Call a "tzick". Arboreal.

1 adult 成鳥：Mai Po 米埔：Oct-07, 07 年 10 月：Wallace Tse 謝鑑超
2 adult 成鳥：Mai Po 米埔：Oct-07, 07 年 10 月：Danny Ho 何國海
3 adult 成鳥：Mai Po 米埔：Oct-07, 07 年 10 月：Wallace Tse 謝鑑超
4 adult 成鳥：Tai Po Kau 大埔滘：Sep-04, 04 年 9 月：Cherry Wong 黃卓研
5 adult 成鳥：Po Toi 蒲台：May-06, 06 年 5 月：Martin Hale 夏敖天
6 adult 成鳥：Mai Po 米埔：Oct-04, 04 年 10 月：Michelle and Peter Wong 江敏兒、黃理沛
7 adult 成鳥：Mai Po 米埔：Oct-07, 07 年 10 月：Freeman Yue 余柏維

春季過境遷徙鳥 Spring Passage Migrant	夏候鳥 Summer Visitor	秋季過境遷徙鳥 Autumn Passage Migrant	冬候鳥 Winter Visitor

常見月份 1 2 **3** **4** **5** 6 7 **8** **9** **10** **11** 12

留鳥 Resident	迷鳥 Vagrant	偶見鳥 Occasional Visitor

2
3
4
5
6
7

栗頭鶲鶯

㊟ lì tóu wēng yīng

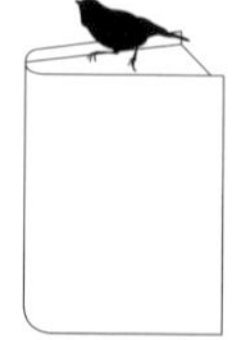

體長 length：9.5cm

Chestnut-crowned Warbler | *Phylloscopus castaniceps*

1

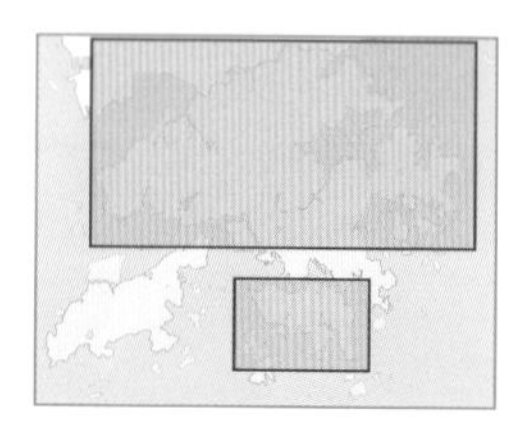

小型橄欖色鶯。冠紋和眉紋紅褐色，側冠紋和過眼線黑色，眼圈白色。上嘴黑色，下嘴淺色，耳羽、頰、頸和胸灰色。上體灰綠色，腰部和腹部鮮明黃色，腳灰色。

Small-sized olive-green warbler. Reddish-brown crown stripe and supercilium, black lateral crown stripes and eye-stripes. White eye-rings. Black upperbill, pale lowerbill. Grey ear coverts, chin, neck and breast. Greyish green upperparts. Bright yellow rump and belly. Grey legs.

1 adult 成鳥：Tai Po Kau 大埔滘：Dec-04, 04 年 12 月：Michelle and Peter Wong 江敏兒、黃理沛
2 adult 成鳥：Tai Po Kau 大埔滘：Dec-04, 04 年 12 月：Michelle and Peter Wong 江敏兒、黃理沛
3 adult 成鳥：Po Toi 蒲台：Nov-07, 07 年 11 月：Allen Chan 陳志雄
4 adult 成鳥：Tai Po Kau 大埔滘：Jan-20, 20 年 1 月：Roman Lo 羅文凱
5 adult 成鳥：Tai Po Kau 大埔滘：Jan-20, 20 年 1 月：Roman Lo 羅文凱
6 adult 成鳥：Tai Po Kau 大埔滘：Jan-20, 20 年 1 月：Roman Lo 羅文凱

春季過境遷徙鳥 Spring Passage Migrant	夏候鳥 Summer Visitor	秋季過境遷徙鳥 Autumn Passage Migrant	冬候鳥 Winter Visitor

常見月份 1 2 3 4 5 6 7 8 9 10 11 12

留鳥 Resident	迷鳥 Vagrant	偶見鳥 Occasional Visitor

2
3
4
5
6

黑眉柳鶯

㊟ hēi méi liǔ yīng

體長 length：10-11cm

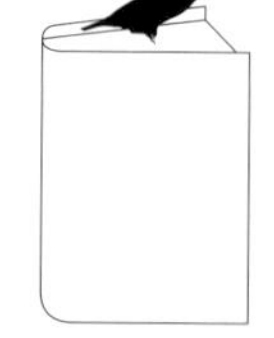

Sulphur-breasted Warbler | *Phylloscopus ricketti*

其他名稱 Other names：黃胸柳鶯

1

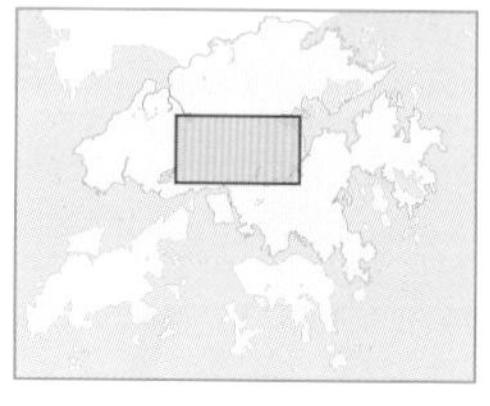

頭黑黃相間，十分搶眼。冠紋和眉紋黃色，側冠紋和過眼線黑綠色，眼圈不明顯，後頸有灰色細紋，上嘴深色，下嘴偏黃。上體綠色，下體鮮明黃色。翼上有兩度翼斑。

Eyecatching black-and-yellow stripes on head. Crown stripe and supercilium yellow. Lateral crown stripe and eye-stripe darkish green. Green upperparts and bright yellow underparts. Upper bill dark colour, lower bill yellowish. Two wing bars on flight feathers.

1 adult 成鳥：Wun Yiu 碗窰：Jan-08, 08 年 1 月：Wallace Tse 謝鑑超
2 adult 成鳥：Tai Po Kau 大埔滘：12-Mar-08, 08 年 3 月 12 日：Ho Wai Chun 何維俊
3 adult 成鳥：Po Toi 蒲台：1-Apr-06, 06 年 4 月 1 日：Michelle and Peter Wong 江敏兒、黃理沛
4 adult 成鳥：Wun Yiu 碗窰：Jan-08, 08 年 1 月：Wallace Tse 謝鑑超
5 adult 成鳥：Tai Po Kau 大埔滘：Dec-18, 18 年 12 月：Roman Lo 羅文凱
6 adult 成鳥：Po Toi 蒲台：1-Apr-06, 06 年 4 月 1 日：Michelle and Peter Wong 江敏兒、黃理沛
7 adult 成鳥：Po Toi 蒲台：1-Apr-06, 06 年 4 月 1 日：Michelle and Peter Wong 江敏兒、黃理沛
8 adult 成鳥：Po Toi 蒲台：1-Apr-06, 06 年 4 月 1 日：Michelle and Peter Wong 江敏兒、黃理沛

	春季過境遷徙鳥 Spring Passage Migrant			夏候鳥 Summer Visitor			秋季過境遷徙鳥 Autumn Passage Migrant			冬候鳥 Winter Visitor		
常見月份	1	2	3	4	5	6	7	8	9	10	11	12
	留鳥 Resident				迷鳥 Vagrant				偶見鳥 Occasional Visitor			

2
3
4
5
6
7
8

古氏[冠紋]柳鶯

㊟ gǔ shì guān wén liǔ yīng

體長 length：10.5-12cm

Hartert's Leaf Warbler | *Phylloscopus goodsoni*

大型而有冠紋的柳鶯。有明顯黃色冠紋，上體呈綠色，下體白色。側冠紋偏黑至深綠色，有黃色闊眉紋。覓食行為特別，類似鳾科鳥類，可以此區別。喜於森林的中及上層活動。和相似的冕柳鶯比較，臀部沒有黃色。

A large "crowned" warbler. Usually recognised by a prominent yellow crown stripe, greenish upperparts with white underparts. Blackish to dark green lateral crown stripes. Broad yellowish supercilium. A special nuthatch-like feeding behaviour is diagnostic. Favours high and middle storeys in woodland. Compared to similar Eastern Crown Warbler, it has no yellow colour on vent.

1 adult, race *goodsoni* 成鳥, *goodsoni* 亞種；Po Toi 蒲台；Sep-08, 08 年 9 月；Isaac Chan 陳家強
2 adult, race *fokiensis* 成鳥, *fokiensis* 亞種；Tai Po Kau 大埔滘；24-Dec-05, 05 年 12 月 24 日；Michelle and Peter Wong 江敏兒、黃理沛
3 adult 成鳥；Tai Po Kau 大埔滘；Feb-07, 07 年 2 月；Henry Lui 呂德恒
4 adult, race *goodsoni* 成鳥, *goodsoni* 亞種；Tai Po Kau 大埔滘；Dec-06, 06 年 12 月；Pippen Ho 何志剛
5 adult, race *fokiensis* 成鳥, *fokiensis* 亞種；Tai Po Kau 大埔滘；22-Feb-07, 07 年 2 月 22 日；Michelle and Peter Wong 江敏兒、黃理沛
6 adult, race *fokiensis* 成鳥, *fokiensis* 亞種；Tai Po Kau 大埔滘；1-Feb-06, 06 年 2 月 1 日；Michelle and Peter Wong 江敏兒、黃理沛

註：根據Olsson *et al.* (2005)，冠紋柳鶯的亞種已於2005年提升為種，包括西冠紋柳鶯 (*P. reguloides*)、冠紋柳鶯 (*P. claudiae*) 及華南冠紋柳鶯 (*P. goodsoni*)。
Note: Blyth's Leat Warbler (*P.reguloides*) has been split into *P. reguloides*, *P. claudiae* and *P. goodsoni* following Olsson *et al.* (2005)

春季過境遷徙鳥 Spring Passage Migrant	夏候鳥 Summer Visitor	秋季過境遷徙鳥 Autumn Passage Migrant	冬候鳥 Winter Visitor

常見月份	1	2	3	4	5	6	7	8	9	10	11	12

留鳥 Resident	迷鳥 Vagrant	偶見鳥 Occasional Visitor

2
3
4
5
6

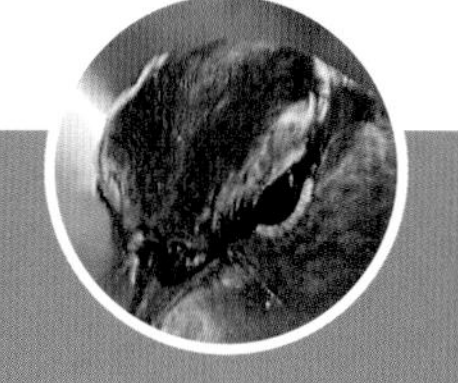

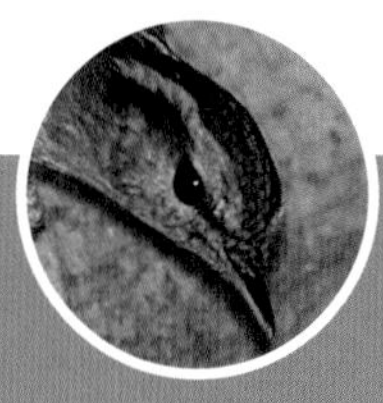

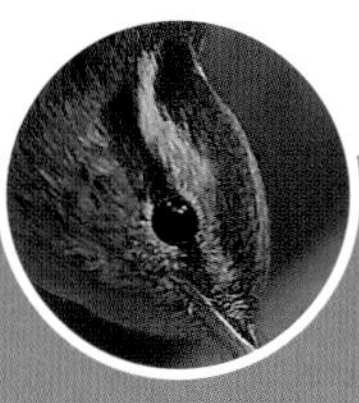

東方大葦鶯

㊟dōng fāng dà wěi yīng

體長 length：18.5cm

Oriental Reed Warbler | *Acrocephalus orientalis*

其他名稱 Other names：大葦鶯, Great Reed Warbler

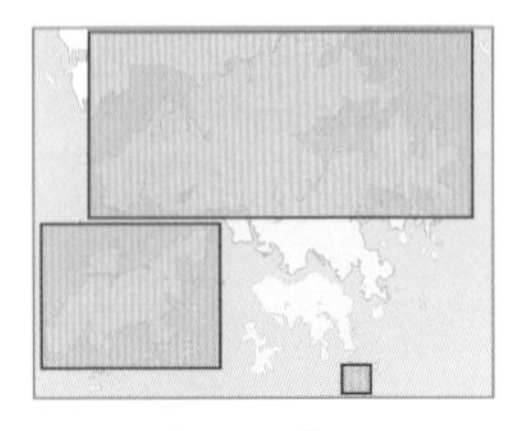

大型葦鶯。上身淺褐色，眉紋明顯淡黃色，腹部白色，脇薑褐色。嘴長黑色，尾長，尾端毛色較淡。會發出一連串沙啞的「charh-charh-charh」叫聲，喜於草地及沼澤活動，但也會在樹冠下出現。

Large-sized warbler. Pale brown upperparts with prominent yellowish supercilium. Bill black and long. Underparts whitish with gingery brown on flanks. Tail long with pale tips. Call is a series of dry "charh-charh-charh". Favours grassy and marshy areas, but appears at under canopy.

1 adult 成鳥：Mai Po 米埔：Sep-07, 07 年 9 月：James Lam 林文華
2 adult 成鳥：Mai Po 米埔：Oct-06, 06 年 10 月：Cherry Wong 黃卓研
3 adult 成鳥：Fung Lok Wai 豐樂圍：Oct-04, 04 年 10 月：Jemi and John Holmes 孔思義、黃亞萍
4 adult 成鳥：Mai Po 米埔：14-Oct-07, 07 年 10 月 14 日：Michelle and Peter Wong 江敏兒、黃理沛
5 adult 成鳥：Mai Po 米埔：15-Oct-07, 07 年 10 月 15 日：Owen Chiang 深藍
6 adult 成鳥：Mai Po 米埔：14-Oct-07, 07 年 10 月 14 日：Michelle and Peter Wong 江敏兒、黃理沛

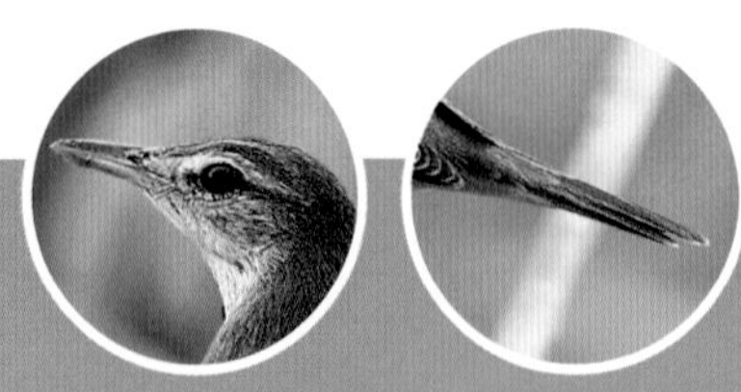

春季過境遷徙鳥 Spring Passage Migrant	夏候鳥 Summer Visitor	秋季過境遷徙鳥 Autumn Passage Migrant	冬候鳥 Winter Visitor

常見月份 1 2 3 4 5 6 7 8 9 10 11 12

留鳥 Resident	迷鳥 Vagrant	偶見鳥 Occasional Visitor

黑眉葦鶯

[普]hēi méi wěi yīng

體長 length：13.5cm

Black-browed Reed Warbler | *Acrocephalus bistrigiceps*

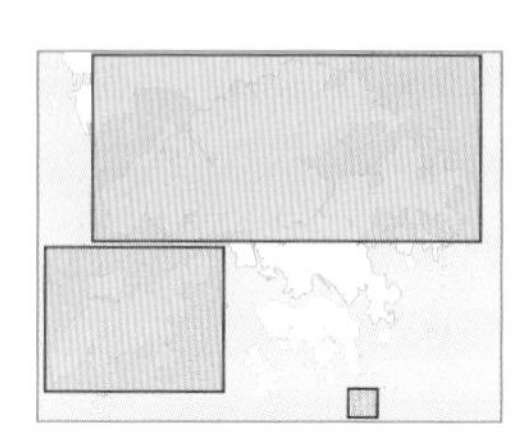

全身濃棕色，沒有斑紋。眉紋白色，上有一道黑紋。飛行時腰部呈栗褐色。腹部白色，脇淡黃色。叫聲與大葦鶯相似但較悦耳，喜在水邊蘆葦和長草之間活動。

A rich-brown warbler with no streaks. Distinctive black stripe above whitish supercilium. In flight, rump slightly chestnut. Underparts white with buff flanks. Call like an Oriental Reed Warbler but more melodious. Prefers reeds and long grass near water.

[1] adult 成鳥：Long Valley 塱原：4-Nov-07, 07 年 11 月 4 日：Andy Cheung 張玉良
[2] Long Valley 塱原：14 Oct-18, 18 年 10 月 14 日：Beetle Cheng 鄭諾銘
[3] adult 成鳥：Tsim Bei Tsui 尖鼻咀：Nov-06, 06 年 11 月：Thomas Chan 陳土飛

春季過境遷徙鳥 Spring Passage Migrant	夏候鳥 Summer Visitor	秋季過境遷徙鳥 Autumn Passage Migrant	冬候鳥 Winter Visitor

1	2	3	4	5	6	7	8	9	10	11	12	常見月份

留鳥 Resident	迷鳥 Vagrant	偶見鳥 Occasional Visitor

鈍翅葦鶯

㊥dùn chì wěi yīng

體長 length：13-14cm

Blunt-winged Warbler | *Acrocephalus concinens*

其他名稱 Other names：鈍翅稻田葦鶯

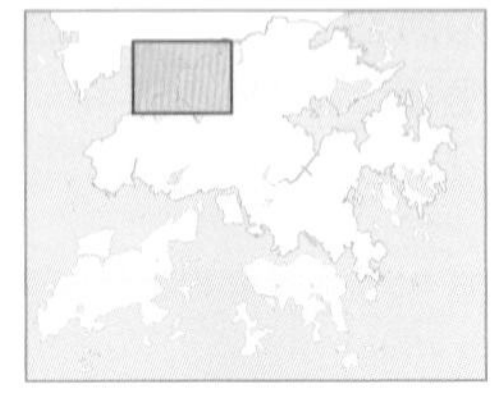

比較其他葦鶯是其短不及眼後白色的眼紋，上體淡棕褐色，圓短的翼，腰和尾上覆羽沾淡棕色，嘴長及下嘴基色淡。下體白色，脇和尾下覆羽沾淡黃褐色。

Short white supercilium compared with other bush warblers. Pale rufous-brown upperparts, short wings, rump and undertail coverts pale brown, long bill and paler lower bill. White underparts, flanks and vent tinted pale brownish yellow.

1 adult 成鳥：Fairview Park 錦繡花園：Feb-09, 09 年 2 月：Martin Hale 夏敖天
2 adult 成鳥：Fairview Park 錦繡花園：Feb-09, 09 年 2 月：Martin Hale 夏敖天
3 adult 成鳥：Fairview Park 錦繡花園：Feb-09, 09 年 2 月：Martin Hale 夏敖天

	春季過境遷徙鳥 Spring Passage Migrant			夏候鳥 Summer Visitor			秋季過境遷徙鳥 Autumn Passage Migrant			冬候鳥 Winter Visitor		
常見月份	1	2	3	4	5	6	7	8	9	10	11	12
	留鳥 Resident				迷鳥 Vagrant				偶見鳥 Occasional Visitor			

遠東葦鶯 ㊟yuǎn dōng wěi yīng

體長 length：13-14.5cm

Manchurian Reed Warbler | *Acrocephalus tangorum*

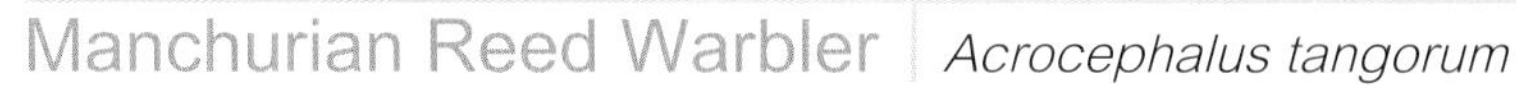

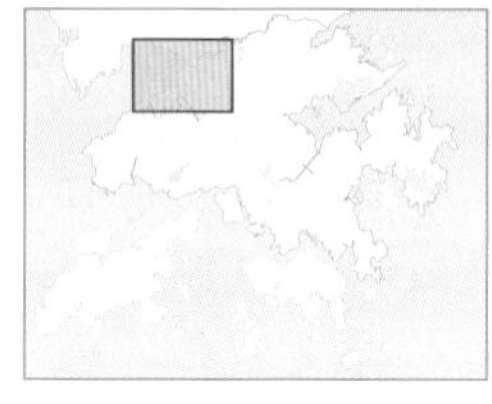

身體大致灰褐色，喉至胸及臀部白色，有白色眉紋、黑褐色貫眼紋及側冠紋。嘴及尾較其他葦鶯長，上嘴深褐色，下嘴粉紅，腳橙褐色。

Mainly greyish-brown body. White from throat to breast. White vent. White supercilia with black eye-stripes and lateral crown-stripes. Bill and tail relatively longer than other reed warblers. Dark brown upper bill with pink lower bill. Brownish-orange legs.

1 adult 成鳥：San Tin 新田：Oct-18, 18 年 10 月：Roman Lo 羅文凱
2 adult 成鳥：Mai Po 米埔：Oct-18, 18 年 10 月：Kwok Tsz Ki 郭子祈
3 adult 成鳥：Mai Po 米埔：Oct-18, 18 年 10 月：Kwok Tsz Ki 郭子祈
4 adult 成鳥：Mai Po 米埔：Oct-18, 18 年 10 月：Roman Lo 羅文凱

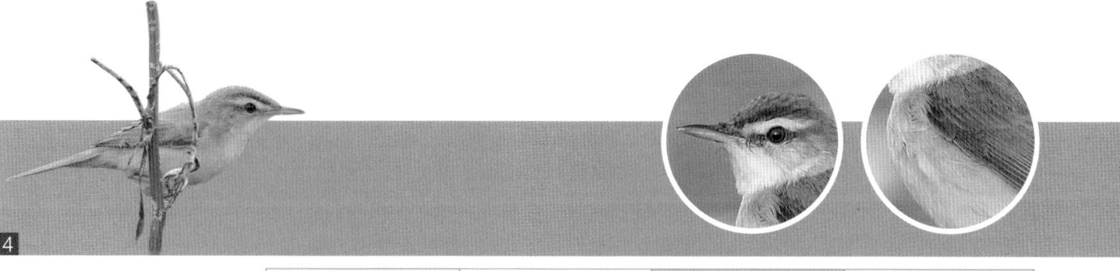

春季過境遷徙鳥 Spring Passage Migrant	夏候鳥 Summer Visitor	秋季過境遷徙鳥 Autumn Passage Migrant	冬候鳥 Winter Visitor

1 2 3 4 5 6 7 8 **9** **10** **11** 12 常見月份

留鳥 Resident	迷鳥 Vagrant	偶見鳥 Occasional Visitor

布氏葦鶯

㊟ bù shì wěi yīng

體長 length：13cm

Blyth's Reed Warbler | *Acrocephalus dumetorum*

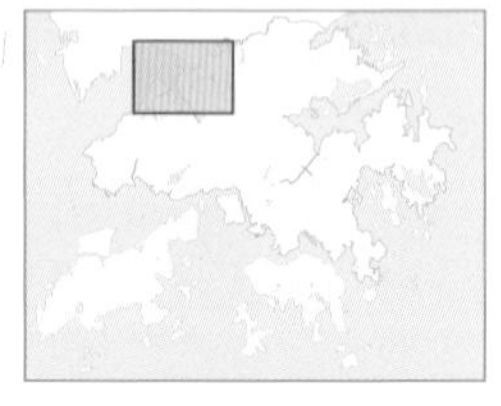

上體均勻淺灰褐色，下體白色，頸側、上胸及脇部沾淡褐色。有白色短眉紋，清晰的淺色眼先。嘴長，偏粉色而嘴端深色，腳近褐色。

Uniformly light greyish-brown upperparts. White underparts. Sides of neck, upper breast and flanks tinged with light brown. Short white supercilia with prominent pale lores. Long pinkish bill darker at tip. Almost brown legs.

[1] adult 成鳥：Sha Tin 沙田：Jan-16, 16 年 1 月：Martin Hale 夏敖天
[2] San Tin 新田：18 Jan-14, 14 年 1 月 18 日：Beetle Cheng 鄭諾銘

春季過境遷徙鳥 Spring Passage Migrant	夏候鳥 Summer Visitor	秋季過境遷徙鳥 Autumn Passage Migrant	冬候鳥 Winter Visitor

常見月份	1	2	3	4	5	6	7	8	9	10	11	12

留鳥 Resident	迷鳥 Vagrant	偶見鳥 Occasional Visitor

厚嘴葦鶯

㊟ hòu zuǐ wěi yīng

體長 length：18-19cm

Thick-billed Warbler | *Arundinax aedon*

其他名稱 Other names：蘆鶯

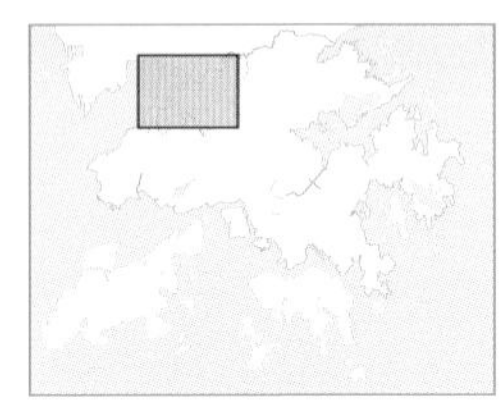

上體至尾部棕紅色大型的葦鶯，極淡色的眼線及無眉紋，嘴粗而短。 頗為長的尾部，下體白，兩脇和尾下覆羽沾淡黃色。

Large-sized warbler. Upperparts to rump and tail reddish-brown, distinctive pale lores and no supercilium, thick and short bill. Long tail, flanks and vent pale brownish-yellow.

1 adult 成鳥；Nov-07, 07 年 11 月；Fung Siu Ping 馮少萍
2 adult 成鳥；Nov-07, 07 年 11 月；Fung Siu Ping 馮少萍

春季過境遷徙鳥 Spring Passage Migrant	夏候鳥 Summer Visitor	秋季過境遷徙鳥 Autumn Passage Migrant	冬候鳥 Winter Visitor

1 2 3 4 5 6 7 8 9 10 11 12 常見月份

留鳥 Resident	迷鳥 Vagrant	偶見鳥 Occasional Visitor

靴籬鶯

[普] xuē lí yīng

體長 length：11-12.5cm

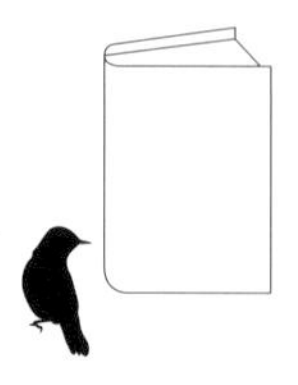

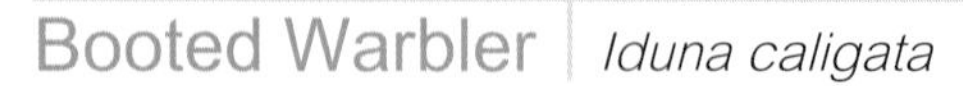

Booted Warbler | *Iduna caligata*

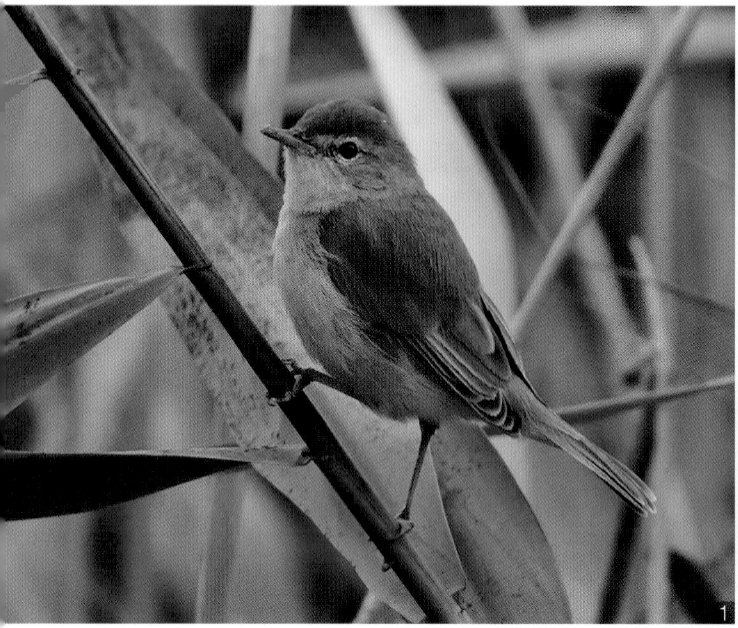

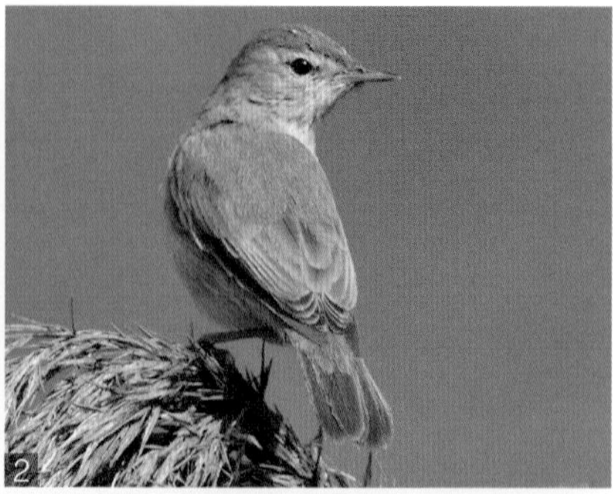

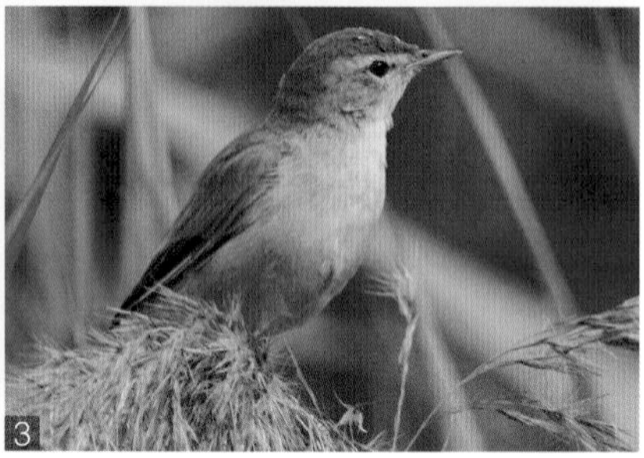

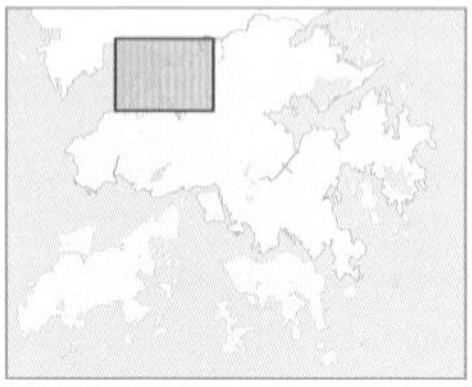

外形似柳鶯但色彩斑紋似葦鶯，上體均勻灰褐色，下體乳白色，脇部及臀部沾淡褐色。白色眉紋既長且寬，有白色眼圈。嘴小，上嘴深褐色，下嘴粉紅色，腳灰色。

Similar shape to leaf warbler but more like reed warbler in colour. Uniformly greyish-brown upperparts with milky white underparts. Flanks and vent tinged with pale brown. Broad and long white supercilia. White eye-rings. Small bill. Dark brown upper bill with pink lower bill. Grey legs.

[1] adult 成鳥；Mai Po 米埔；Dec-15, 15 年 12 月；Kinni Ho 何建業
[2] adult 成鳥；Mai Po 米埔；Dec-15, 15 年 12 月；Sam Chan 陳巨輝
[3] adult 成鳥；Mai Po 米埔；Dec-15, 15 年 12 月；Sam Chan 陳巨輝
[4] adult 成鳥；Mai Po 米埔；Dec-15, 15 年 12 月；Sam Chan 陳巨輝

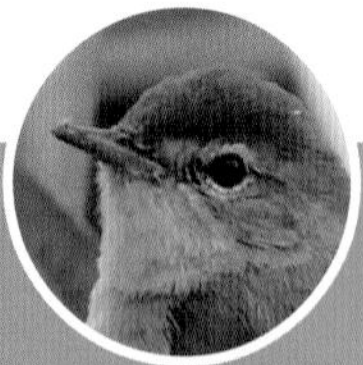

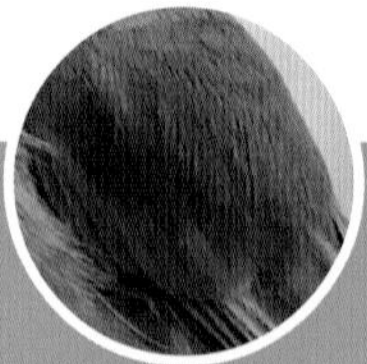

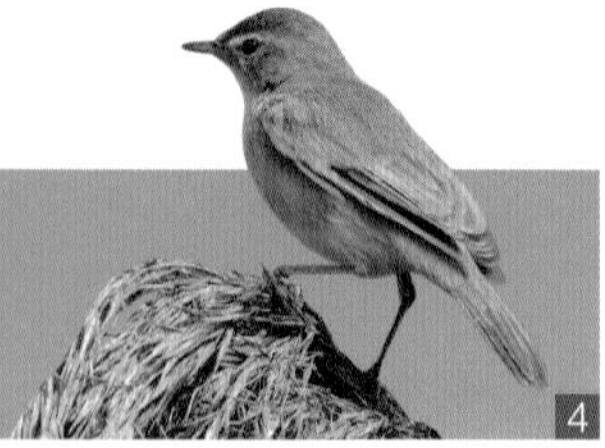

春季過境遷徙鳥 Spring Passage Migrant	夏候鳥 Summer Visitor	秋季過境遷徙鳥 Autumn Passage Migrant	冬候鳥 Winter Visitor

常見月份 1 2 3 4 5 6 7 8 9 10 11 12

留鳥 Resident	迷鳥 Vagrant	偶見鳥 Occasional Visitor

賽氏籬鶯

[普] sài shì lí yīng

體長 length：11.5-13cm

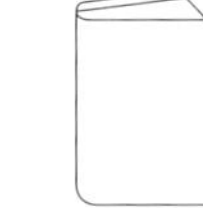

Sykes's Warbler | *Iduna rama*

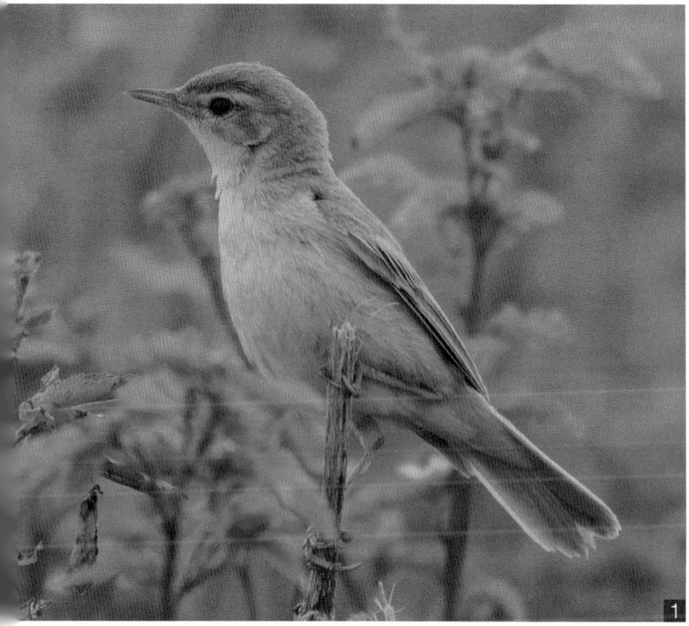
1

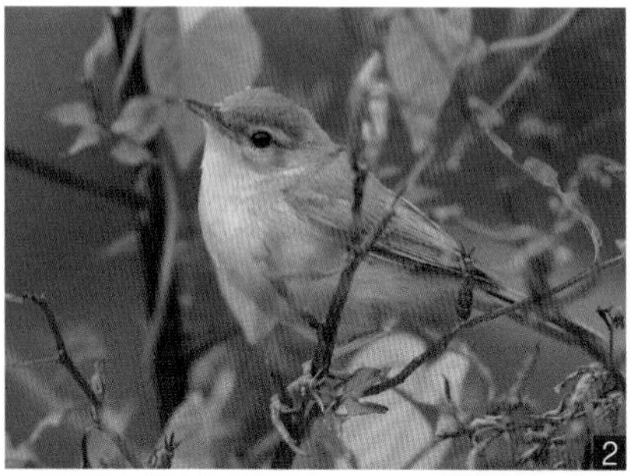
2

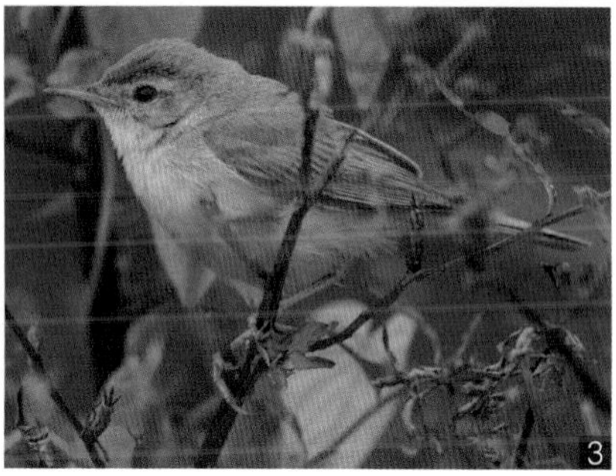
3

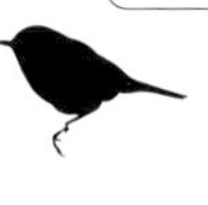

上體灰褐色，下體白色，有白色短眉紋。嘴較大，上嘴深褐色，下嘴粉紅色，腳偏粉色。

Greyish-brown upperparts. White underparts. Short and white supercilia. Bill relatively big. Dark brown upper bill with pink lower bill. Pinkish legs.

1 adult 成鳥：San Tin 新田：Oct-18, 18 年 10 月：Kwok Tsz Ki 郭子祈
2 adult 成鳥：San Tin 新田：Oct-18, 18 年 10 月：Kwok Tsz Ki 郭子祈
3 adult 成鳥：San Tin 新田：Oct-18, 18 年 10 月：Kwok Tsz Ki 郭子祈

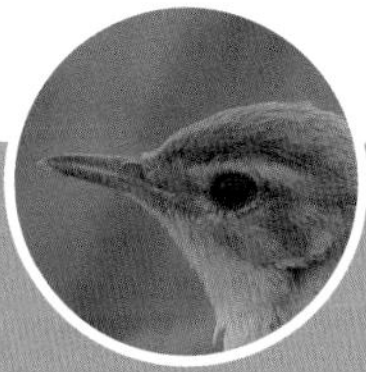

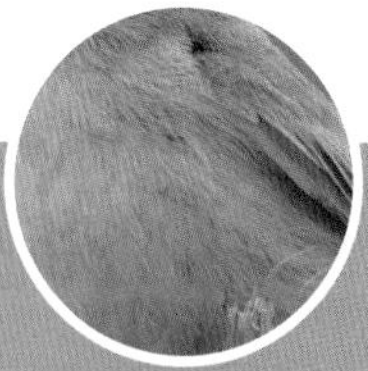

春季過境遷徙鳥 Spring Passage Migrant	夏候鳥 Summer Visitor	秋季過境遷徙鳥 Autumn Passage Migrant	冬候鳥 Winter Visitor

1 2 3 4 5 6 7 8 9 10 11 12 常見月份

留鳥 Resident	迷鳥 Vagrant	偶見鳥 Occasional Visitor

小蝗鶯

㊟ xiǎo huáng yīng

體長 length：13-14cm

Pallas's Grasshopper Warbler | *Helopsaltes certhiola*

其他名稱 Other names：Rusty-rumped Warbler

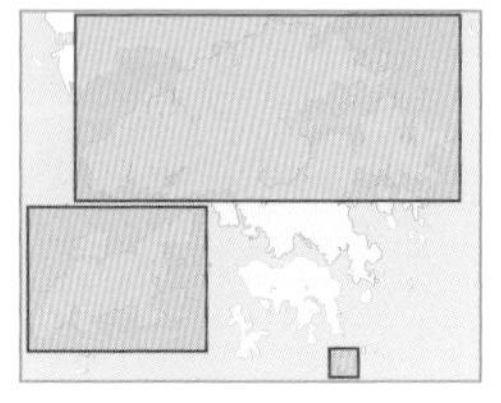

小型蝗鶯，上體大致深褐色，有黑色縱紋。頭頂偏灰帶深色縱紋，有淡褐色眼眉。腰部鮮栗色，縱紋較少，尾羽末端白色。下體淡皮黃色，胸和脇帶有縱紋。喜歡在蘆葦、矮樹叢或近水的草地活動。

Small-sized *locustella* warbler. Upperparts mainly dark brown with black streaks. It has greyish crown with dark streaks, pale brown supercilium. Rump is chestnut-brown in colour with fewer streaks. White tips at tail feathers. Warm buff underparts, with streaks on breast and flanks. Favours reedbeds, shrubland or grassland near water.

1 Kam Tin 錦田：Tam Yiu Leung 譚耀良
2 adult 成鳥：Lai Chi Kok 荔枝角：Dec-07, 07 年 12 月：Michelle and Peter Wong 江敏兒、黃理沛
3 1st winter 第一年冬天：Lai Chi Kok 荔枝角：Dec-07, 07 年 12 月：Michelle and Peter Wong 江敏兒、黃理沛
4 adult 成鳥：Lai Chi Kok 荔枝角：Nov-07, 07 年 11 月：Bill Man 文權溢
5 adult 成鳥：Lai Chi Kok 荔枝角：Nov-07, 07 年 11 月：Andy Kwok 郭匯昌
6 1st winter 第一年冬天：Lai Chi Kok 荔枝角：Nov-07, 07 年 11 月：Law Kam Man 羅錦文
7 1st winter 第一年冬天：Lai Chi Kok 荔枝角：Dec-07, 07 年 12 月：Michelle and Peter Wong 江敏兒、黃理沛
8 1st winter 第一年冬天：Lai Chi Kok 荔枝角：Dec-07, 07 年 12 月：Michelle and Peter Wong 江敏兒、黃理沛

春季過境遷徙鳥 Spring Passage Migrant	夏候鳥 Summer Visitor	秋季過境遷徙鳥 Autumn Passage Migrant	冬候鳥 Winter Visitor

常見月份 1 2 3 4 5 6 7 8 9 10 11 12

留鳥 Resident	迷鳥 Vagrant	偶見鳥 Occasional Visitor

史氏蝗鶯 [普] shǐ shì huáng yīng

體長 length：16-17cm

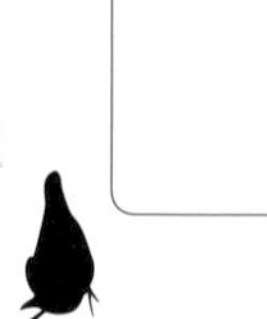

Styan's Grasshopper Warbler | *Helopsaltes pleskei*

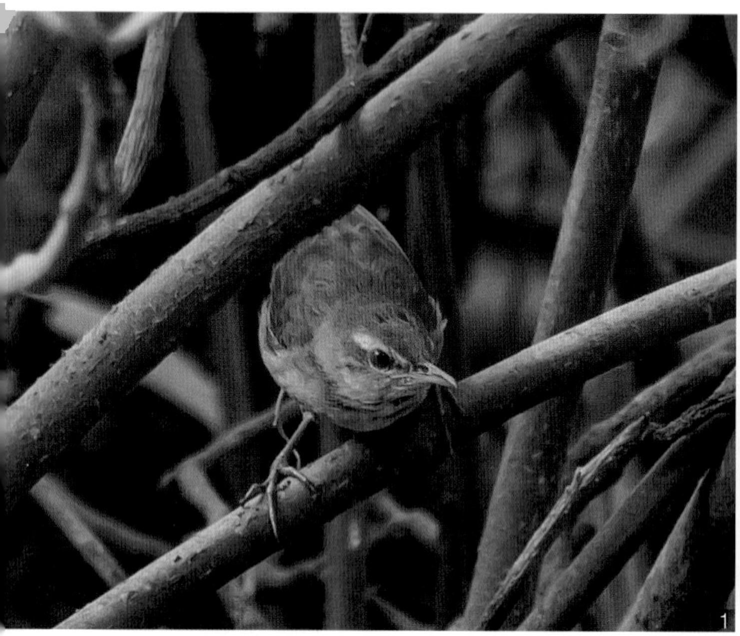

1

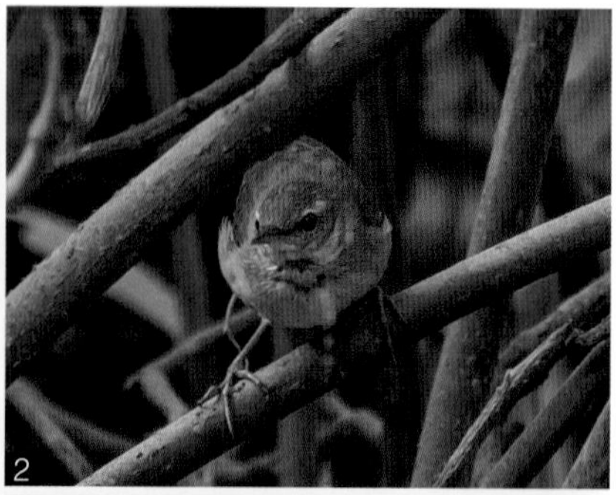

2

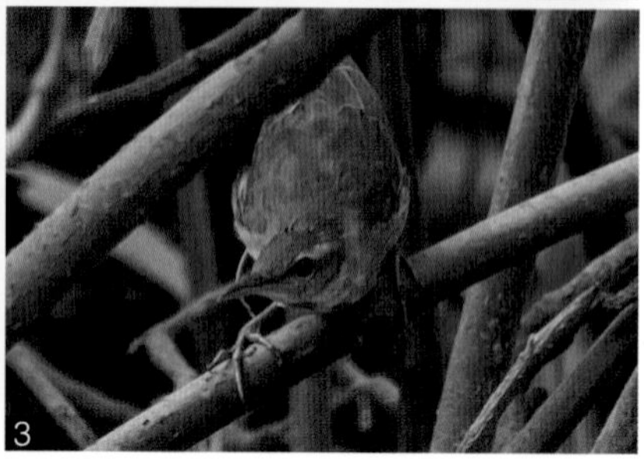

3

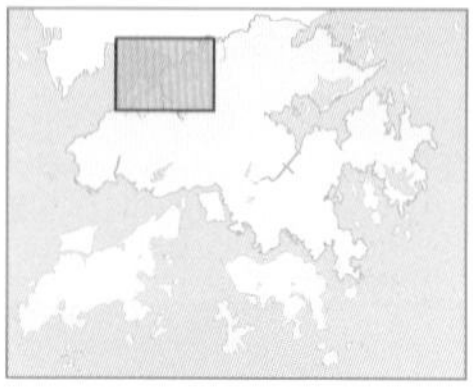

上體灰橄欖褐色，尾羽端有白斑，下體污白色，胸部、脇部、臀部沾淡青褐色。上嘴深褐色，下嘴粉紅色，腳粉紅色。

Olive-brown upperparts. White spots at tail end. Dirty-white underparts. Breast, flanks and rump tinged with light greenish-brown. Upper bill dark brown with pink lower bill. Pink legs.

[1] adult 成鳥；Mai Po 米埔；Mar-19, 19 年 3 月；Kenneth Lam 林釗
[2] adult 成鳥；Mai Po 米埔；Mar-19, 19 年 3 月；Kenneth Lam 林釗
[3] adult 成鳥；Mai Po 米埔；Mar-19, 19 年 3 月；Kenneth Lam 林釗

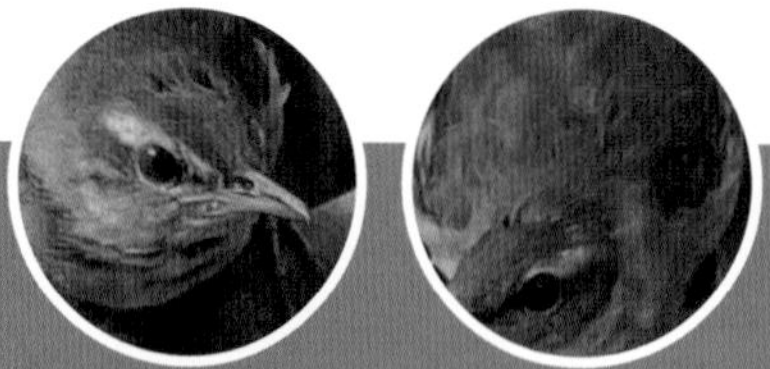

春季過境遷徙鳥 Spring Passage Migrant	夏候鳥 Summer Visitor	秋季過境遷徙鳥 Autumn Passage Migrant	冬候鳥 Winter Visitor

常見月份 1 2 3 4 5 6 7 8 9 10 11 12

留鳥 Resident	迷鳥 Vagrant	偶見鳥 Occasional Visitor

矛斑蝗鶯 [普]máo bān huáng yīng

體長 length：12cm

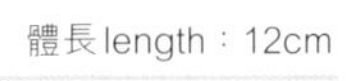

Lanceolated Warbler | *Locustella lanceolata*

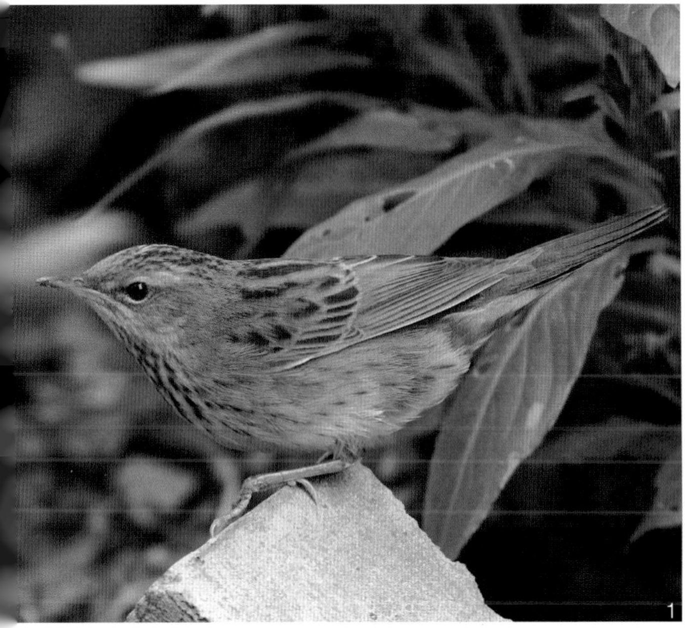
1

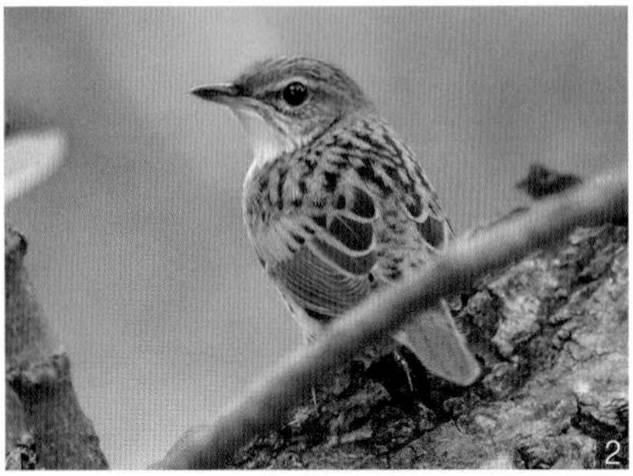
2

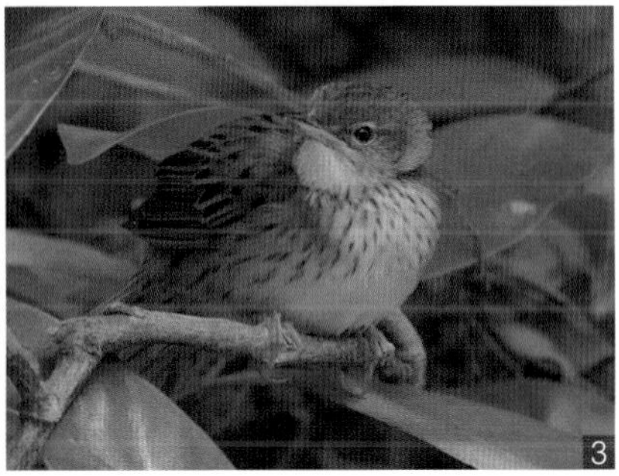
3

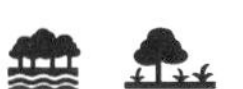

上體褐色且具黑色粗縱紋，下體淡褐色帶黑色縱紋，有淡褐色不顯眼眉紋。上嘴褐色，下嘴帶黃色，腳粉色。

Brown upperparts with thick black streaks. Paler underparts with black streaks. Vague pale brown supercilia. Brown upper bill with yellowish lower bill. Pink legs.

[1] adult 成鳥：Telford Garden 德福花園：Oct-18, 18 年 10 月：Leo Sit 薛國華
[2] adult 成鳥：Mai Po 米埔：Oct-18, 18 年 10 月：Wong Wong Yung 王煌容
[3] adult 成鳥：Shek Pik 石壁：Oct-18, 18 年 10 月：Kwok Tsz Ki 郭子祈
[4] adult 成鳥：Tam Kon Chau Road 担竿洲路：Oct-18, 18 年 10 月：Captain Wong 黃倫昌

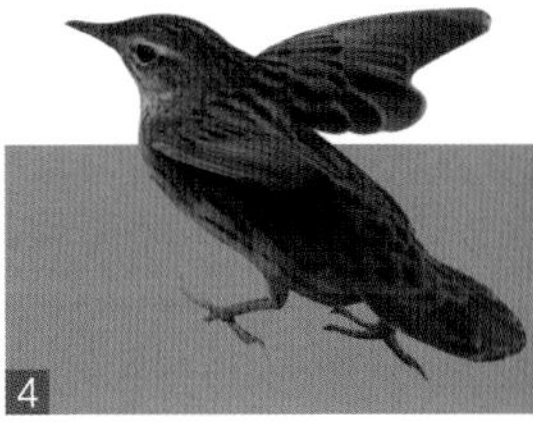
4

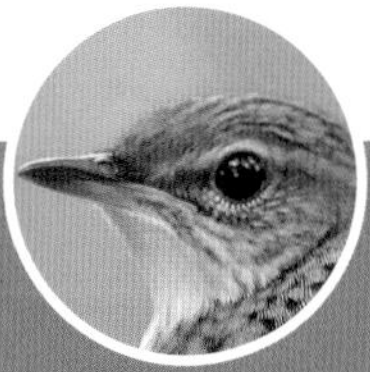

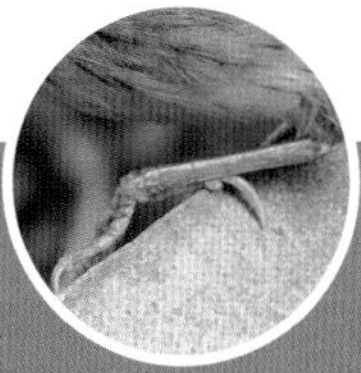

春季過境遷徙鳥 Spring Passage Migrant	夏候鳥 Summer Visitor	秋季過境遷徙鳥 Autumn Passage Migrant	冬候鳥 Winter Visitor

1	2	3	4	5	6	7	8	9	10	11	12	常見月份

留鳥 Resident	迷鳥 Vagrant	偶見鳥 Occasional Visitor

北短翅鶯

[普] běi duǎn chì yīng

體長 length：12cm

Baikal Bush Warbler | *Locustella davidi*

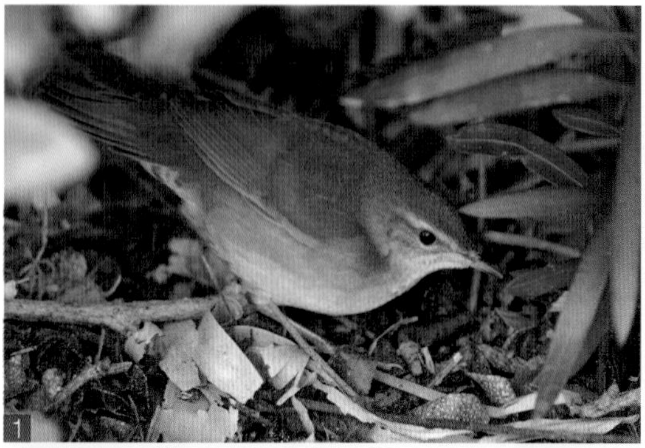

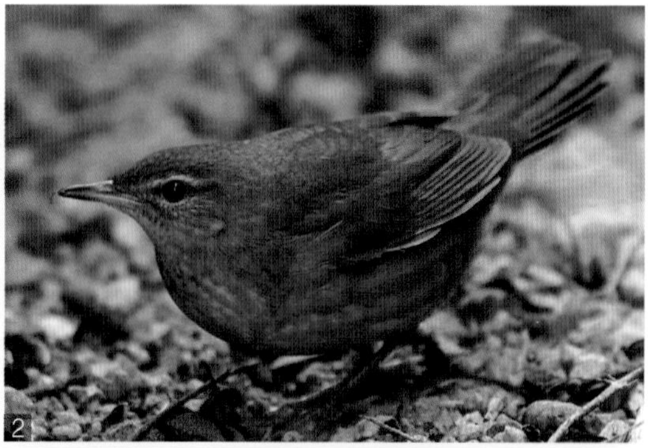

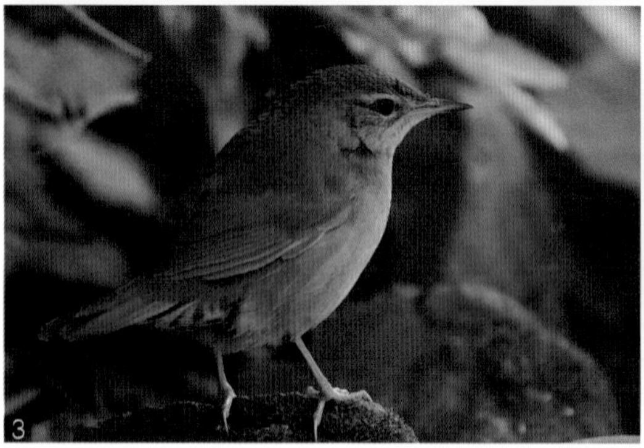

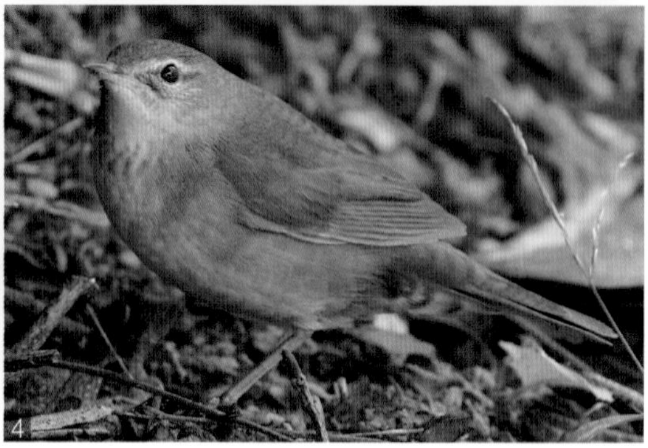

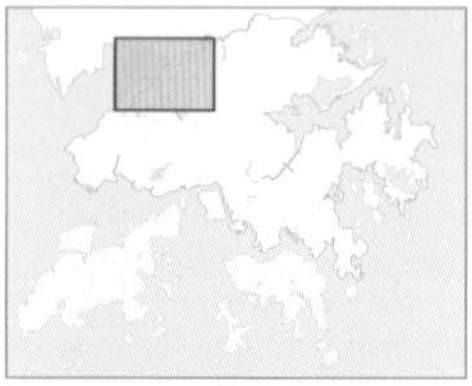

上體灰褐色，下體灰色，胸部有黑色小斑點。嘴黑色，腳粉色。

Greyish-brown upperparts with grey underparts. Black spots on breast. Black bill. Pale legs.

[1] adult 成鳥；Kowloon Bay 九龍灣；Feb-17, 17 年 2 月；Jemi and John Holmes 孔思義、黃亞萍
[2] adult 成鳥；Telford Garden 德福花園；Jan-17, 17 年 1 月；Kinni Ho 何建業
[3] adult 成鳥；Telford Garden 德福花園；Jan-17, 17 年 1 月；Kinni Ho 何建業
[4] adult 成鳥；Kowloon Bay 九龍灣；Feb-17, 17 年 2 月；Jemi and John Holmes 孔思義、黃亞萍
[5] adult 成鳥；Telford Garden 德福花園；Jan-17, 17 年 1 月；Kinni Ho 何建業

春季過境遷徙鳥 Spring Passage Migrant	夏候鳥 Summer Visitor	秋季過境遷徙鳥 Autumn Passage Migrant	冬候鳥 Winter Visitor

常見月份	1	2	3	4	5	6	7	8	9	10	11	12

留鳥 Resident	迷鳥 Vagrant	偶見鳥 Occasional Visitor

高山短翅鶯

[普] gāo shān duǎn chì yīng

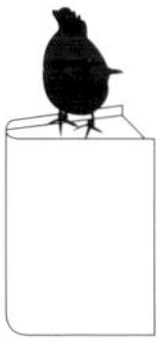

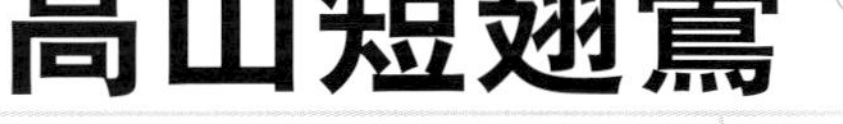

體長 length：13cm

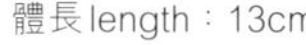

Russet Bush Warbler | *Locustella mandelli*

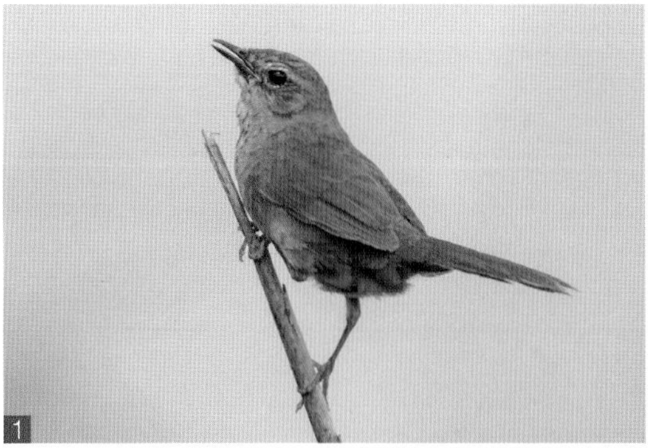
1

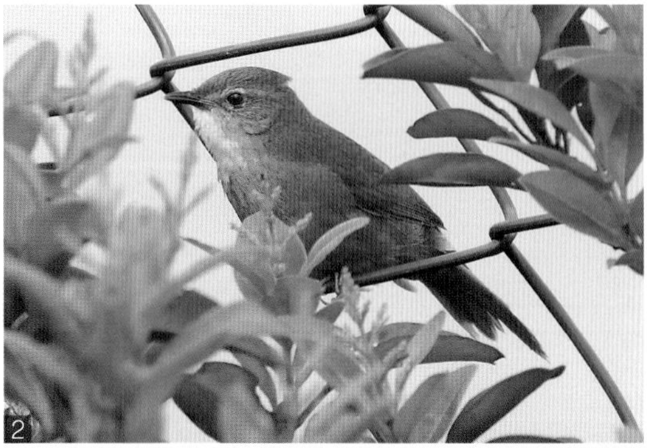
2

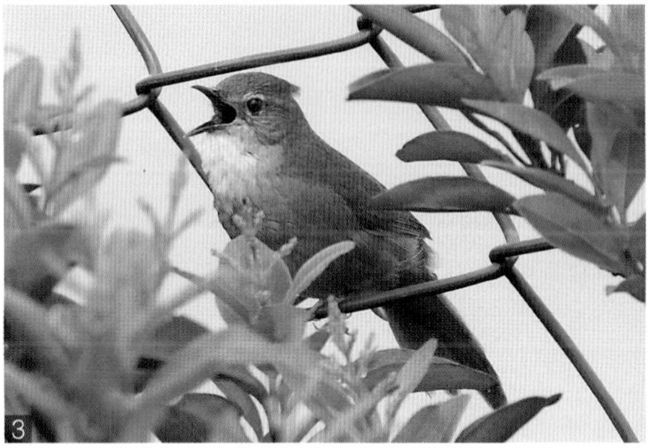
3

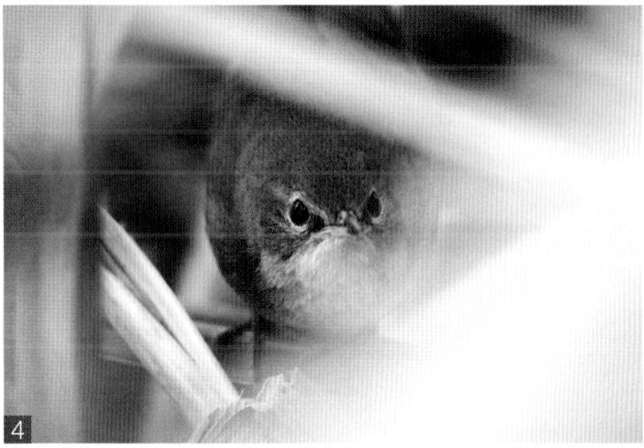
4

🌳

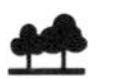

中等體型深褐色的鶯，非常隱秘。上體橄欖褐色，尾頗長及尖，頦和喉部白色並具深色縱紋，冬天時深色紋消失，下體其餘部分白色。叫聲「zee-bit」「zee-bit」「zee-bit」很易辨認。

Medium-sized dark brown warbler. Extremely skulking. Olive-brown upperparts with a longish pointed tail. Chin and throat white with dark streaks is disappeared in winter. Rest of the underparts white. Repeatedly call as "zee-bit" "zee-bit" "zee-bit".

[1] adult 成鳥：Tai Mo Shan 大帽山：Jun-16, 16 年 6 月：Matthew Kwan 關朗曦
[2] adult 成鳥：Tai Mo Shan 大帽山：Apr-18, 18 年 4 月：Kenneth Lam 林釗
[3] adult 成鳥：Tai Mo Shan 大帽山：Apr-18, 18 年 4 月：Kenneth Lam 林釗
[4] adult 成鳥：Po Toi 蒲台：Nov-07, 07 年 11 月：Geoff Welch
[5] adult 成鳥：Po Toi 蒲台：Nov-07, 07 年 11 月：Geoff Welch

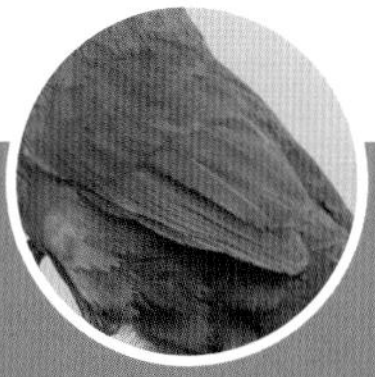

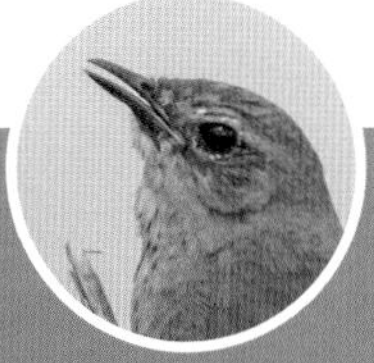

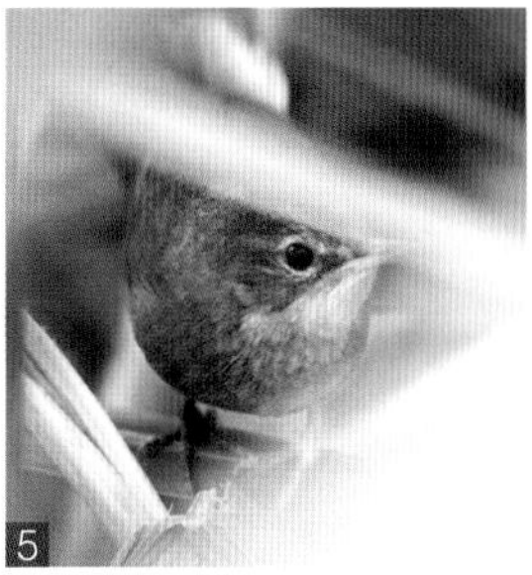
5

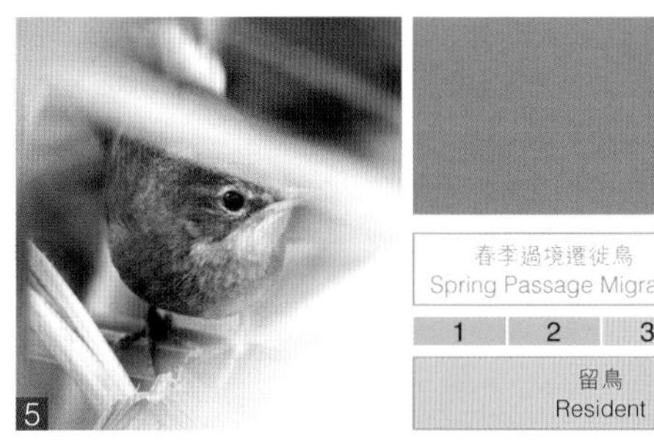

春季過境遷徙鳥 Spring Passage Migrant	夏候鳥 Summer Visitor	秋季過境遷徙鳥 Autumn Passage Migrant	冬候鳥 Winter Visitor

1	2	3	4	5	6	7	8	9	10	11	12	常見月份

留鳥 Resident	迷鳥 Vagrant	偶見鳥 Occasional Visitor

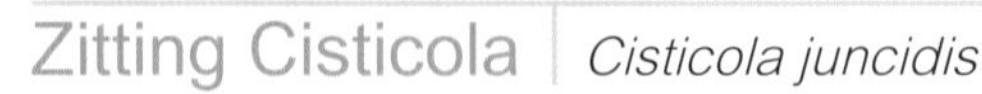

棕扇尾鶯

[普] zōng shàn wěi yīng

體長 length：10cm

Zitting Cisticola | *Cisticola juncidis*

其他名稱 Other names：Fantail Warbler

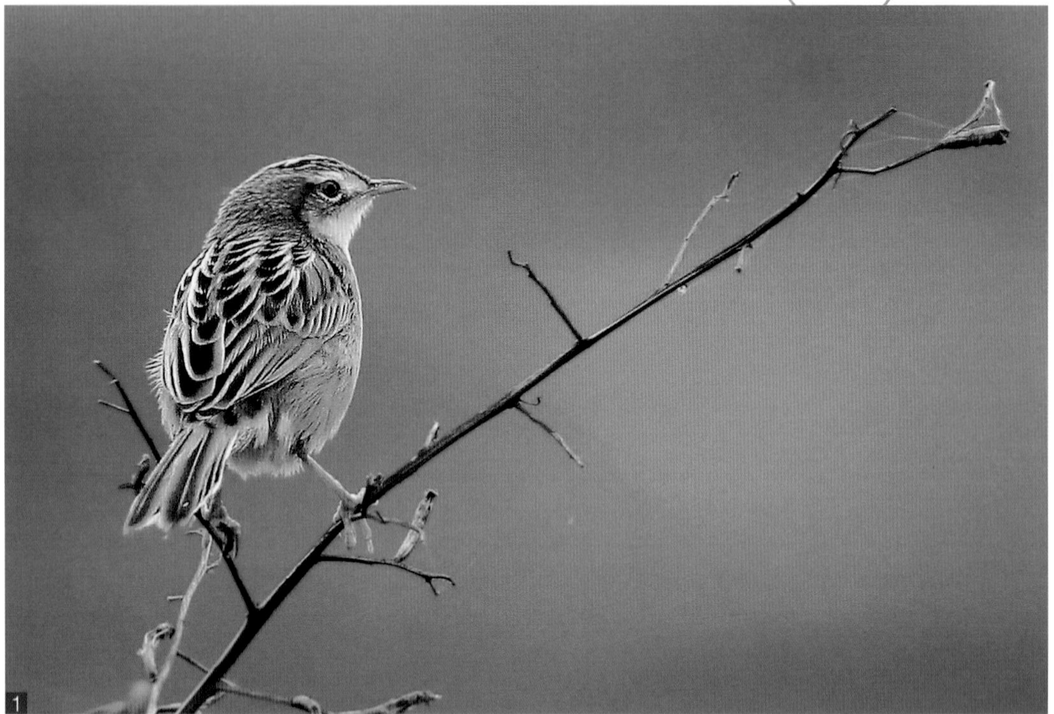

小型鶯，尾短呈扇狀，頭頂及上背有黑色條紋。頭及腰均為褐色，喉至腹部白色，脇黃褐色。尾羽褐色，有黑色寬橫帶，末端白色，飛行時可見。通常一兩隻出現，叫聲「dsip-dsip-dsip」，喜於水邊草叢頂出沒。

A small-sized warbler with short fan-shaped tail. Black streaks on crown and mantle. Brown head and rump. White throat to belly with yellowish brown flanks. Tail brown with thick black sub-terminal bar and white tips, visible in flight. Usually seen in singles or twos. Call a "dsip-dsip-dsip". Favours top of grassy stalks near water.

1 adult 成鳥：Long Valley 塱原：Dec-06, 06 年 12 月：James Lam 林文華
2 adult 成鳥：Long Valley 塱原：Feb-07, 07 年 2 月：Winnie Wong and Sammy Sam 森美與雲妮
3 adult 成鳥：Long Valley 塱原：Mar-08, 08 年 3 月：Ken Fung 馮漢城
4 adult 成鳥：Long Valley 塱原：Feb-07, 07 年 2 月：Cherry Wong 黃卓研
5 1st winter 第一年冬天：Long Valley 塱原：Oct-04, 04 年 10 月：Henry Lui 呂德恆
6 1st winter 第一年冬天：Long Valley 塱原：Oct-04, 04 年 10 月：Henry Lui 呂德恆
7 adult 成鳥：Long Valley 塱原：Dec-07, 07 年 12 月：Cherry Wong 黃卓研

春季過境遷徙鳥 Spring Passage Migrant			夏候鳥 Summer Visitor			秋季過境遷徙鳥 Autumn Passage Migrant			冬候鳥 Winter Visitor		
常見月份 1	2	3	4	5	6	7	8	9	10	11	12

留鳥 Resident	迷鳥 Vagrant	偶見鳥 Occasional Visitor

2
3
4
5
6

7

金頭扇尾鶯

[普] jīn tóu shàn wěi yīng

體長 length：9cm

Golden-headed Cisticola | *Cisticola exilis*

其他名稱 Other names：黃頭扇尾鶯

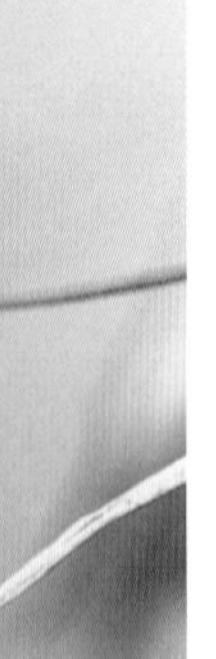

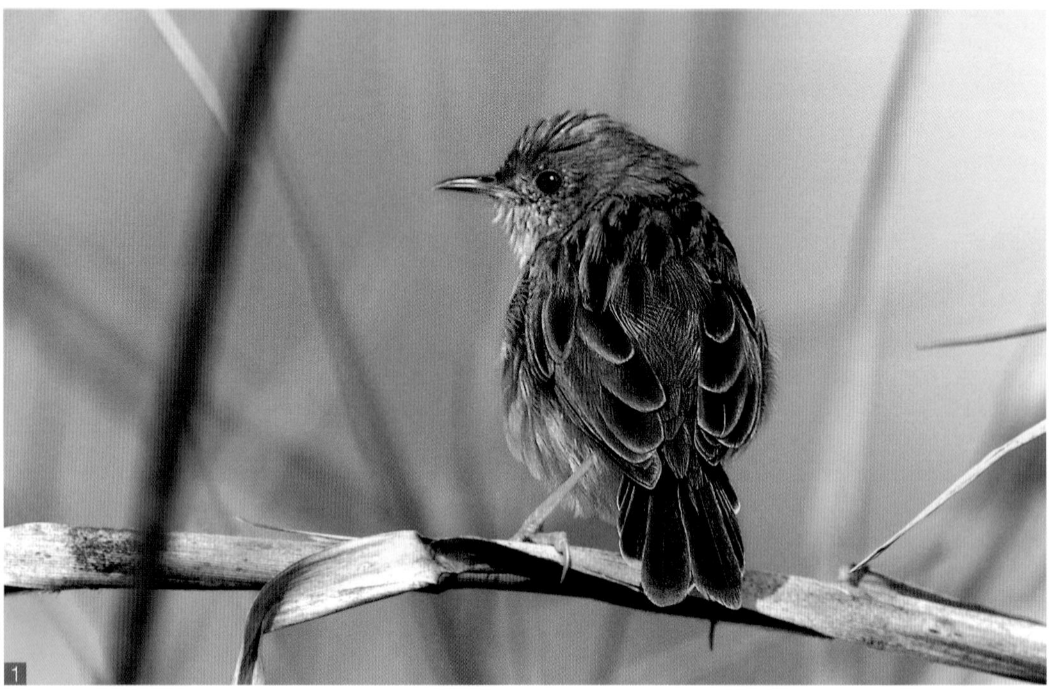

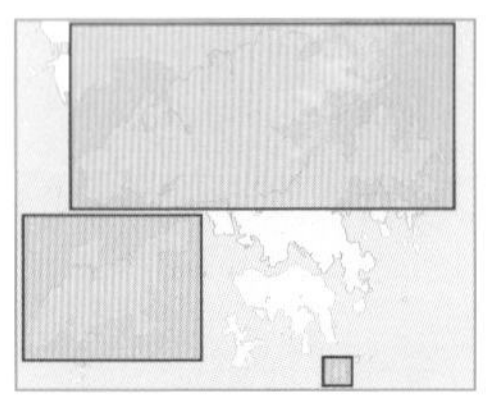

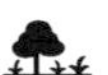

看似鷦鶯，具有長尾，以及明顯淡灰色耳羽和白喉。上身褐色，頭部深灰色及明顯白色條紋。上背、翼及尾羽深褐色，有淡褐色羽緣。下體淡黃褐色。尾部於冬季時較繁殖季節時長，羽尖具有淡色點。叫聲為具有鼻音而長的「斯…」聲，喜歡在長有長草的開闊原野出沒。

Resembles a prinia, with long tail. Greyish-white ear coverts and white throat. Upperparts brownish, head darkish grey with narrow light stripes. Wings to tail dark brown, with light brown fringes. Underparts light brownish-yellow. The tail is longer at winter compared to breeding season, with light spots at tail tip. Monosyllabic nasal call "tzee". Favours open country with long grasses.

1 breeding 繁殖羽：Long Valley 塱原：Mar-08, 08 年 3 月：Lee Kai Hong 李啟康
2 non-breeding adult 非繁殖羽成鳥：Long Valley 塱原：Dec-07, 07 年 12 月：James Lam 林文華
3 non-breeding adult 非繁殖羽成鳥：Kai Shan 雞山：Dec-07, 07 年 12 月：Lee Kai Hong 李啟康
4 non-breeding adult 非繁殖羽成鳥：Long Valley 塱原：Mar-08, 08 年 3 月：Martin Hale 夏敖天
5 breeding 繁殖羽：Long Valley 塱原：Mar-08, 08 年 3 月：Wallace Tse 謝鑑超
6 non-breeding adult, race *courtoisi* 非繁殖羽成鳥，*courtoisi* 亞種：Long Valley 塱原：Dec-07, 07 年 12 月：Eling Lee 李佩玲
7 non-breeding adult 非繁殖羽成鳥：Kai Shan 雞山：Dec-07, 07 年 12 月：Lee Kai Hong 李啟康
8 breeding 繁殖羽：Long Valley 塱原：Mar-08, 08 年 3 月：Lee Kai Hong 李啟康

	春季過境遷徙鳥 Spring Passage Migrant			夏候鳥 Summer Visitor			秋季過境遷徙鳥 Autumn Passage Migrant			冬候鳥 Winter Visitor	
常見月份	1	2	3	4	5	6	7	8	9	10	11	12
	留鳥 Resident				迷鳥 Vagrant				偶見鳥 Occasional Visitor			

2
4
5
6
3
7

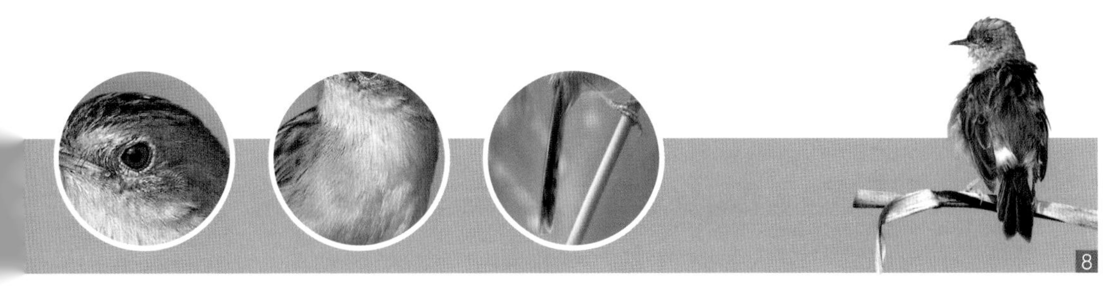
8

黃腹鷦鶯

[普]huáng fù jiāo yīng
[粵]鷦：音招

體長 length：12-14cm

Yellow-bellied Prinia | *Prinia flaviventris*

其他名稱 Other names：灰頭鷦鶯

小型鶯，體短尾長。頭灰色，眼前黃色眉紋伸延至眼上。上體褐色，腹部白色，脇較黃。叫聲似貓的「喵喵」聲，繁殖期有急促輕快的歌聲。喜於草叢間出沒，但也會在其他有植物的生境出現。

Small-sized warbler with relatively short body and long tail. Grey head with yellowish supercilium extending to top of eye. Brown upperparts. White underparts with yellowish flanks. Call like a cat-mew, also a rollicking song when breeding. Favours grassland but also occurs in any vegetated habitat.

1 breeding 繁殖羽；Long Valley 塱原；21-Nov-08, 08 年 11 月 21 日；Sung Yik Hei 宋亦希
2 breeding 繁殖羽；Tsim Bei Tsui 尖鼻咀；Mar-04, 04 年 3 月；Cherry Wong 黃卓研
3 non-breeding adult 非繁殖羽成鳥；Long Valley 塱原；Dec-07, 07 年 12 月；Cherry Wong 黃卓研
4 breeding 繁殖羽；Mai Po 米埔；Apr-03, 03 年 4 月；Henry Lui 呂德恒
5 breeding 繁殖羽；Long Valley 塱原；Apr-08, 08 年 4 月；Michael Schmitz
6 breeding 繁殖羽；Mai Po 米埔；Mar-07, 07 年 3 月；Henry Lui 呂德恒

春季過境遷徙鳥 Spring Passage Migrant	夏候鳥 Summer Visitor	秋季過境遷徙鳥 Autumn Passage Migrant	冬候鳥 Winter Visitor

常見月份 1 2 3 4 5 6 7 8 9 10 11 12

留鳥 Resident	迷鳥 Vagrant	偶見鳥 Occasional Visitor

2
3
4
5

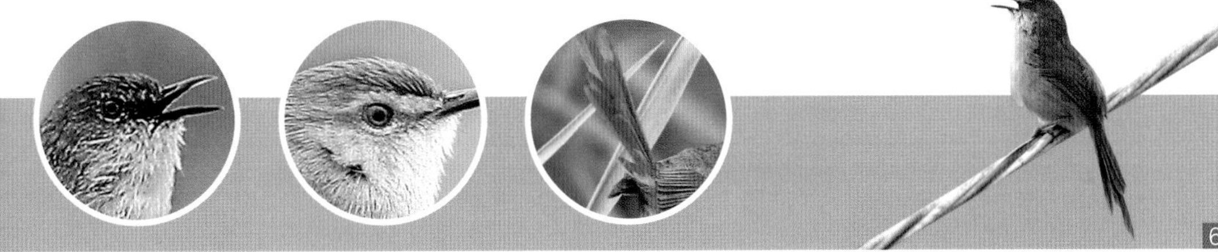
6

純色鷦鶯

㊢ chún sè jiāo yīng
[粵] 鷦：音招

體長 length：11cm

Plain Prinia *Prinia inornata*

其他名稱 Other names：褐頭鷦鶯 / 純色山鷦鶯

1

外貌和黃腹鷦鶯相若，但頭偏褐，眼前有黃色短眉紋。全身較黃腹鷦鶯褐色，嘴較粗，尾亦較長，底部末端顏色較淡。叫聲為「tee-tee-tee」，好像絞動魚桿時發出的聲響。喜在近水的開闊田野及草坡活動。

Resembles Yellow-bellied Prinia but has brown head. Short yellowish supercilium in front of eye. Compared to Yellow-bellied Prinia, more brown in colour, has thicker bill, longer tail and also pale tips on underside. Call a "tee-tee-tee", like sound of winding a fishing reel. Prefers open country near water and grassy hillsides.

1 breeding, race *extensicauda* 繁殖羽, *extensicauda* 亞種；Tsim Bei Tsui 尖鼻咀；Mar-07, 07 年 3 月；Frankie Chu 朱錦滿
2 non-breeding adult, race *extensicauda* 非繁殖羽成鳥, *extensicauda* 亞種；Mai Po 米埔；Jan-07, 07 年 1 月；Bill Man 文權溢
3 non-breeding adult, race *extensicauda* 非繁殖羽成鳥, *extensicauda* 亞種；Fung Lok Wai 豐樂圍；Oct-04, 04 年 10 月；Jemi and John Holmes 孔思義、黃亞萍
4 breeding, race *extensicauda* 繁殖羽, *extensicauda* 亞種；Mai Po 米埔；Mar-05, 05 年 3 月；Henry Lui 呂德恒
5 non-breeding adult, race *extensicauda* 非繁殖羽成鳥, *extensicauda* 亞種；Long Valley 塱原；Sep-06, 06 年 9 月；Cherry Wong 黃卓研
6 juvenile, race *extensicauda* 幼鳥, *extensicauda* 亞種；Mai Po 米埔；Sep-08, 08 年 9 月；Kami Hui 許淑君
7 non-breeding adult, race *extensicauda* 非繁殖羽成鳥, *extensicauda* 亞種；Long Valley 塱原；Aug-08, 08 年 8 月；Ken Fung 馮漢城
8 non-breeding adult, race *extensicauda* 非繁殖羽成鳥, *extensicauda* 亞種；Tsim Bei Tsui 尖鼻咀；Oct-04, 04 年 10 月；Cherry Wong 黃卓研

	春季過境遷徙鳥 Spring Passage Migrant			夏候鳥 Summer Visitor			秋季過境遷徙鳥 Autumn Passage Migrant			冬候鳥 Winter Visitor		
常見月份	1	2	3	4	5	6	7	8	9	10	11	12
	留鳥 Resident				迷鳥 Vagrant				偶見鳥 Occasional Visitor			

2
5
6
3
4
7
8

長尾縫葉鶯

[普] cháng wěi féng yè yīng

體長 length：10-14cm

Common Tailorbird | *Orthotomus sutorius*

其他名稱 Other names：Long-tailed Tailorbird

1

小型鶯，尾長嘴長，前額和頭頂紅褐色，背部至尾部橄欖綠色，下體白色，喉部有時可見到黑紋，特別在鳴叫時。叫聲是獨特而響亮的「即－即－」聲，不斷重複。喜在林中下層植被活動。

Small-sized warbler with long bill and tail. Reddish-brown forehead and crown. Olive-green mantle to tail. White underparts. Throat may show some blackish streaks, especially when calling. Call a distinctive loud and repeating "chink-chink-chink". Prefers lower storey in forest.

[1] adult male, race *longicauda* 雄成鳥, *longicauda* 亞種：Kwai Chung 葵涌：Mar-07, 07 年 3 月：Martin Hale 夏敖天
[2] adult female, race *longicauda* 雌成鳥, *longicauda* 亞種：Tai Po Kau 大埔滘：Dec-06, 06 年 12 月：Cherry Wong 黃卓研
[3] adult male, race *longicauda* 雄成鳥, *longicauda* 亞種：Cheung Chau 長洲：Dec-04, 04 年 12 月：Henry Lui 呂德恆
[4] adult female, race *longicauda* 雌成鳥, *longicauda* 亞種：Kowloon Park 九龍公園：Nov-06, 06 年 11 月：James Lam 林文華
[5] adult male, race *longicauda* 雄成鳥, *longicauda* 亞種：Long Valley 塱原：Dec-07, 07 年 12 月：Cherry Wong 黃卓研
[6] adult probably male, race *longicauda* 成鳥, *longicauda* 亞種, 很可能是雄鳥：Kowloon Park 九龍公園：Sep-07, 07 年 9 月：Wong Shui Chi 黃瑞芝
[7] adult female, race *longicauda* 雌成鳥, *longicauda* 亞種：Cheung Chau 長洲：Feb-06, 06 年 2 月：Henry Lui 呂德恆

	春季過境遷徙鳥 Spring Passage Migrant			夏候鳥 Summer Visitor			秋季過境遷徙鳥 Autumn Passage Migrant			冬候鳥 Winter Visitor		
常見月份	1	2	3	4	5	6	7	8	9	10	11	12
	留鳥 Resident				迷鳥 Vagrant				偶見鳥 Occasional Visitor			

2
4
5
3
6
7

棕頸鉤嘴鶥

㊟ zōng jǐng gōu zuǐ méi

體長 length：16-19cm

Streak-breasted Scimitar Babbler | *Pomatorhinus ruficollis*

1

2

3

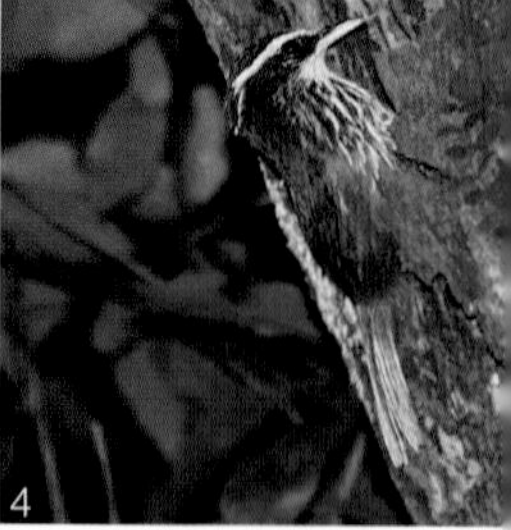

4

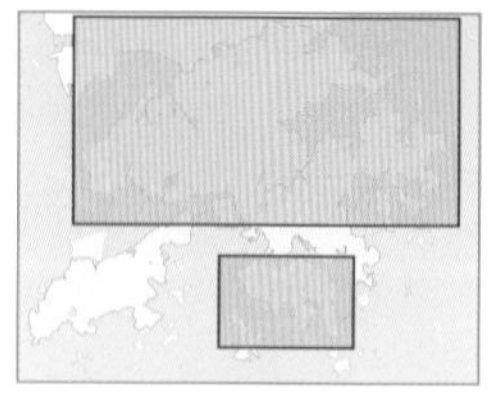

全身大致褐色，有明顯白色眼眉，和臉上黑色斑紋成強烈對比。嘴長黃色，末端微向下彎；喉部和胸部白色，有褐色縱紋。上體褐色，後頸紅褐色，腹部白色。常以小群出沒，常混在鳥浪中。叫聲為獨特的「doo-doo-which」三聲。

Overall brown. Prominent white eyebrow contrasts with blackish marking on face. Long, yellow decurved bill. Throat white with brown streaks on breast. Upperparts brown and reddish brown on nape. Belly white. Usually in small flocks, often found in bird waves of other birds. Distinctive three-note call "doo-doo-which".

1 Tai Po Kau 大埔滘：May-04, 04 年 5 月：Cherry Wong 黃卓研
2 Tai Po Kau 大埔滘：Jan-04, 04 年 1 月：Michelle and Peter Wong 江敏兒、黃理沛
3 Tai Po Kau 大埔滘：Feb-07, 07 年 2 月：Henry Lui 呂德恆
4 Tai Po Kau 大埔滘：Jan-05, 05 年 1 月：Cherry Wong 黃卓研
5 Tai Po Kau 大埔滘：Aug-08, 08 年 8 月：Ken Fung 馮漢城

5

	春季過境遷徙鳥 Spring Passage Migrant			夏候鳥 Summer Visitor			秋季過境遷徙鳥 Autumn Passage Migrant			冬候鳥 Winter Visitor		
常見月份	1	2	3	4	5	6	7	8	9	10	11	12
	留鳥 Resident				迷鳥 Vagrant				偶見鳥 Occasional Visitor			

紅頭穗鶥

[普] hóng tóu suì méi

體長 length：12cm

Rufous-capped Babbler | *Stachyridopsis ruficeps*

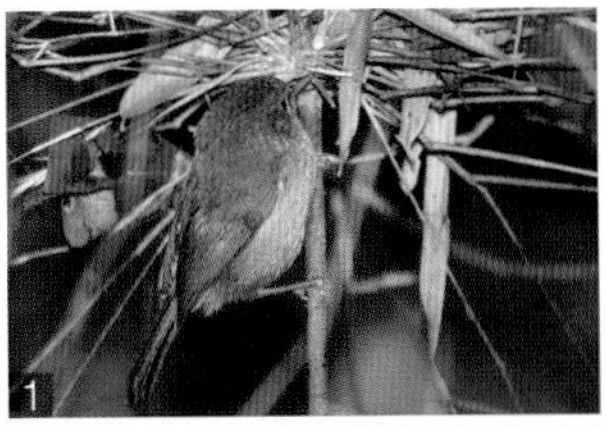
1

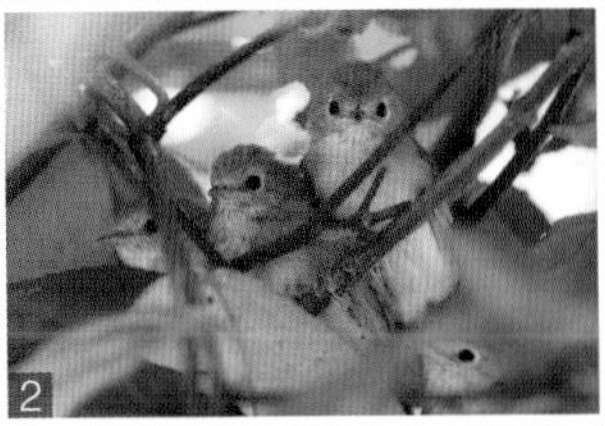
2

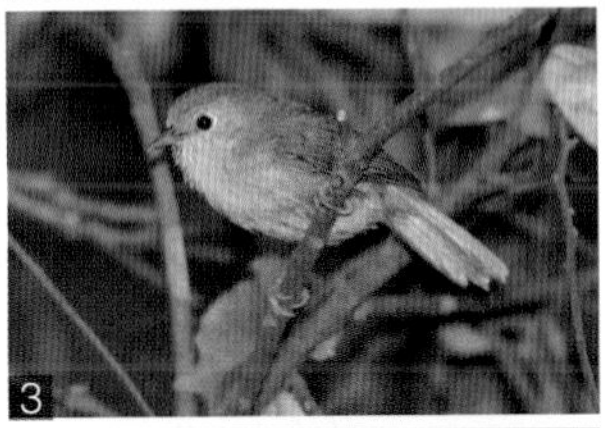
3

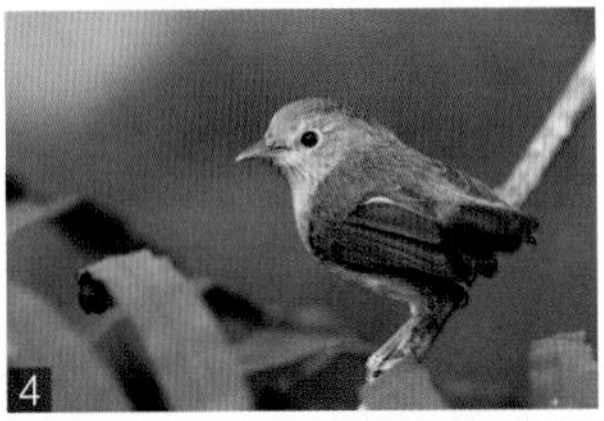
4

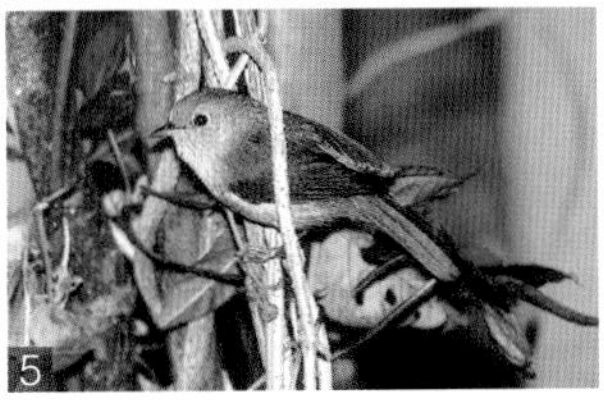
5

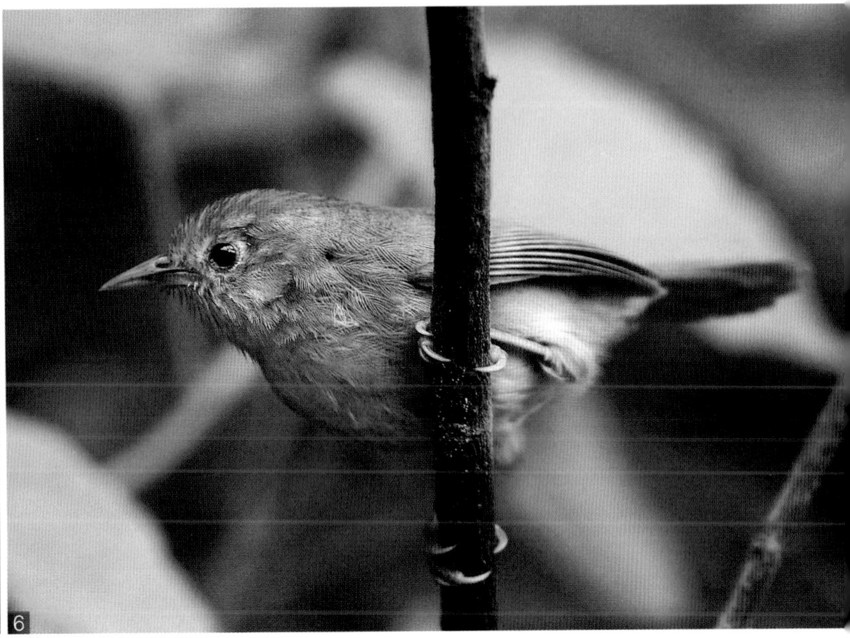
6

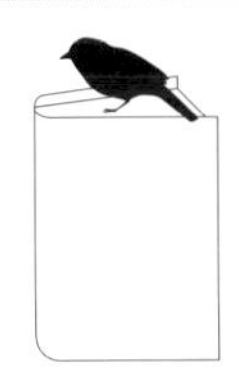

細小穗鶥和常見的長尾縫葉鶯相似，但稍健碩，嘴較短而成三角形。頭頂紅褐色、喉淡黃色，背部橄欖褐色。有獨特而嘹亮的「do-do-do-do」叫聲。常於樹林下層近河溪旁的叢林出沒。

Small, olive-green babbler. Alike Common Tailorbird but with slightly bigger and stout body, more triangular bill. Crown reddish-brown and throat pale yellow. Mantle olive brown. Call is a musical prolonged "do-do-do-do". Favours forest undergrowth near streams.

[1] Tai Po Kau 大埔滘：Feb-05, 05 年 2 月：Cherry Wong 黃卓研
[2] Tai Po Kau 大埔滘：May-07, 07 年 5 月：Wallace Tse 謝鑑超
[3] Tai Po Kau 大埔滘：Apr-04, 04 年 4 月：Henry Lui 呂德恒
[4] Tai Po Kau 大埔滘：Nov-08, 08 年 11 月：Wallace Tse 謝鑑超
[5] Tai Po Kau 大埔滘：Jan-03, 03 年 1 月：Michelle & Peter Wong 江敏兒、黃理沛
[6] Tai Po Kau 大埔滘：Dec-08, 08 年 12 月：Andy Kwok 郭匯昌

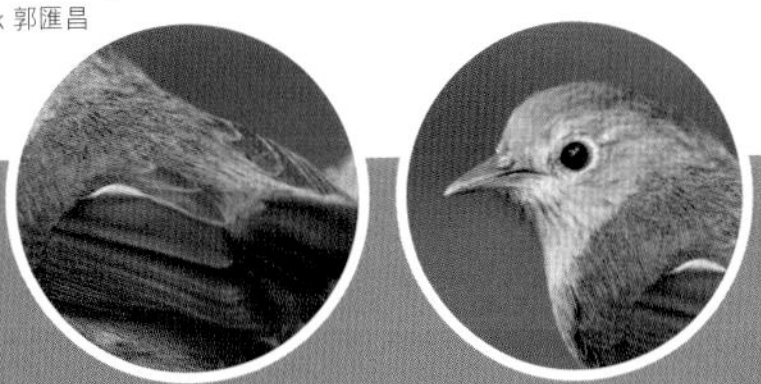

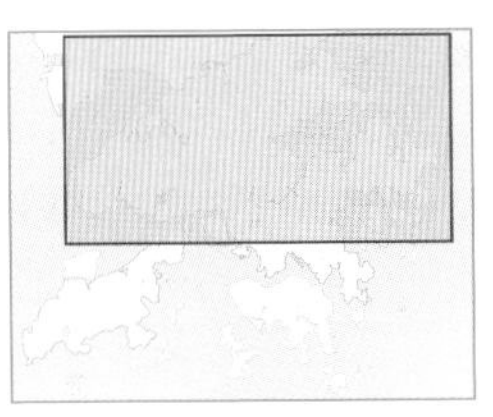

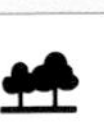

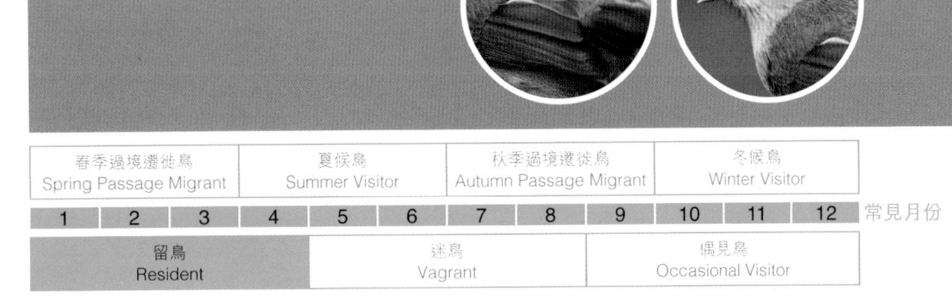

春季過境遷徙鳥 Spring Passage Migrant	夏候鳥 Summer Visitor	秋季過境遷徙鳥 Autumn Passage Migrant	冬候鳥 Winter Visitor

常見月份: 1 2 3 4 5 6 7 8 9 10 11 12

留鳥 Resident	迷鳥 Vagrant	偶見鳥 Occasional Visitor

黑眉雀鶥

㊟ hēi méi què méi

體長 length：12.5-14cm

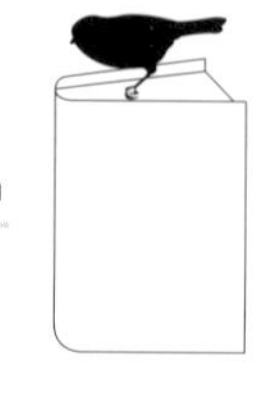

Huet's Fulvetta *Alcippe hueti*

其他名稱 Other names：白眶雀鶥

1

2

3

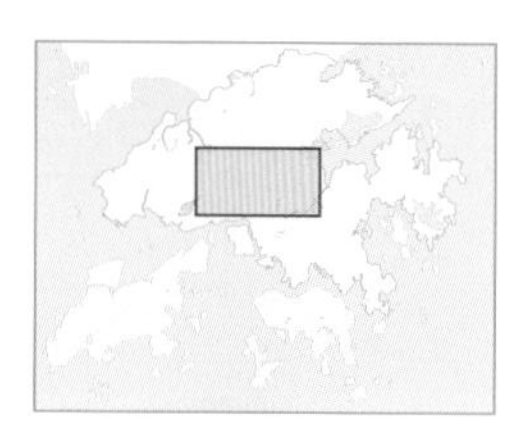

頭部灰色，有明顯白眼圈，嘴黑色。上體至尾部褐色，喉及下體淡黃褐色，沾有灰斑，腳粉紅色。

Small forest babbler. Greyish head with prominent white eye-ring. Black bill. Upperparts to tail brown. Throat to underparts pale yellowish brown, with grey patches. Pink legs.

1 Tai Po Kau 大埔滘：Aug-07, 07 年 8 月：Allen Chan 陳志雄
2 Tai Po Kau 大埔滘：Aug-07, 07 年 8 月：Allen Chan 陳志雄
3 Tai Po Kau 大埔滘：Apr-04, 04 年 4 月：Henry Lui 呂德恆
4 Tai Po Kau 大埔滘：Dec-06, 06 年 12 月：Wallace Tse 謝鑑超
5 Tai Po Kau 大埔滘：Dec-06, 06 年 12 月：Wallace Tse 謝鑑超
6 Tai Po Kau 大埔滘：Dec-06, 06 年 12 月：Henry Lui 呂德恆
7 Tai Po Kau 大埔滘：Feb-07, 07 年 2 月：Wallace Tse 謝鑑超
8 Tai Po Kau 大埔滘：Nov-08, 08 年 11 月：Wallace Tse 謝鑑超

	春季過境遷徙鳥 Spring Passage Migrant			夏候鳥 Summer Visitor			秋季過境遷徙鳥 Autumn Passage Migrant			冬候鳥 Winter Visitor		
常見月份	1	2	3	4	5	6	7	8	9	10	11	12
	留鳥 Resident				迷鳥 Vagrant				偶見鳥 Occasional Visitor			

大草鶯 ㊟dà cǎo yīng

體長 length：16-18cm

Chinese Grassbird | *Graminicola striatus*

其他名稱 Other names：Rufous-rumped Grassbird, Large Grass Warbler

1

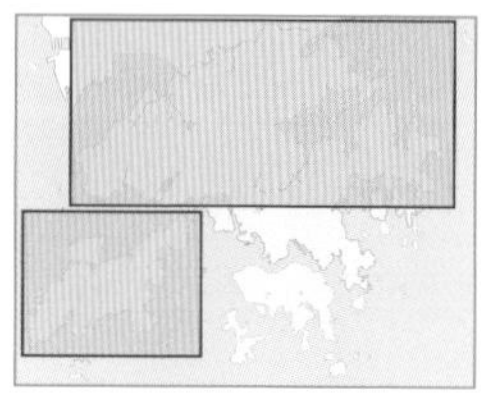

全身大致褐色，頭圓，嘴粗壯，尾長而翼短，看起來有如一隻體形健碩的鷦鶯。後頸羽毛有黑白條紋，腹部白色，脇及尾下覆羽淡黃色。叫聲為乾涸沙啞的「charh-charh-charh」。喜在高山草地活動，但冬天時也會降到低地棲息。

Mainly brown. Round head, stout bill, short wings with long tail, resembling a robust prinia. Black-and-white streaks on nape. Whitish underparts with buff flanks and vent. Call a dry and harsh "charh-charh-charh". Prefers grassland at high altitude, descends to lower levels in winter.

1 Robin's Nest 紅花嶺：18-May-09, 09 年 5 月 18日：Martin Hale 夏敖天
2 Tai Mo Shan 大帽山：Jul-00, 00 年 7 月：Martin Hale 夏敖天
3 Tai Mo Shan 大帽山：Jun-09, 09 年 6 月：Michelle and Peter Wong 江敏兒、黃理沛
4 Martin Hale 夏敖天
5 Robin's Nest 紅花嶺：18-May-09, 09 年 5 月 18日：Martin Hale 夏敖天

	春季過境遷徙鳥 Spring Passage Migrant			夏候鳥 Summer Visitor			秋季過境遷徙鳥 Autumn Passage Migrant			冬候鳥 Winter Visitor		
常見月份	1	2	3	4	5	6	7	8	9	10	11	12
	留鳥 Resident				迷鳥 Vagrant				偶見鳥 Occasional Visitor			

2
4
3
5

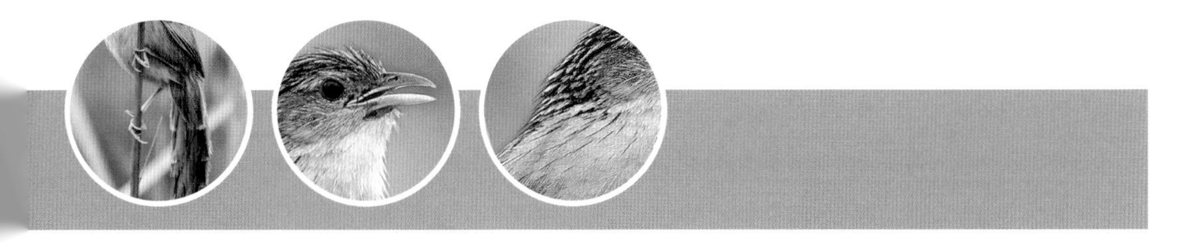

藍翅希鶥

[普] lán chì xī mēi

體長 length：14-15.5cm

Blue-winged Minla | *Actinodura cyanouroptera*

1

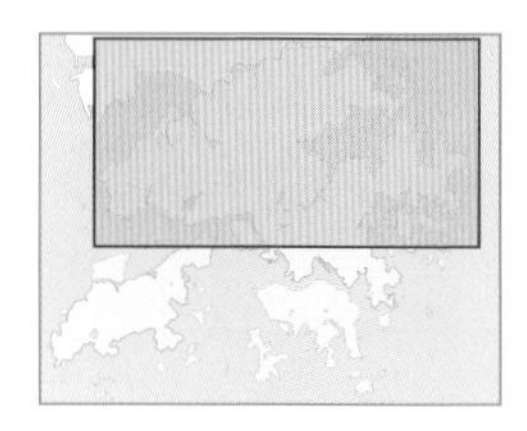

大致灰色，頭部及翼上有藍色色斑，但只有在光線充足的情況下才能看到。眉淺色，上方具有黑線。尾長，尾下銀白色，兩側邊緣黑色。常成小群出現，叫聲有多個音節，聲調哀怨。喜在樹間活動，甚少在地上走動。

A grey babbler. Blue head and wing panels can only be seen under good light. Black line above pale eyebrow. Long tail with distinctive silvery white undertail with thin black margin. Usually in small flocks. Call plaintive with several notes. Arboreal, rarely descends to ground.

1 adult 成鳥：Tai Po Kau 大埔滘：Michelle & Peter Wong 江敏兒、黃理沛
2 adult 成鳥：Tai Po Kau 大埔滘：Feb-08, 08 年 2 月：Gary Chow 周家禮
3 adult 成鳥：Tai Po Kau 大埔滘：Jan-09, 09 年 1 月：Cherry Wong 黃卓研
4 adult 成鳥：Tai Po Kau 大埔滘：Jan-08, 08 年 1 月：Owen Chiang 深藍
5 adult 成鳥：Tai Po Kau 大埔滘：Mar-09, 09 年 3 月：Cherry Wong 黃卓研
6 adult 成鳥：Tai Po Kau 大埔滘：Jan-08, 08 年 1 月：Andy Kwok 郭匯昌
7 adult 成鳥：Tai Po Kau 大埔滘：Jan-09, 09 年 1 月：Cherry Wong 黃卓研

春季過境遷徙鳥 Spring Passage Migrant	夏候鳥 Summer Visitor	秋季過境遷徙鳥 Autumn Passage Migrant	冬候鳥 Winter Visitor

常見月份 | 1 | 2 | 3 | 4 | 5 | 6 | 7 | 8 | 9 | 10 | 11 | 12 |

留鳥 Resident	迷鳥 Vagrant	偶見鳥 Occasional Visitor

2
3
4
5
6
7

紅嘴相思鳥

[普]hóng zuǐ xiāng sī niǎo

體長length：14-15cm

Red-billed Leiothrix | *Leiothrix lutea*

其他名稱 Other names：Pekin Robin, Red-billed Robin

1

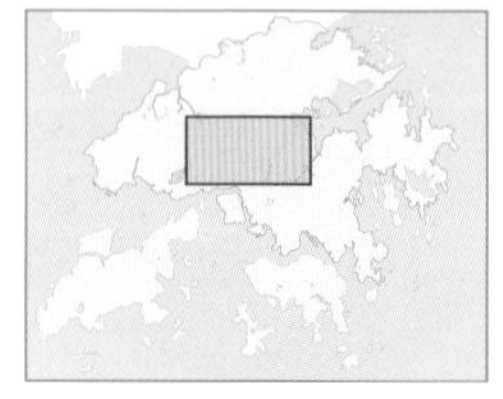

外形獨特。嘴紅而較粗，嘴尖偏黃。上體橄欖綠，下體黃色，喉及胸部帶紅色。叫聲通常較嘈吵。經常小群出現，喜棲身於茂密竹林中。

Very distinctive. Thick red bill and yellowish tip. Upperparts olive green. Underparts yellow with red tinge on throat and breast. Wing panels red and yellow. Call usually a noisy chatter. Often in small flocks. Prefers dense bamboo thickets.

1 adult male 雄成鳥：Kowloon Park 九龍公園：Dec-07, 07 年 12 月：Chan Kai Wai 陳佳瑋
2 adult female 雌成鳥：Kowloon Park 九龍公園：Sep-06, 06 年 9 月：Joyce Tang 鄧玉蓮
3 adult female 雌成鳥：Kowloon Park 九龍公園：Aug-03, 03 年 8 月：Cherry Wong 黃卓研
4 adult male 雄成鳥：Tai Po Kau 大埔滘：Jan-07, 07 年 1 月：Henry Lui 呂德恆
5 adult male 雄成鳥：Tai Po Kau 大埔滘：Jan-07, 07 年 1 月：Henry Lui 呂德恆
6 adult female 雌成鳥：Tai Po Kau 大埔滘：Aug-07, 07 年 8 月：Wallace Tse 謝鑑超
7 adult male 雄成鳥：Kowloon Park 九龍公園：19-Nov-07, 07 年 11 月 19 日：Owen Chiang 深藍

	春季過境遷徙鳥 Spring Passage Migrant			夏候鳥 Summer Visitor			秋季過境遷徙鳥 Autumn Passage Migrant			冬候鳥 Winter Visitor		
常見月份	1	2	3	4	5	6	7	8	9	10	11	12
	留鳥 Resident				迷鳥 Vagrant				偶見鳥 Occasional Visitor			

2
3
4
5
6
7

銀耳相思鳥

㊟ yín ěr xiāng sī niǎo

體長 length：15.5-17cm

Silver-eared Mesia | *Leiothrix argentauris*

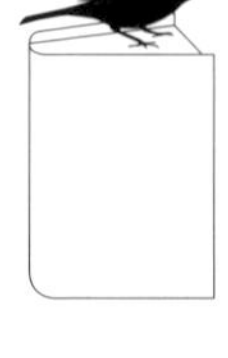

1

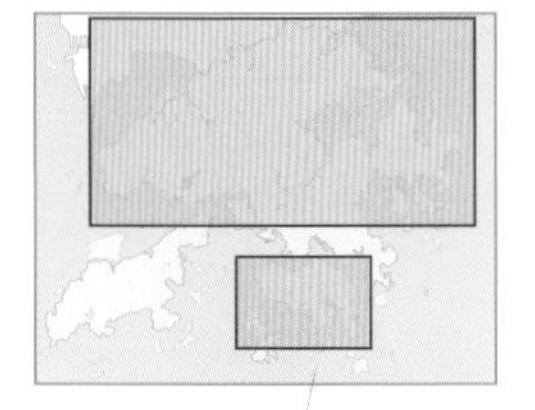

顏色獨特且鮮艷奪目。頭黑色而有銀灰色耳羽，喉及胸呈亮麗的橙紅色，翼上有紅色和黃色斑。叫聲變化多端，但較多為四聲口哨。常與紅嘴相思鳥為伍，喜在樹林下層茂密的灌草叢中活動。

A distinctively colourful babbler. Black head with silvery ear coverts. Orange-red throat and breast. Wing panels red and yellow. Calls vary but often a four-note whistle. Usually in mixed flocks with Red-billed Leiothrix. Prefers dense undergrowth in forest.

1 adult, race *ricketti* 成鳥, *ricketti* 亞種：Tai Po Kau 大埔滘：Dec-07, 07 年 12 月：Martin Hale 夏敖天
2 adult, race *ricketti* 成鳥, *ricketti* 亞種：Tai Po Kau 大埔滘：Nov-08, 08 年 11 月：Andy Kwok 郭匯昌
3 adult female, race *galbana* 雌成鳥, *galbana* 亞種：Lam Tsuen 林村：Jan-06, 06 年 1 月：Jemi and John Holmes 孔思義、黃亞萍
4 adult with juvenile, race *ricketti* 成鳥和幼鳥, *ricketti* 亞種：Tai Po Kau 大埔滘：19-May-07, 07 年 5 月 19 日：Michelle and Peter Wong 江敏兒、黃理沛
5 adult male, race *galbana* 雄成鳥, *galbana* 亞種：Tai Po Kau 大埔滘：Feb-08, 08 年 2 月：Ken Fung 馮漢城
6 adult, race *ricketti* 成鳥, *ricketti* 亞種：Tai Po Kau 大埔滘：Feb-04, 04 年 2 月：Henry Lui 呂德恒

	春季過境遷徙鳥 Spring Passage Migrant			夏候鳥 Summer Visitor			秋季過境遷徙鳥 Autumn Passage Migrant			冬候鳥 Winter Visitor		
常見月份	1	2	3	4	5	6	7	8	9	10	11	12
	留鳥 Resident				迷鳥 Vagrant				偶見鳥 Occasional Visitor			

2
3
4
5
6

畫眉 [普]huà méi

體長 length：21-24cm

Chinese Hwamei | *Garrulax canorus*

其他名稱 Other names：Hwamei

全身褐色有細緻黑紋，眼圈及眼眉明顯白色，嘴和腳黃色。不易觀察。由於叫聲悅耳多變，在本地為十分普遍的籠養鳥。

Brown body with fine black streaks. Prominent white eyering and eyebrow. Yellow bill and legs. More often heard than seen. Melodious whistling calls account for it's popularity as a cagebird.

1 adult 成鳥：Kowloon Park 九龍公園：Dec-07, 07 年 12 月：Chan Kai Wai 陳佳瑋
2 adult 成鳥：Cheung Chau 長洲：Nov-04, 04 年 11 月：Henry Lui 呂德恆
3 adult 成鳥：Cheung Chau 長洲：Dec-04, 04 年 12 月：Henry Lui 呂德恆
4 adult 成鳥：Kowloon Park 九龍公園：Mar-04, 04 年 3 月：Cherry Wong 黃卓研
5 adult 成鳥：Kowloon Park 九龍公園：Nov-03, 03 年 11 月：Cherry Wong 黃卓研
6 adult 成鳥：Kowloon Park 九龍公園：Aug-08, 08 年 8 月：Kitty Koo 古愛婉

	春季過境遷徙鳥 Spring Passage Migrant			夏候鳥 Summer Visitor			秋季過境遷徙鳥 Autumn Passage Migrant			冬候鳥 Winter Visitor		
常見月份	1	2	3	4	5	6	7	8	9	10	11	12
	留鳥 Resident				迷鳥 Vagrant				偶見鳥 Occasional Visitor			

2
3
4
5

6

黑喉噪鶥

㊟ hēi hóu zào méi

體長 length：23-30cm

Black-throated Laughingthrush | *Pterorhinus chinensis*

1

2

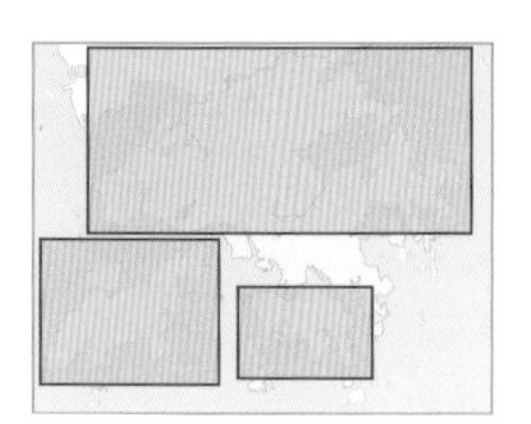

深色的中型噪鶥，上體灰黑色，下體較淺色。臉及喉黑色，面頰是鮮明的白色。常小群出現，叫聲嘹亮，如口哨般「胡壺一胡壺一」，有時會模仿其他雀鳥的鳴聲。

A dark medium-sized laughingthrush. Dark grey upperparts and paler underparts. Black face and throat. Prominent white cheeks. Usually in small flocks. Loud whistling call, like musical note "do - sol". Imitates other bird calls.

1 adult 成鳥：Tso Kung Tam 曹公潭：23-Feb-08, 08 年 2 月 23 日：Owen Chiang 深藍
2 adult 成鳥：Hong Kong Park 香港公園：20-Jan-08, 08 年 1 月 20 日：Anita Lee 李雅婷
3 adult 成鳥：Lung Fu Shan 龍虎山：Mar-08, 08 年 3 月：Herman Ip 葉紀江
4 adult 成鳥：Hong Kong Park 香港公園：21-Dec-07, 07 年 12 月 21 日：Owen Chiang 深藍
5 adult 成鳥：Lung Fu Shan 龍虎山：15-Jun-08, 08 年 6 月 15 日：Christina Chan 陳燕明
6 adult 成鳥：Hong Kong Park 香港公園：20-Jan-08, 08 年 1 月 20 日：Anita Lee 李雅婷
7 adult 成鳥：Lung Fu Shan 龍虎山：Jul-08, 08 年 7 月：Herman Ip 葉紀江

	春季過境遷徙鳥 Spring Passage Migrant			夏候鳥 Summer Visitor			秋季過境遷徙鳥 Autumn Passage Migrant			冬候鳥 Winter Visitor		
常見月份	1	2	3	4	5	6	7	8	9	10	11	12
	留鳥 Resident				迷鳥 Vagrant				偶見鳥 Occasional Visitor			

3
4
5
6
7

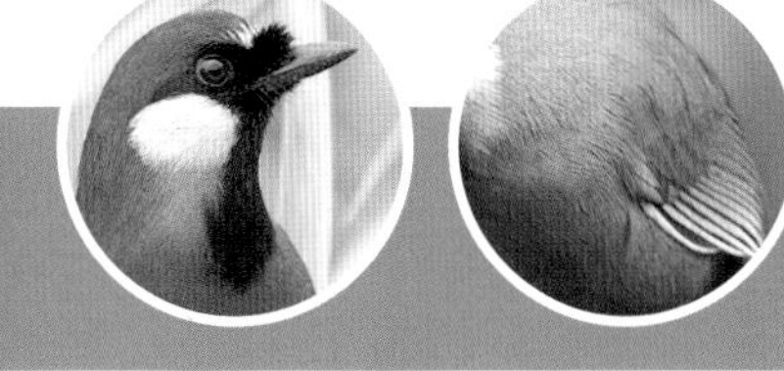

黑領噪鶥

㊟ hēi lǐng zào méi

體長 length：26.5-34.5cm

Greater Necklaced Laughingthrush | *Pterorhinus pectoralis*

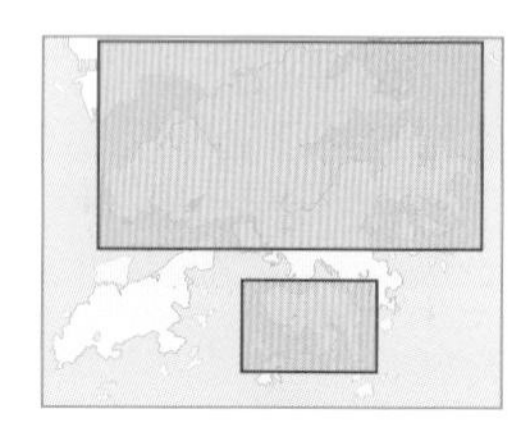

體型較大的噪鶥。背部鮮栗褐色，眼後及耳羽有黑色斑紋伸延到頸邊。香港有兩個亞種，*picticollis* 亞種常於新界中部出沒，頸上斑紋灰色；另一亞種 *pectoralis* 較多於市區公園中見到，頸上斑紋黑色，來自西部地區，多為逸鳥。性合群，叫聲嘹亮。

Large-sized laughingthrush. Upperparts chestnut brown. Black marking behind eyes and from ear coverts down to both sides of neck. Two races occur in Hong Kong. South China race *picticollis* (with grey markings on neck) usually occurs in central New Territories. Individuals with darker markings on neck belong to a west China race *pectoralis* and most are escapes, usually recorded in urban parks. Gregarious. Loud calls.

1 adult, race *picticollis* 成鳥, *picticollis* 亞種：Lung Fu Shan 龍虎山：Aug-08, 08 年 8 月：Herman Ip 葉紀江
2 adult, race *picticollis* 成鳥, *picticollis* 亞種：Shatin Pass 沙田坳：5-Apr-06, 06 年 4 月 5 日：Michelle and Peter Wong 江敏兒、黃理沛
3 adult, race *picticollis* 成鳥, *picticollis* 亞種：Wu Kau Tang 烏蛟騰：Feb-04, 04 年 2 月：Jemi and John Holmes 孔思義、黃亞萍
4 adult, race *picticollis* 成鳥, *picticollis* 亞種：Lung Fu Shan 龍虎山：Jul-08, 08 年 7 月：Herman Ip 葉紀江
5 adult, race *picticollis* 成鳥, *picticollis* 亞種：Lung Fu Shan 龍虎山：Jul-08, 08 年 7 月：Herman Ip 葉紀江
6 juvenile 幼鳥：Lung Fu Shan 龍虎山：Apr-07, 07 年 4 月：Cheng Nok Ming 鄭諾銘
7 adult, race *picticollis* 成鳥, *picticollis* 亞種：Siu Lek Yuen 小瀝源：Mar-08, 08 年 3 月：Ken Fung 馮漢城
8 adult, race *picticollis* 成鳥, *picticollis* 亞種：Shatin Pass 沙田坳：24-Dec-05, 05 年 12 月 24 日：Michelle and Peter Wong 江敏兒、黃理沛

	春季過境遷徙鳥 Spring Passage Migrant			夏候鳥 Summer Visitor			秋季過境遷徙鳥 Autumn Passage Migrant			冬候鳥 Winter Visitor		
常見月份	1	2	3	4	5	6	7	8	9	10	11	12
	留鳥 Resident				迷鳥 Vagrant				偶見鳥 Occasional Visitor			

2
3
4
5
6
7

8

矛紋草鶥

㊟máo wén cǎo méi

體長length：22.5-29.5cm

Chinese Babax | *Pterorhinus lanceolatus*

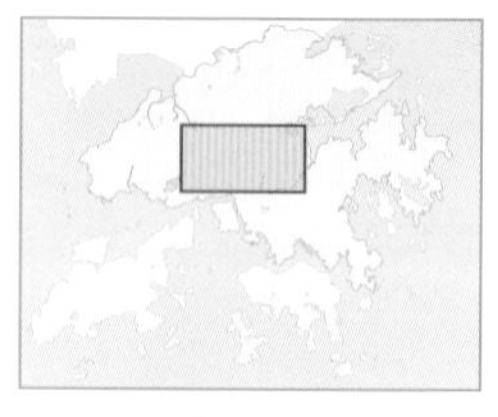

身軀大致灰褐色，滿佈栗褐色縱紋，喉至胸部白色。黑色嘴稍向下彎，有灰黑眼圈與黃色虹膜對比鮮明。尾有淺色窄橫斑，腳粉褐色。2005年後再無野化種群記錄。

Mainly greyish-brown body with chestnut-brown streaks. White from throat to breast. Black curved bill. Greyish-black eye-rings contrast with yellow irises. Pale narrow bands on tail. Pale brown legs. No feral population after 2005.

1 adult 成鳥：Tai Mo Shan 大帽山：May-11, 11 年 5 月：Lee Yat Ming 李逸明
2 adult 成鳥：Tai Mo Shan 大帽山：May-11, 11 年 5 月：Kinni Ho 何建業
3 adult 成鳥：Tai Mo Shan 大帽山：May-11, 11 年 5 月：Lee Yat Ming 李逸明
4 adult 成鳥：Tai Mo Shan 大帽山：May-11, 11 年 5 月：Lee Yat Ming 李逸明
5 adult 成鳥：Tai Mo Shan 大帽山：May-11, 11 年 5 月：Lee Yat Ming 李逸明
6 adult 成鳥：Tai Mo Shan 大帽山：May-11, 11 年 5 月：Lee Yat Ming 李逸明
7 adult 成鳥：Tai Mo Shan 大帽山：May-11, 11 年 5 月：Lee Yat Ming 李逸明
8 adult 成鳥：Tai Mo Shan 大帽山：May-11, 11 年 5 月：Lee Yat Ming 李逸明

春季過境遷徙鳥 Spring Passage Migrant	夏候鳥 Summer Visitor	秋季過境遷徙鳥 Autumn Passage Migrant	冬候鳥 Winter Visitor

常見月份 1 2 3 4 5 6 7 8 9 10 11 12

留鳥 Resident	迷鳥 Vagrant	偶見鳥 Occasional Visitor

3
4
5
6
7
8

白頰噪鶥

㊟ bái jiá zào méi
㊟ 頰：音夾

體長 length：22-24cm

White-browed Laughingthrush | *Pterorhinus sannio*

1

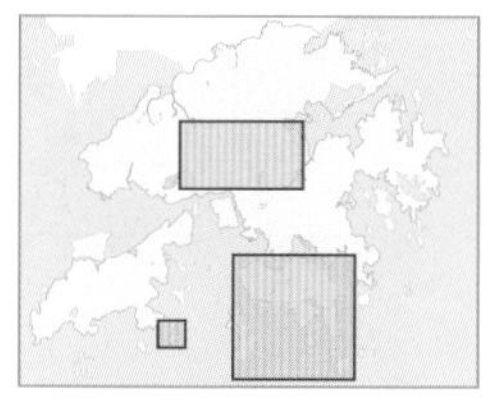

全身褐色，和畫眉相似，眼眉和面頰均為白色。喜成小群出沒，叫聲類似黑臉噪鶥的「標－標－」聲，但有顫音。

Brown body. Similar to Chinese Hwamei but with distinctive white eye brow and cheeks. Usually in small flocks. Call a loud "bill-bill-" similar to Masked Laughingthrush, but appeared vibrating.

1 adult 成鳥：HK Zoological and Botanical Gardens 香港動植物公園：Apr-07, 07 年 4 月：Cherry Wong 黃卓研
2 adult 成鳥：Cheung Chau 長洲：Nov-04, 04 年 11 月：Henry Lui 呂德恒
3 adult 成鳥：HK Zoological and Botanical Gardens 香港動植物公園：Apr-07, 07 年 4 月：Cherry Wong 黃卓研
4 adult 成鳥：HK Zoological and Botanical Gardens 香港動植物公園：Apr-07, 07 年 4 月：Cherry Wong 黃卓研
5 adult 成鳥：Cheung Chau 長洲：Dec-04, 04 年 12 月：Henry Lui 呂德恒
6 adult 成鳥：HK Zoological and Botanical Gardens 香港動植物公園：Mar-03, 03 年 3 月：Henry Lui 呂德恒
7 adult 成鳥：Hong Kong Park 香港公園：May-08, 08 年 5 月：Eling Lee 李佩玲

	春季過境遷徙鳥 Spring Passage Migrant			夏候鳥 Summer Visitor			秋季過境遷徙鳥 Autumn Passage Migrant			冬候鳥 Winter Visitor		
常見月份	1	2	3	4	5	6	7	8	9	10	11	12
	留鳥 Resident				迷鳥 Vagrant				偶見鳥 Occasional Visitor			

2
3
4
5
6
7

黑臉噪鶥

㊟ hēi liǎn zào méi

體長 length：28-31.5cm

Masked Laughingthrush | *Pterorhinus perspicillatus*

其他名稱 Other names：Black-faced Laughingthrush

1

香港最常見的噪鶥。背灰褐色，頭較灰，臉部有黑色面罩，尾部深褐色，尾下覆羽紅棕色。常成小群出沒，叫聲為嘈吵的「標一標一」聲。

The most common Laughingthrush in Hong Kong. Body greyish brown. Head greyer with black mask on the face and rufous undertail coverts. Tail dark brown. Usually in small flocks. Call is a loud, scolding "bill-bill-".

1 adult 成鳥：Mai Po 米埔：Feb-07, 07 年 2 月：Cherry Wong 黃卓研
2 adult 成鳥：Cheung Chau 長洲：Nov-04, 04 年 11 月：Henry Lui 呂德恆
3 juvenile 幼鳥：Kowloon Park 九龍公園：Oct-03, 03 年 10 月：Cherry Wong 黃卓研
4 adult 成鳥：HK Zoological and Botanical Gardens 香港動植物公園：Nov-04, 04 年 11 月：Henry Lui 呂德恆
5 adult 成鳥：Hong Kong Park 香港公園：Jul-08, 08 年 7 月：Kami Hui 許淑君
6 adult 成鳥：Kam Tin 錦田：Mar-07, 07 年 3 月：Martin Hale 夏敖天
7 adult 成鳥：Hong Kong Park 香港公園：Jul-08, 08 年 7 月：Kami Hui 許淑君

	春季過境遷徙鳥 Spring Passage Migrant			夏候鳥 Summer Visitor			秋季過境遷徙鳥 Autumn Passage Migrant			冬候鳥 Winter Visitor		
常見月份	1	2	3	4	5	6	7	8	9	10	11	12
	留鳥 Resident				迷鳥 Vagrant				偶見鳥 Occasional Visitor			

2
3
4
5
6
7

棕頭鴉雀 [普] zōng tóu yā què

體長 length：11-12.5cm

Vinous-throated Parrotbill | *Sinosuthora webbiana*

其他名稱 Other names：粉紅鸚嘴

1

⛰

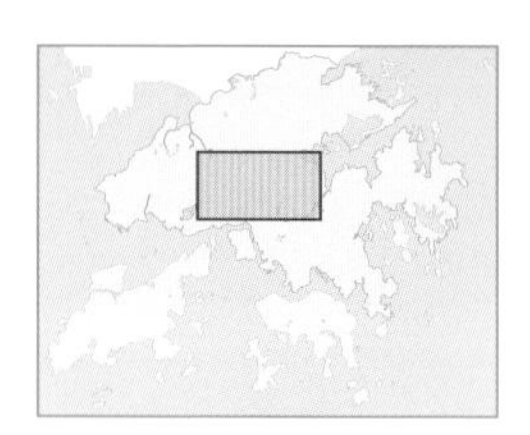

全身紅褐色，羽毛蓬鬆，尾長。頭大而圓，嘴部細小。背至尾部褐色，翼紅褐色。下體淡褐色。在大帽山處有小種群。

Mainly reddish-brown, feathers look shaggy, long tail. Head round and big, bill small and short. Mantle to tail brownish, wings reddish-brown. Underparts greyish brown. A small population can be found at Tai Mo Shan.

[1] race *suffusus*, *suffusus* 亞種：Tai Mo Shan 大帽山：Jun-04, 04 年 6 月：Henry Lui 呂德恒
[2] race *suffusus*, *suffusus* 亞種：Tai Mo Shan 大帽山：Jun-04, 04 年 6 月：Henry Lui 呂德恒
[3] race *suffusus*, *suffusus* 亞種：Tai Mo Shan 大帽山：Jun-04, 04 年 6 月：Henry Lui 呂德恒
[4] race *suffusus*, *suffusus* 亞種：Tai Mo Shan 大帽山：May-04, 04 年 5 月：Michelle and Peter Wong 江敏兒、黃理沛
[5] race *suffusus*, *suffusus* 亞種：Tai Mo Shan 大帽山：Jun-04, 04 年 6 月：Henry Lui 呂德恒
[6] juvenile 幼鳥：Tai Mo Shan 大帽山：Jul-09, 09 年 7 月：Michelle and Peter Wong 江敏兒、黃理沛
[7] race *suffusus*, *suffusus* 亞種：Tai Mo Shan 大帽山：Jun-04, 04 年 6 月：Cherry Wong 黃卓研

	春季過境遷徙鳥 Spring Passage Migrant			夏候鳥 Summer Visitor			秋季過境遷徙鳥 Autumn Passage Migrant			冬候鳥 Winter Visitor		
常見月份	1	2	3	4	5	6	7	8	9	10	11	12
	留鳥 Resident				迷鳥 Vagrant				偶見鳥 Occasional Visitor			

2
3
4
5
6
7

白喉林鶯 ㊗ bái hóu lín yīng

體長length：12.5-14cm

Lesser Whitethroat | *Sylvia curruca*

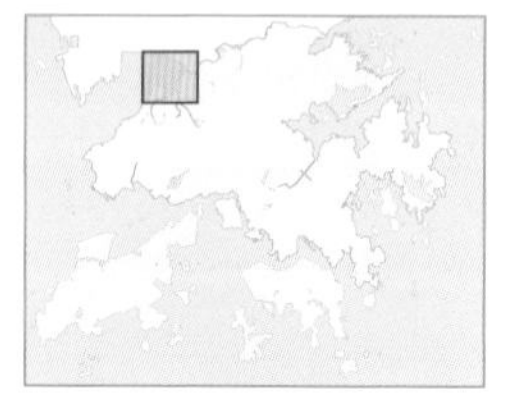

身型纖細，頭頂帶灰色，上體褐色。嘴短，尾長而末端呈方形。喉及下體白色，胸兩側及脇偏淡黃色，眼先及耳羽深色。腳長，嘴及腳均為黑色。喜開闊原野，多接近地面活動。

Slim-bodied. Crown greyish. Upperparts brown. Short bill and square-ended long tail. Throat and underparts white. Buffish on side of breast and flanks. Dark lores and ear coverts. Long legs, legs and bill are black. Prefers open country and keeps close to ground.

1 adult, race *margelanica* 成鳥, *margelanica* 亞種：Lai Chi Kok 荔枝角：Jan-08, 08 年 1 月：Martin Hale 夏敖天
2 adult, race *margelanica* 成鳥, *margelanica* 亞種：Lai Chi Kok 荔枝角：Jan-08, 08 年 1 月：Martin Hale 夏敖天
3 race *blythi*, *blythi* 亞種：Long Valley 塱原：Yu Yat Tung 余日東

春季過境遷徙鳥 Spring Passage Migrant			夏候鳥 Summer Visitor			秋季過境遷徙鳥 Autumn Passage Migrant			冬候鳥 Winter Visitor		
常見月份 1	2	3	4	5	6	7	8	9	10	11	12

留鳥 Resident	迷鳥 Vagrant	偶見鳥 Occasional Visitor

紅脇繡眼鳥

㊅hóng xié xiù yǎn niǎo
㊊脇：音脅

體長length：10.5-11.5cm

Chestnut-flanked White-eye | *Zosterops erythropleurus*

體型細小的繡眼鳥。頭、上體及尾部綠色，眼圈白色。喉和臀部黃色，胸和腹部白色，脇部栗色。有時混在暗綠繡眼鳥群中，不過顏色較暗啞，喉部黃色較小，下嘴顏色較淡。

Small-sized white-eye. Head, upperparts and tail green, eye-ring is white, bill and legs are black. Throat and vent yellow, breast and belly white. Chestnut flanks. Sometimes mixed in flocks of Swinhoe's White-eyes, but their colour is generally duller, size of yellow throat is smaller, and lower bill is paler.

1 adult 成鳥：Tai Po Kau 大埔滘：Dec-06, 06 年 12 月：Henry Lui 呂德恆
2 adult 成鳥：Tai Po Kau 大埔滘：Dec-06, 06 年 12 月：Henry Lui 呂德恆

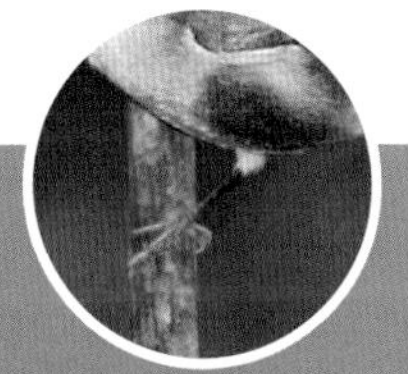

春季過境遷徙鳥 Spring Passage Migrant	夏候鳥 Summer Visitor	秋季過境遷徙鳥 Autumn Passage Migrant	冬候鳥 Winter Visitor

1	2	3	4	5	6	7	8	9	10	11	12	常見月份

留鳥 Resident	迷鳥 Vagrant	偶見鳥 Occasional Visitor

栗耳鳳鶥 [普]lì ěr fèng méi

體長length：14-15cm

Indochinese Yuhina | *Yuhina torqueola*

其他名稱 Other names：栗頭鳳鶥

有明顯灰色短冠羽和後枕，面頰深褐色，並連接着紅褐色、帶有白色縱紋的領圈。上體、背部至尾部灰褐色，喉至腹部、尾下覆羽灰白色，尾羽末端有白點。喜在森林中層至樹冠處活動。

Prominent small grey crown and nape, dark brown cheeks join collar-ring which is reddish-brown with white streaks. Upperparts, mantle to tail greyish brown. Throat to belly greyish white. White tips on tail feathers. Favours canopy to mid-storey in forest.

1 race *torqueola*, *torqueola* 亞種：Tai Po Kau 大埔滘：Dec-06, 06 年 12 月：Law Kam Man 羅錦文
2 race *torqueola*, *torqueola* 亞種：Tai Po Kau 大埔滘：19-Dec-06, 06 年 12 月 19 日：Owen Chiang 深藍
3 race *torqueola*, *torqueola* 亞種：Shatin Pass 沙田坳：Jul-04, 04 年 7 月：Michelle and Peter Wong 江敏兒、黃理沛
4 race *torqueola*, *torqueola* 亞種：Ng Tung Chai 梧桐寨：Dec-08, 08 年 12 月：Cheng Nok Ming 鄭諾銘
5 race *torqueola*, *torqueola* 亞種：Tai Po Kau 大埔滘：Dec-06, 06 年 12 月：Law Kam Man 羅錦文

	春季過境遷徙鳥 Spring Passage Migrant			夏候鳥 Summer Visitor			秋季過境遷徙鳥 Autumn Passage Migrant			冬候鳥 Winter Visitor		
常見月份	1	2	3	4	5	6	7	8	9	10	11	12
	留鳥 Resident				迷鳥 Vagrant				偶見鳥 Occasional Visitor			

暗綠繡眼鳥 [普] àn lǜ xiù yǎn niǎo

體長 length：10-11.5cm

Swinhoe's White-eye | *Zosterops simplex*

其他名稱 Other names：相思，Japanese White-eye

1

體型細小。頭、上身及尾部綠色，有明顯的白眼圈，嘴和腳黑色。喉和臀部黃色，胸和腹部白色。常成群一起活動，叫聲為輕柔的「tzee-tzee」聲。

Small-sized. Head, upperparts and tail green, with prominent white eyering, black bill and legs. Throat and vent yellow, breast and belly white. Call is a soft "tzee". Wanders and feeds in flocks.

[1] Wanchai 灣仔；Aug-06, 06 年 8 月；Law Kam Man 羅錦文
[2] Shek Kong 石崗；Jan-07, 07 年 1 月；Andy Kwok 郭匯昌
[3] Shek Kong 石崗；Jan-07, 07 年 1 月；Martin Hale 夏敖天
[4] Shek Kong 石崗；Feb-07, 07 年 2 月；Michelle and Peter Wong 江敏兒、黃理沛
[5] juvenile 幼鳥；Siu Lek Yuen 小瀝源；Apr-08, 08 年 4 月；Ken Fung 馮漢城
[6] Mai Po 米埔；Oct-08, 08 年 10 月；Cherry Wong 黃卓研
[7] Cheung Chau 長洲；Nov-03, 03 年 11 月；Henry Lui 呂德恆

	春季過境遷徙鳥 Spring Passage Migrant			夏候鳥 Summer Visitor			秋季過境遷徙鳥 Autumn Passage Migrant			冬候鳥 Winter Visitor		
常見月份	1	2	3	4	5	6	7	8	9	10	11	12
	留鳥 Resident				迷鳥 Vagrant				偶見鳥 Occasional Visitor			

2
3
4
5
6

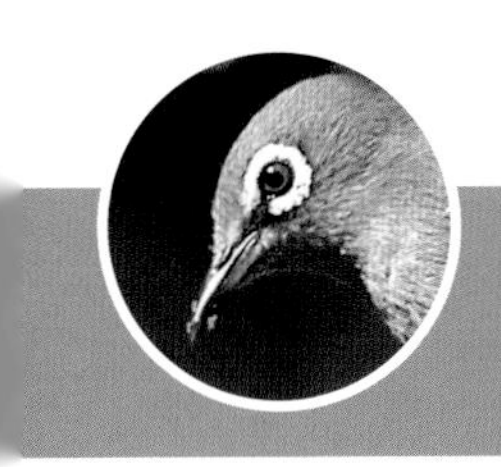

7

絨額鳾

㊟róng é shī
㊟鳾：音司

體長 length：12.5cm

Velvet-fronted Nuthatch | *Sitta frontalis*

體型細小。頭部和上體淺紫藍色，虹膜黃色，前額至眼先有黑斑，面頰淡粉紅色。嘴紅色，幼鳥的嘴黑色。喉白色，下體淡黃褐色。常沿樹幹或樹枝上下爬行，有時會頭下腳上或腹部朝天倒懸。

Small-sized. Head and upperparts pale purplish blue, yellowish iris, red eye-ring. Forehead to lores black, cheeks pale pink. The bill is red, but dark in juveniles. White throat, underparts buffy yellow. Roves along tree trunks and branches, sometimes upside-down.

1 adult male 雄成鳥：Chinese University of HK 香港中文大學：May-07, 07 年 5 月：Eling Lee 李佩玲
2 adult male 雄成鳥：Tai Po Kau 大埔滘：Nov-04, 04 年 11 月：Jemi and John Holmes 孔思義、黃亞萍
3 adult male 雄成鳥：Tai Po Kau 大埔滘：Nov-04, 04 年 11 月：Jemi and John Holmes 孔思義、黃亞萍
4 adult male 雄成鳥：Tai Po Kau 大埔滘：Dec-08, 08 年 12 月：Andy Kwok 郭匯昌
5 adult male 雄成鳥：Tai Po Kau 大埔滘：Nov-04, 04 年 11 月：Jemi and John Holmes 孔思義、黃亞萍

	春季過境遷徙鳥 Spring Passage Migrant			夏候鳥 Summer Visitor			秋季過境遷徙鳥 Autumn Passage Migrant			冬候鳥 Winter Visitor		
常見月份	1	2	3	4	5	6	7	8	9	10	11	12
	留鳥 Resident				迷鳥 Vagrant				偶見鳥 Occasional Visitor			

八哥 [普]bā gē

體長length：25cm

Crested Myna | *Acridotheres cristatellus*

全身黑色而有光澤，虹膜橙黃色，嘴的上部和頭頂之間有長冠羽。嘴黃色，腳淡粉紅色。飛行時，翼底有明顯大白斑；靜止或站立時，兩旁有兩片白斑。

Shiny black, with yellow-orange iris and long tufts above the bill. Bill yellow, legs pale pink. A distinctive white patch under each wing in flight. Two small white marks on the folded wing at rest.

[1] adult 成鳥：Po Toi 蒲台：May-07, 07 年 5 月：Sammy Sam and Winnie Wong 森美與雲妮
[2] adult 成鳥：Long Valley 塱原：Dec-04, 04 年 12 月：Henry Lui 呂德恒
[3] adult 成鳥：Kowloon Park 九龍公園：Apr-03, 03 年 4 月：Cherry Wong 黃卓研
[4] adult 成鳥：Po Toi 蒲台：May-07, 07 年 5 月：Sammy Sam and Winnie Wong 森美與雲妮
[5] adult 成鳥：Po Toi 蒲台：Apr-07, 07 年 4 月：Wallace Tse 謝鑑超

春季過境遷徙鳥 Spring Passage Migrant			夏候鳥 Summer Visitor			秋季過境遷徙鳥 Autumn Passage Migrant			冬候鳥 Winter Visitor			
1	2	3	4	5	6	7	8	9	10	11	12	常見月份

留鳥 Resident	迷鳥 Vagrant	偶見鳥 Occasional Visitor

家八哥 ㊟ jiā bā gē

體長 length：25cm

Common Myna | *Acridotheres tristis*

1

頭黑色，眼綠色，眼眶及周圍黃色。嘴和腳黃色，上體和下體深褐色，飛羽和尾羽黑色。腰白色。有時會與八哥一起活動。

Black head with yellow patch around green eye. Bill and legs yellow, upperparts and underparts dark brown, flight feathers and tail black. White rump. Sometime mixes with Crested Myna.

1 adult 成鳥：Shek Kong 石崗：Feb-08, 08 年 2 月：Kam Wing Lok 甘永樂
2 adult 成鳥：Long Valley 塱原：Apr-07, 07 年 4 月：Sammy Sam and Winnie Wong 森美與雲妮
3 adult 成鳥：Kam Tin 錦田：4-Jun-06, 06 年 6 月 4 日：Michelle and Peter Wong 江敏兒、黃理沛
4 adult 成鳥：Kam Tin 錦田：Aug-05, 05 年 8 月：Cherry Wong 黃卓研
5 adult 成鳥：Mai Po 米埔：Feb-07, 07 年 2 月：Cherry Wong 黃卓研
6 adult 成鳥：Mai Po 米埔：Mar-20, 20 年 3 月：Henry Lui 呂德恆
7 adult 成鳥：Kowloon Tsai Park 九龍仔公園：Aug-03, 03 年 8 月：Henry Lui 呂德恆

	春季過境遷徙鳥 Spring Passage Migrant			夏候鳥 Summer Visitor			秋季過境遷徙鳥 Autumn Passage Migrant			冬候鳥 Winter Visitor		
常見月份	1	2	3	4	5	6	7	8	9	10	11	12
	留鳥 Resident			迷鳥 Vagrant				偶見鳥 Occasional Visitor				

絲光椋鳥

㊥ sī guāng liáng niǎo
㊕ 椋：音良

體長 length：22cm

Red-billed Starling | *Spodiopsar sericeus*

其他名稱 Other names：Silky Starling

頭部灰白色，看來像絲絨。嘴和腳鮮紅色。上體淡褐灰色，翼和尾黑色而帶藍綠色金屬光澤。飛行時初級飛羽底部有白斑。冬天時成大群出現。

Head greyish white and silky, with distinctive red bill and legs. Upperparts pale brownish grey, wings and tail black with glossy green. White wing patches in flight. Occurs in large flocks in winter.

1 adult male 雄成鳥：Mai Po 米埔：Nov-08, 08 年 11 月：Pippen Ho 何志剛
2 adult 成鳥：Nam Sang Wai 南生圍：Dec-07, 07 年 12 月：Aka Ho
3 adult female 雌成鳥：Mai Po 米埔：Jun-07, 07 年 6 月：Cherry Wong 黃卓研
4 adult female 雌成鳥：Pui O 貝澳：Mar-08, 08 年 3 月：Jacky Chan 陳家華
5 adult male 雄成鳥：Shatin Central Park 沙田中央公園：Dec-08, 08 年 12 月：Christine and Samuel Ma 馬志榮・蔡美蓮
6 adult female 雌成鳥：Pui O 貝澳：Feb-08, 08 年 2 月：Andy Kwok 郭匯昌

春季過境遷徙鳥 Spring Passage Migrant			夏候鳥 Summer Visitor			秋季過境遷徙鳥 Autumn Passage Migrant			冬候鳥 Winter Visitor		
1	2	3	4	5	6	7	8	9	10	11	12

常見月份

留鳥 Resident | 迷鳥 Vagrant | 偶見鳥 Occasional Visitor

灰椋鳥

㊀普 huī liáng niǎo
㊀粵 椋：音良

體長 length：22cm

White-cheeked Starling | *Spodiopsar cineraceus*

1

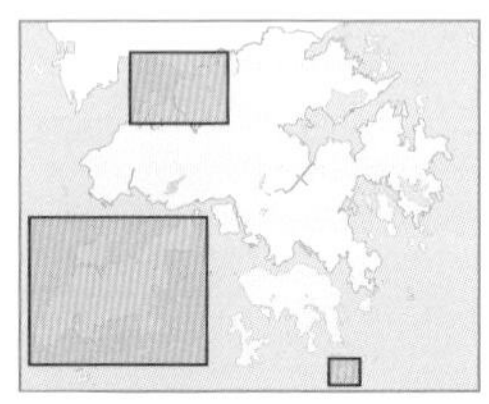

頭黑色，眼部周圍至臉頰有大片白斑，嘴和腳橙黃色。上體和翅膀深褐色，腰部明顯白色，尾端有白點。喉至上胸黑色，下體淡黃褐色。有時大群出現。

Black head with a large white patch around the eyes and the cheeks. Bill and legs orange-yellow. Upperparts and wings dark brown, rump white, white tips on tail feathers. Black throat and upper breast. Underparts pale buff. Can occur in large flocks.

1 adult male 雄成鳥：Mai Po 米埔：Jan-04, 04 年 1 月：Henry Lui 呂德恒
2 adult male 雄成鳥：Shatin Central Park 沙田中央公園：14-Feb-09, 09 年 2 月 14 日：Cheng Wai Keung 鄭偉強
3 adult male 雄成鳥：Mai Po 米埔：Nov-08, 08 年 11 月：Cheng Nok Ming 鄭諾銘
4 adult male 雄成鳥：Mai Po 米埔：Nov-08, 08 年 11 月：Cheng Nok Ming 鄭諾銘
5 adult female 雌成鳥：Kam Tin 錦田：Mar-08, 08 年 3 月：Aka Ho
6 adult male 雄成鳥：Tsim Bei Tsui 尖鼻咀：Feb-05, 05 年 2 月：Cherry Wong 黃卓研

春季過境遷徙鳥 Spring Passage Migrant			夏候鳥 Summer Visitor			秋季過境遷徙鳥 Autumn Passage Migrant			冬候鳥 Winter Visitor		
1	2	3	4	5	6	7	8	9	10	11	12
留鳥 Resident				迷鳥 Vagrant				偶見鳥 Occasional Visitor			

常見月份

2
3
4
5

6

黑領椋鳥

㊟hēi lǐng liáng niǎo
㊟椋：音良

體長 length：28cm

Black-collared Starling | *Gracupica nigricollis*

其他名稱 Other names：Black-necked Starling

大型椋鳥，頭部白色，眼部周圍皮膚黃色。有明顯黑色領帶，上體和翼深褐或黑色，羽毛邊緣白色。嘴黑色，腳淡粉紅色。幼鳥沒有黑色領帶，整體褐色。叫聲嘈吵，常成群活動。

Large starling. White head, with yellow bare skin around the eye. Prominent black collar, upperparts and wings dark brown or black with white tips. Juvenile has no black collar and is browner overall. Noisy and gregarious.

1 adult 成鳥：Long Valley 塱原：22-Nov-08, 08 年 11 月 22 日：Isaac Chan 陳家強
2 juvenile 幼鳥：Long Valley 塱原：Aug-07, 07 年 8 月：Sammy Sam and Winnie Wong 森美與雲妮
3 adult 成鳥：Fo Tan 火炭：Dec-07, 07 年 12 月：Christine and Samuel Ma 馬志榮、蔡美蓮
4 adult 成鳥：Long Valley 塱原：Dec-04, 04 年 12 月：Henry Lui 呂德恒
5 juvenile 幼鳥：Kowloon Park 九龍公園：Jun-07, 07 年 6 月：Bill Man 文權溢
6 adult 成鳥：Kowloon Park 九龍公園：Mar-05, 05 年 3 月：Henry Lui 呂德恒

	春季過境遷徙鳥 Spring Passage Migrant			夏候鳥 Summer Visitor			秋季過境遷徙鳥 Autumn Passage Migrant			冬候鳥 Winter Visitor		
常見月份	1	2	3	4	5	6	7	8	9	10	11	12
	留鳥 Resident				迷鳥 Vagrant				偶見鳥 Occasional Visitor			

北椋鳥

普 běi liáng niǎo
粵 椋：音良

體長 length：17cm

Daurian Starling | *Agropsar sturninus*

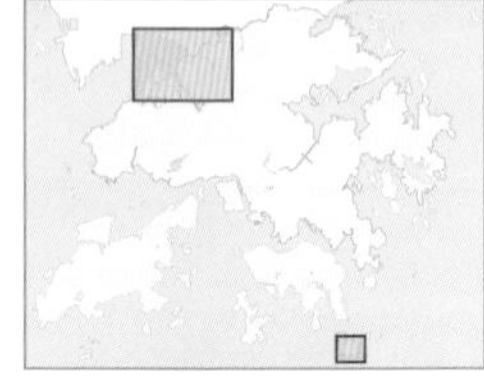

頭部深灰色，眼睛深色。上背和肩部深灰，有時沾紫褐色，翼上有一片大翼斑，腰黃褐色，下體淡灰色。

Dark grey head and dark eyes. Upperparts and scapulars dark grey, sometimes purplish brown. Distinctive large white wingbars. Rump brownish yellow, underparts pale grey.

1 female 雌鳥；Long Valley 塱原；Martin Hale 夏敖天

2 juvenile male moulting into adult plumage 雄幼鳥轉換成羽；Po Toi 蒲台；Sep-08, 08 年 9 月；Sammy Sam and Winnie Wong 森美與雲妮

3 adult female 雌成鳥；Tai Wai 大圍；15 Apr-13, 13年4月15日；Beetle Cheng 鄭諾銘

4 adult male 雄成鳥；Po Toi 蒲台；23-Apr-07, 07 年 4 月 23 日；Geoff Welch

春季過境遷徙鳥 Spring Passage Migrant	夏候鳥 Summer Visitor	秋季過境遷徙鳥 Autumn Passage Migrant	冬候鳥 Winter Visitor

常見月份 1 2 3 4 5 6 7 8 9 10 11 12

留鳥 Resident	迷鳥 Vagrant	偶見鳥 Occasional Visitor

栗頰椋鳥

[普]lì jiá liáng niǎo
[粵]頰椋：音夾良

體長 length：17cm

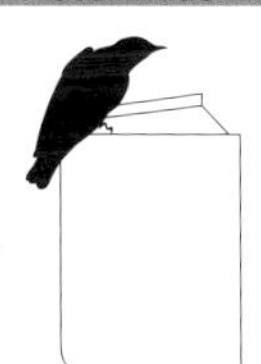

Chestnut-cheeked Starling | *Agropsar philippensis*

其他名稱 Other names：紫背椋鳥

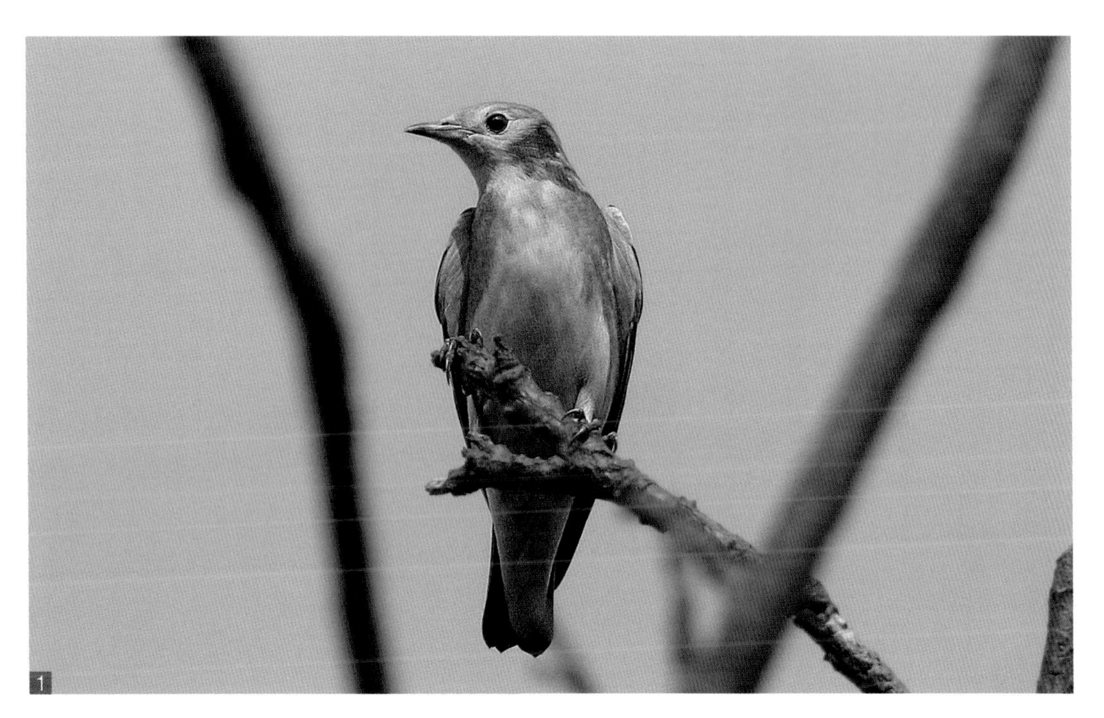

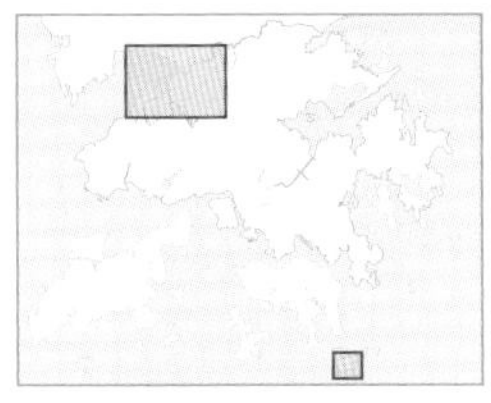

雄性頭白色，面部及頭部有明顯紅褐色的斑。翅膀上有明顯白帶，腰部帶栗色。雌性全身帶淺褐色。

Male has white head and prominent reddish brown cheeks. White wing bar and chestnut rump. Female: pale brown overall.

1 adult male 雄成鳥：Mai Po 米埔：Oct-05, 05 年 10 月：Cherry Wong 黃卓研
2 adult male 雄成鳥：Mai Po 米埔：Oct-05, 05 年 10 月：Cherry Wong 黃卓研

春季過境遷徙鳥 Spring Passage Migrant	夏候鳥 Summer Visitor	秋季過境遷徙鳥 Autumn Passage Migrant	冬候鳥 Winter Visitor

1 2 3 **4** 5 6 7 8 9 **10** 11 12 常見月份

留鳥 Resident	迷鳥 Vagrant	偶見鳥 Occasional Visitor

灰背椋鳥

[普]huī bèi liáng niǎo
[粵]椋：音良

體長 length：17cm

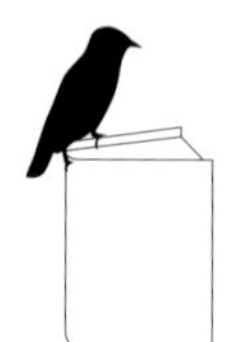

White-shouldered Starling | *Sturnia sinensis*

其他名稱 Other names：Chinese Starling

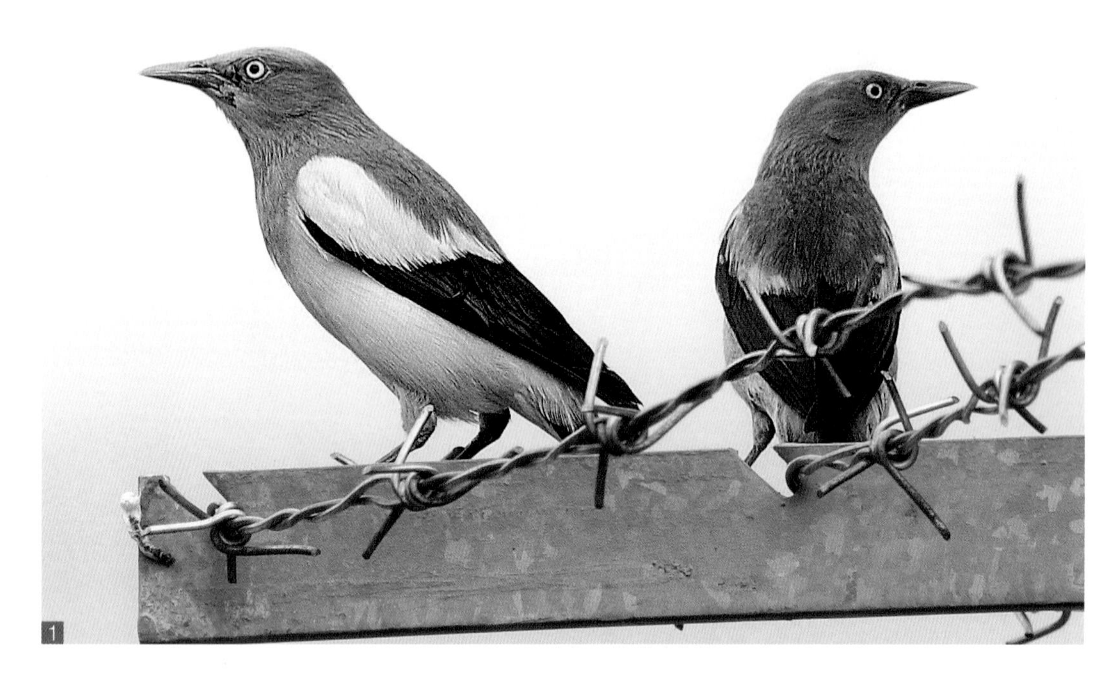

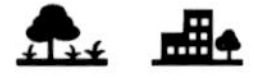

頭部灰色，有白眼圈，嘴和腳灰色。上體灰色，肩部有一片大白斑，飛羽黑色，腰白色，下體白色。

Grey head with white eye-rings. Bill and legs grey. Upperparts grey, with prominent white scapulars. Flight feathers black, rump white. Underparts white.

1 adult male and female 雄成鳥和雌成鳥：Kam Tin 錦田：Jun-08, 08 年 6 月：Herman Ip 葉紀江
2 adult female 雌成鳥：Kam Tin 錦田：Jun-07, 07 年 6 月：Cherry Wong 黃卓研
3 adult male 雄成鳥：Mai Po 米埔：May-07, 07 年 5 月：Bill Man 文權溢
4 adult male 雄成鳥：Kam Tin 錦田：May-04, 04 年 5 月：Martin Hale 夏敖天
5 adult male 雄成鳥：Kam Tin 錦田：May-04, 04 年 5 月：Martin Hale 夏敖天
6 adult male 雄成鳥：Nam Sang Wai 南生圍：Apr-06, 06 年 4 月：Jemi and John Holmes 孔思義、黃亞萍
7 adult female 雌成鳥：Kam Tin 錦田：Jun-08, 08 年 6 月：Herman Ip 葉紀江
8 adult male 雄成鳥：Kam Tin 錦田：Jun-08, 08 年 6 月：Herman Ip 葉紀江

	春季過境遷徙鳥 Spring Passage Migrant			夏候鳥 Summer Visitor			秋季過境遷徙鳥 Autumn Passage Migrant			冬候鳥 Winter Visitor		
常見月份	1	2	3	4	5	6	7	8	9	10	11	12
	留鳥 Resident				迷鳥 Vagrant				偶見鳥 Occasional Visitor			

2
3
4
5
6
7

8

灰頭椋鳥

[普]huī tóu liáng niǎo
[粵]椋：音良

體長 length：20cm

Chestnut-tailed Starling | *Sturnia malabaricus*

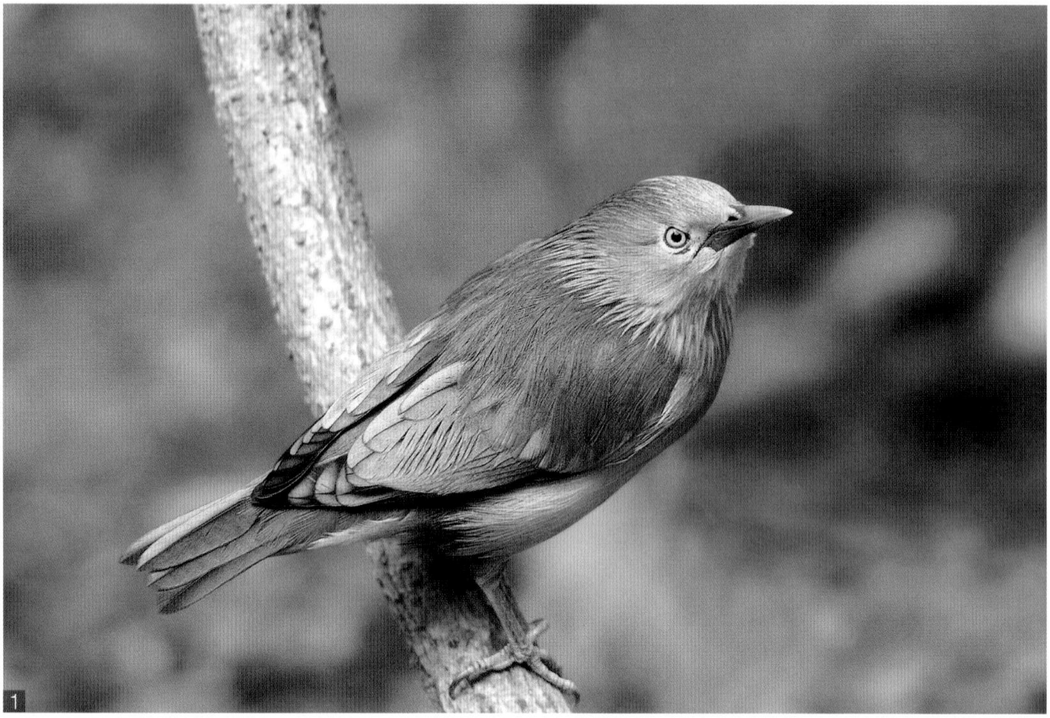
1

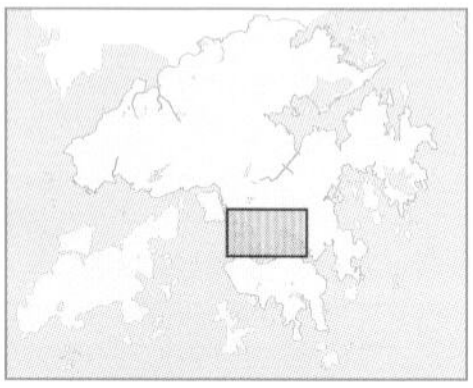

🌊 🏙

頭灰色，虹膜白色，嘴黃色，嘴基藍色，腳黃色。上體深灰色，下體淡橙灰色，尾羽外側栗色。

Grey head with white iris. Bill yellow with blue base. Legs yellow. Upperparts dark grey, underparts pale orangey-grey. Black tail with chestnut colour on outer tail feathers.

[1] adult, race *nemoricola* 成鳥, *nemoricola* 亞種；Kowloon Park 九龍公園；Dec-06, 06 年 12 月；James Lam 林文華
[2] adult, race *nemoricola* 成鳥, *nemoricola* 亞種；Kowloon Park 九龍公園；Feb-09, 09 年 2 月；Danny Ho 何國海
[3] adult, race *nemoricola* 成鳥, *nemoricola* 亞種；Kowloon Park 九龍公園；Jan-07, 07 年 1 月；Allen Chan 陳志雄
[4] adult male and female, race *nemoricola* 雄成鳥和雌成鳥, *nemoricola* 亞種；Kowloon Park 九龍公園；Mar-04, 04 年 3 月；Cherry Wong 黃卓研
[5] adult, race *nemoricola* 成鳥, *nemoricola* 亞種；Kowloon Park 九龍公園；Feb-09, 09 年 2 月；Danny Ho 何國海
[6] adult, race *nemoricola* 成鳥, *nemoricola* 亞種；Kowloon Park 九龍公園；Sep-07, 07 年 9 月；Man Kuen Yat Bill 文權溢
[7] adult, race *nemoricola* 成鳥, *nemoricola* 亞種；Kowloon Park 九龍公園；Mar-04, 04 年 3 月；Cherry Wong 黃卓研

	春季過境遷徙鳥 Spring Passage Migrant			夏候鳥 Summer Visitor			秋季過境遷徙鳥 Autumn Passage Migrant			冬候鳥 Winter Visitor		
常見月份	1	2	3	4	5	6	7	8	9	10	11	12
	留鳥 Resident				迷鳥 Vagrant				偶見鳥 Occasional Visitor			

2 3 4 5 6

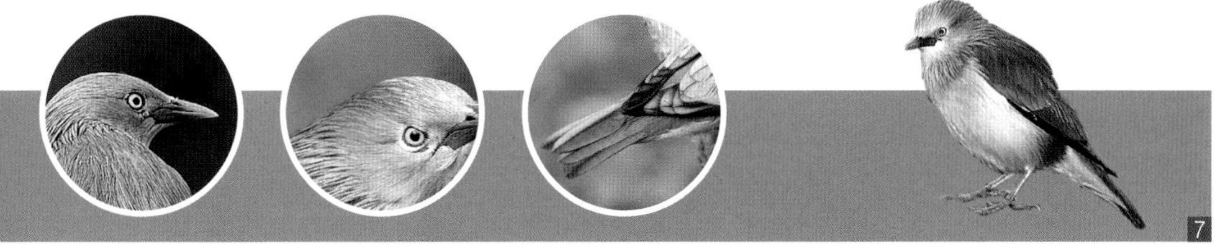

7

粉紅椋鳥

㊟ fěn hóng liáng niǎo
㊟ 椋：音良

體長 length：21cm

Rosy Starling | *Pastor roseus*

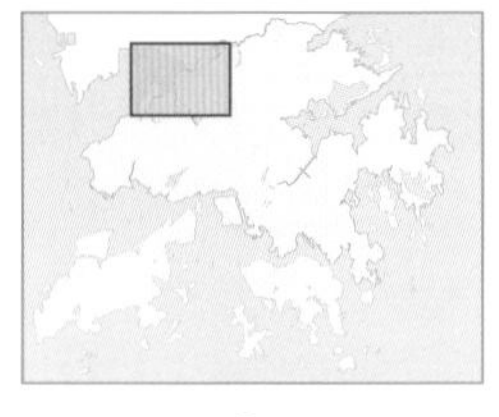

雄性成鳥有黑色頭部、翅膀、尾下覆羽和尾部，其他部分包括上體至腰、胸和腹部均為粉紅色。雌性顏色較為暗淡。幼鳥全身褐色，嘴和腳粉紅色，嘴基淡黃色。

Adult male has black head, wings, undertail coverts and tail. Other parts including upperparts to rump, breast and belly are pink in colour. Female has paler plumage. Juvenile is brownish in general, with pinkish bill and legs and yellowish bill base.

1 immature 未成年鳥；Long Valley 塱原；6-Oct-02, 02 年 10 月 6 日；Doris Chu 朱詠兒
2 immature 未成年鳥；Long Valley 塱原；Oct-02, 02 年 10 月；Michelle and Peter Wong 江敏兒、黃理沛
3 immature 未成年鳥；Long Valley 塱原；Oct-02, 02 年 10 月；Michelle and Peter Wong 江敏兒、黃理沛
4 immature 未成年鳥；Cyberport 數碼港；26 Feb-13, 13 年 2 月 26 日；Beetle Cheng 鄭諾銘

春季過境遷徙鳥 Spring Passage Migrant			夏候鳥 Summer Visitor			秋季過境遷徙鳥 Autumn Passage Migrant			冬候鳥 Winter Visitor		
常見月份 1	2	3	4	5	6	7	8	9	10	11	12

留鳥 Resident	迷鳥 Vagrant	偶見鳥 Occasional Visitor

紫翅椋鳥

㊟ zǐ chì liáng niǎo
㊟ 椋：音良

體長 length：21cm

Common Starling | *Sturnus vulgaris*

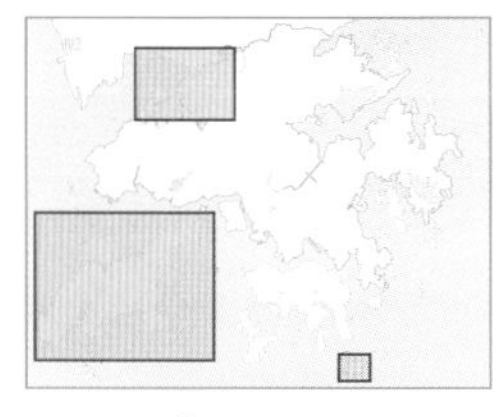

全身黑色，有藍綠色和紫色光澤，冬季時有濃密的淡黃斑點。繁殖期嘴黃色，尖而長。幼鳥顏色較淡，嘴黑色，羽毛褐色，喉和下體淡褐色。

Overall plumage black with glossy greenish-blue and purple. Dense pale yellow spots and pointed brown bill in winter. Yellow bill in breeding plumage. Juvenile is paler and browner overall.

[1] non-breeding adult 非繁殖羽成鳥：Pui O 貝澳：Dec-08, 08 年 12 月：Ng Lin Yau 吳璉宥
[2] non-breeding adult 非繁殖羽成鳥：Pui O 貝澳：Dec-08, 08 年 12 月：Ng Lin Yau 吳璉宥
[3] 1st winter 第一年冬天：Sai Kung 西貢：Dec-08, 08 年 12 月：Andy Kwok 郭匯昌
[4] non-breeding adult 非繁殖羽成鳥：Kam Tin 錦田：Dec-04, 04 年 12 月：Cherry Wong 黃卓研

春季過境遷徙鳥 Spring Passage Migrant			夏候鳥 Summer Visitor			秋季過境遷徙鳥 Autumn Passage Migrant			冬候鳥 Winter Visitor			
1	2	3	4	5	6	7	8	9	10	11	12	常見月份
留鳥 Resident				迷鳥 Vagrant				偶見鳥 Occasional Visitor				

鷯哥

[普] liáo gē
[粵] 鷯：音了

體長 length：27-37cm

Common Hill Myna | *Gracula religiosa*

1

全身光亮的黑色羽毛，顯眼的白色翼斑，還有獨特的橘黃色的肉垂，嘴和腳黃色，非常突出。所有記錄被判斷為逸鳥。

Shiny black body with yellowish wattles, distinctive white wing patch, large orange bill and yellowish legs, unmistakable. All records are considered as birds esacped from captivity.

[1] adult 成鳥：Hong Kong Park 香港公園：4-Mar-08, 08 年 3 月 4 日：Anita Lee 李雅婷
[2] adult 成鳥：Hong Kong Park 香港公園：4-Mar-08, 08 年 3 月 4 日：Anita Lee 李雅婷

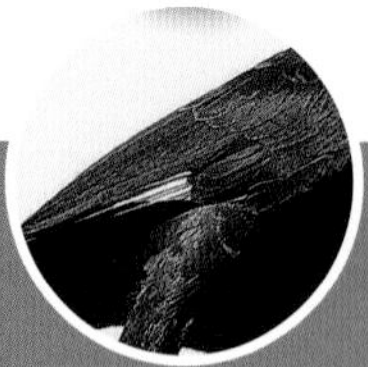

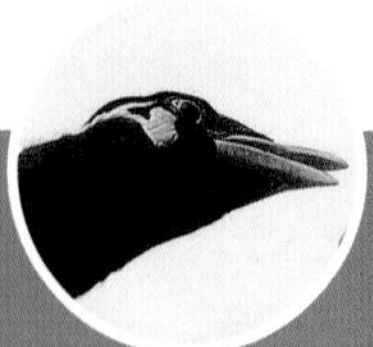

2

春季過境遷徙鳥 Spring Passage Migrant	夏候鳥 Summer Visitor	秋季過境遷徙鳥 Autumn Passage Migrant	冬候鳥 Winter Visitor

常見月份	1	2	3	4	5	6	7	8	9	10	11	12

留鳥 Resident	迷鳥 Vagrant	偶見鳥 Occasional Visitor

白眉地鶇

[普] bái méi dù dōng
[粵]鶇：音東

體長 length：20.5-23cm

Siberian Thrush | *Geokichla sibirica*

有顯眼的白色幼長眉紋，嘴黑色，腳黃色，飛行時黑色翼底有兩條粗白帶。雄鳥石板灰黑或灰藍色，下體有小量白色；雌鳥和幼鳥上體大致深褐色，胸部有橫紋。

An obvious long and thin white supercilium. Black bill and yellow legs. The black underwings with two thick white bands seen in flight. Male is overall blackish blue or blackish grey except some whites at underparts. Female and juvenile are overall brown with bars on breast.

1 2nd year male 第二年雄鳥；Kadoorie Farm 嘉道理農場；Oct-04, 04 年 10 月；Jemi and John Holmes 孔思義、黃亞萍
2 adult male 雄成鳥；Tai Po Kau 大埔滘；Feb-18, 18 年 2 月；Henry Lui 呂德恒
3 adult male 雄成鳥；Tai Po Kau 大埔滘；Feb-18, 18 年 2 月；Henry Lui 呂德恒

春季過境遷徙鳥 Spring Passage Migrant			夏候鳥 Summer Visitor			秋季過境遷徙鳥 Autumn Passage Migrant			冬候鳥 Winter Visitor			
1	2	3	4	5	6	7	8	9	10	11	12	常見月份

留鳥 Resident	迷鳥 Vagrant	偶見鳥 Occasional Visitor

橙頭地鶇

[普]chéng tóu dì dōng
[粵]鶇：音東

體長 length：20-23cm

Orange-headed Thrush | *Geokichla citrina*

橙灰二色非常獨特。嘴黑色，頭、胸至腹部橙色，上體及翼灰色，翼斑及臀部白色。*melli* 亞種面頰上有兩條黑紋。雌鳥和雄鳥相似，但上體較橄欖灰色。未成年鳥和雌鳥相似，上體有細紋及鱗狀紋。

Unmistakable combination of orange and grey. Black bill, orange head, breast and belly. White wing bar and vent. The race *melli* has two black bars on paler face. Female resembles male but upperparts are olive-brown. Immature has scaly pattern on upperparts.

1 adult male 雄成鳥：Kap Lung 甲龍：Oct-08, 08 年 10 月：Koel Ko 高偉琛
2 adult female 雌成鳥：Lantau Island 大嶼山：Owen Chiang 深藍
3 adult female 雌成鳥：Tuen Mun 屯門：Feb-07, 07 年 2 月：Andy Kwok 郭匯昌
4 adult male 雄成鳥：Chung Mei 涌尾：8-Feb-04, 04 年 2 月 8 日：Michelle and Peter Wong 江敏兒、黃理沛
5 1st winter female 第一年冬天雌鳥：Tai Po Kau 大埔滘：Sep-07, 07 年 9 月：Henry Lui 呂德恒
6 adult female 雌成鳥：Tuen Mun 屯門：Feb-07, 07 年 2 月：Lee Kai Hong 李啟康

	春季過境遷徙鳥 Spring Passage Migrant			夏候鳥 Summer Visitor			秋季過境遷徙鳥 Autumn Passage Migrant			冬候鳥 Winter Visitor		
常見月份	1	2	3	4	5	6	7	8	9	10	11	12
	留鳥 Resident				迷鳥 Vagrant				偶見鳥 Occasional Visitor			

2
3
4
5

6

懷氏地鶇

㊟ huái shì dì dōng
㊟ 鶇：音東

體長 length：24-30cm

White's Thrush | *Zoothera aurea*

1

嘴及腳淡黃色，較其他鶇大，全身金褐色而有黑色鱗狀斑紋，腹部斑紋較稀疏。尾外緣有明顯白點。常單獨出現。

Pale yellow bill and legs. Large and striking golden-brown thrush with distinctive black crescent-shaped markings above and below. The markings are less heavy on the belly. Conspicuous white tips to dark outer tail feathers. Usually solitary.

1 Chung Mei 涌尾：Feb-04, 04 年 2 月：Jemi and John Holmes 孔思義・黃亞萍
2 Shing Mun 城門水塘：Dec-06, 06 年 12 月：Allen Chan 陳志雄
3 Shing Mun 城門水塘：Dec-06, 06 年 12 月：Wallace Tse 謝鑑超
4 Shing Mun 城門水塘：Dec-06, 06 年 12 月：Allen Chan 陳志雄
5 Kwai Chung 葵涌：Feb-07, 07 年 2 月：Cherry Wong 黃卓研
6 Pok Fu Lam 薄扶林：Feb-08, 08 年 2 月：Herman Ip 葉紀江
7 Wonderland Villas 華景山莊：Dec-06, 06 年 12 月：Matthew and TH Kwan 關朗曦、關子凱

	春季過境遷徙鳥 Spring Passage Migrant			夏候鳥 Summer Visitor			秋季過境遷徙鳥 Autumn Passage Migrant			冬候鳥 Winter Visitor		
常見月份	1	2	3	4	5	6	7	8	9	10	11	12
	留鳥 Resident				迷鳥 Vagrant				偶見鳥 Occasional Visitor			

灰背鶇

㊟huī bèi dōng
㊟鶇：音東

體長 length：20-23cm

Grey-backed Thrush | *Turdus hortulorum*

1

嘴及腳淡色。雄鳥上體及胸為灰色，喉及腹白色，脇橙色。雌鳥及未成年鳥上體褐色，胸前有黑色斑紋。常在地上翻起落葉覓食，比較怕人。

Pale bill and legs. Male bird has distinctive grey upperparts and breast. Throat and belly are white. Flanks orange. Female and immature birds have brownish upperparts, black spots on breast and orange flanks. Feeds on the ground by overturning fallen leaves. Common in winter, but very shy.

1 adult male 雄成鳥：Shek Kong 石崗：Mar-07, 07 年 3 月：Martin Hale 夏敖天
2 adult male 雄成鳥：Mai Po 米埔：Mar-08, 08 年 3 月：Danny Ho 何國海
3 adult female 雌成鳥：Pui O 貝澳：Feb-08, 08 年 2 月：Andy Kwok 郭匯昌
4 adult male 雄成鳥：Shek Kong 石崗：Dec-08, 08 年 12 月：Ken Fung 馮漢城
5 1st winter female 第一年冬天雌鳥：Shing Mun 城門水塘：27-Jan-09, 09 年 1 月 27 日：Owen Chiang 深藍
6 1st winter male 第一年冬天雄鳥：Shek Kong 石崗：18-Jan-09, 09 年 1 月 18 日：Jacky Chan 陳家華

春季過境遷徙鳥 Spring Passage Migrant			夏候鳥 Summer Visitor			秋季過境遷徙鳥 Autumn Passage Migrant			冬候鳥 Winter Visitor		
1	2	3	4	5	6	7	8	9	10	11	12

常見月份

留鳥 Resident	迷鳥 Vagrant	偶見鳥 Occasional Visitor

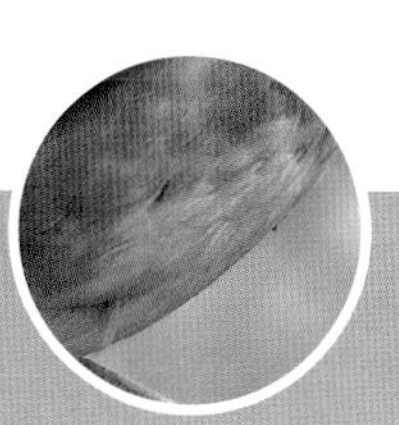

烏灰鶇

㊥ wū huī dōng
㊣ 鶇：音東

體長 length：21-22cm

Japanese Thrush | *Turdus cardis*

1

嘴及腳黃色。雄鳥上體、喉和胸部黑色，腹部白色帶黑色斑點。第一年雄鳥上體、喉和胸部灰藍色，像灰背鶇。雌鳥體形較灰背鶇小，脇部橙色較少但斑點較多。

Yellow bill and legs. Male has unmistakable black upperparts throat and breast, white underparts with black spots. First year male has dark blue-grey upperparts, throat and breast, similar to Grey-backed Thrush. Female resembles female Grey-backed Thrush but is smaller and generally has less orange and more spots on flanks.

1 adult male 雄成鳥：Shek Kong 石崗：Jan-07, 07 年 1 月：Owen Chiang 深藍
2 1st winter female 第一年冬天雌鳥：The University of HK 香港大學：Apr-06, 06 年 4 月：Cherry Wong 黃卓研
3 adult female 雌成鳥：Kowloon Park 九龍公園：Apr-08, 08 年 4 月：Cherry Wong 黃卓研
4 adult male 雄成鳥：Shing Mun 城門水塘：Dec-06, 06 年 12 月：Allen Chan 陳志雄
5 adult female 雌成鳥：The University of HK 香港大學：25-Mar-06, 06 年 3 月 25 日：Michelle and Peter Wong 江敏兒、黃理沛
6 male 雄鳥：The University of HK 香港大學：25-Mar-06, 06 年 3 月 25 日：Michelle and Peter Wong 江敏兒、黃理沛
7 adult male 雄成鳥：Pui O 貝澳：Feb-08, 08 年 2 月：Andy Kwok 郭匯昌
8 adult male 雄成鳥：Lamma Island 南丫島：Apr-05, 05 年 4 月：Cherry Wong 黃卓研

春季過境遷徙鳥 Spring Passage Migrant			夏候鳥 Summer Visitor			秋季過境遷徙鳥 Autumn Passage Migrant			冬候鳥 Winter Visitor		
常見月份 1	2	3	4	5	6	7	8	9	10	11	12
留鳥 Resident				迷鳥 Vagrant				偶見鳥 Occasional Visitor			

2
3
4
5
6
7

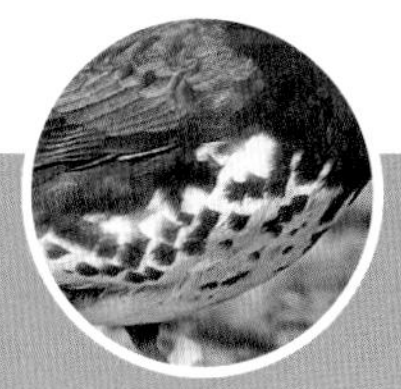

8

烏鶇

㊟wū dōng
㊟鶇：音東

體長 length：28-29cm

Chinese Blackbird | *Turdus mandarinus*

其他名稱 Other names：Blackbird

上體黑色至深褐色，下體較淡。嘴由黃至褐色都有，眼圈可能不明顯，腳黑色。常小群出沒於樹上，叫聲為哀怨的「dweep」。

Black to brownish-black upperparts, underparts slightly paler. Bill varies from yellow to brownish, the eyering can be indistinct. Black legs. Often in flocks among trees. Call is a plaintive "dweep".

1 male 雄鳥：Chinese University of HK 香港中文大學：Feb-05, 05 年 2 月：Cherry Wong 黃卓研
2 male 雄鳥：Long Valley 塱原：Nov-04, 04 年 11 月：Henry Lui 呂德恒
3 male 雄鳥：Pui O 貝澳：Feb-08, 08 年 2 月：Cherry Wong 黃卓研
4 female 雌鳥：Wun Yiu 碗窰：18-Mar-06, 06 年 3 月 18 日：Owen Chiang 深藍
5 male 雄鳥：Kowloon Park 九龍公園：Jan-06, 06 年 1 月：Cherry Wong 黃卓研
6 female 雌鳥：Wun Yiu 碗窰：Jan-03, 03 年 1 月：Henry Lui 呂德恒
7 female 雌鳥：Kowloon City Park 九龍城公園：Jan-08, 08 年 1 月：Ng Lin Yau 吳璉宥
8 female 雌鳥：Wun Yiu 碗窰：Jan-03, 03 年 1 月：Henry Lui 呂德恒

春季過境遷徙鳥 Spring Passage Migrant			夏候鳥 Summer Visitor			秋季過境遷徙鳥 Autumn Passage Migrant			冬候鳥 Winter Visitor		
1	2	3	4	5	6	7	8	9	10	11	12

常見月份

留鳥 Resident	迷鳥 Vagrant	偶見鳥 Occasional Visitor

2
3
4
5
6
7

8

白眉鶇

㊟ bái méi dōng
㊟ 鶇：音東

體長 length：21-23cm

Eyebrowed Thrush | *Turdus obscurus*

頭灰色，有白色幼長眼眉，眼下有白色橫紋，雌鳥喉部淡色。上體褐色，胸部和脇部橙褐色，腹部白色。

Male has grey head with long narrow white supercilium and a short white stripe below eye. Female is pale throated. Brown upperparts, brownish-orange breast and flank, and white belly.

1 adult male 雄成鳥：Pui O 貝澳：Mar-08, 08 年 3 月：Michelle and Peter Wong 江敏兒、黃理沛
2 adult male 雄成鳥：Pui O 貝澳：Mar-08, 08 年 3 月：Ken Fung 馮漢城
3 1st winter female 第一年冬天雌鳥：Victoria Peak 維多利亞山頂：May-07, 07 年 5 月：Law Kam Man 羅錦文
4 adult male 雄成鳥：Pui O 貝澳：Mar-08, 08 年 3 月：Michelle and Peter Wong 江敏兒、黃理沛
5 adult male 雄成鳥：Pui O 貝澳：Mar-08, 08 年 3 月：Cherry Wong 黃卓研
6 1st winter male 第一年冬天雄鳥：Pui O 貝澳：Feb-08, 08 年 2 月：Andy Kwok 郭匯昌

春季過境遷徙鳥 Spring Passage Migrant			夏候鳥 Summer Visitor			秋季過境遷徙鳥 Autumn Passage Migrant			冬候鳥 Winter Visitor		
1	2	3	4	5	6	7	8	9	10	11	12

常見月份

留鳥 Resident	迷鳥 Vagrant	偶見鳥 Occasional Visitor

3
4
2
5

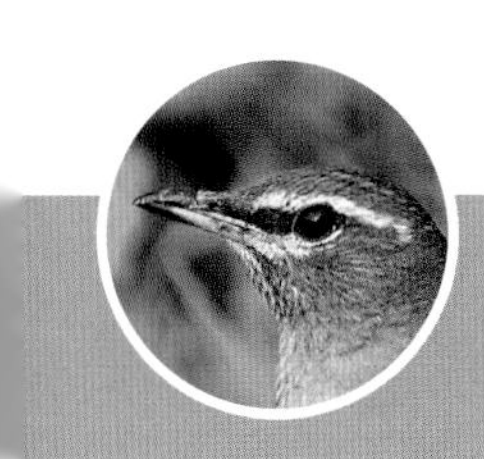

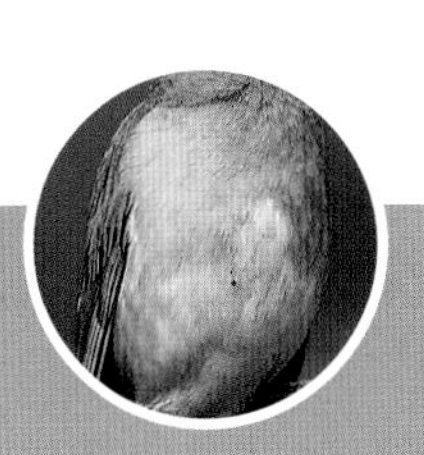

6

白腹鶇

㊟ bái fù dōng
㊟ 鶇：音東

體長 length：22-23cm

Pale Thrush | *Turdus pallidus*

頭部灰褐色，喉顏色較淡並具有幼紋，眼圈黃色，上嘴深色，下嘴淡橙色。上體及尾部褐色，外側尾羽具有明顯白點。下體淡灰褐色，脇部沾橙色。雌鳥具有淡眉，耳羽、頰和喉部淡色具有斑紋。

Head greyish brown, pale throat with narrow streaks. Yellowish eye-ring, dark upperbill, light orange lowerbill. Upperparts and tail brownish, with conspicuous white spot on outertail feather. Underparts light greyish brown, tinted with orange colour. Female has light eyebrow and ear coverts, with streaked cheek and throat.

1 adult male 雄成鳥：Shing Mun 城門水塘：Dec-06, 06 年 12 月：Owen Chiang 深藍
2 1st winter male 第一年冬天雄鳥：Tso Kung Tam 曹公潭：15-Jan-07, 07 年 1 月 15 日：Christina Chan 陳燕明
3 1st winter female 第一年冬天雌鳥：Po Toi 蒲台：Dec-08, 08 年 12 月：Isaac Chan 陳家強
4 1st winter male 第一年冬天雄鳥：Po Toi 蒲台：Jan-08, 08 年 1 月：Andy Kwok 郭匯昌
5 adult male 雄成鳥：Po Toi 蒲台：Mar-08, 08 年 3 月：Jasper Lee 李君哲
6 adult male 雄成鳥：Po Toi 蒲台：Mar-08, 08 年 3 月：Jasper Lee 李君哲

春季過境遷徙鳥 Spring Passage Migrant			夏候鳥 Summer Visitor			秋季過境遷徙鳥 Autumn Passage Migrant			冬候鳥 Winter Visitor		
1	2	3	4	5	6	7	8	9	10	11	12

常見月份

留鳥 Resident	迷鳥 Vagrant	偶見鳥 Occasional Visitor

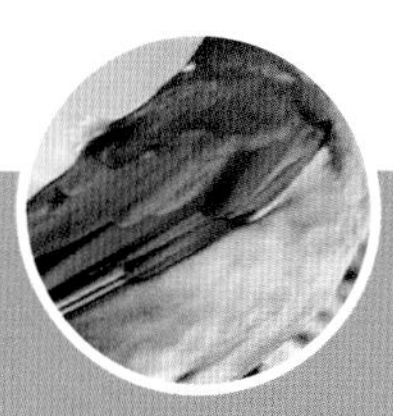

赤胸鶇

㊟chì xiōng dōng
㊟鶇：音東

體長length：23-24cm

Brown-headed Thrush | *Turdus chrysolaus*

其他名稱Other names：赤腹鶇

1

頭部及喉部褐色，嘴基淡橙褐色，腳粉紅色。上體及尾部深褐色，飛羽具有淡色羽緣。胸及脇明顯橙色，腹部至尾下白色，尾下覆羽具有淡色斑紋。雌鳥及幼鳥眼眉淡褐色，喉部白色，具有幼細直紋。

Brownish head and throat, bill base light brownish orange, pink legs. Upperparts and tail dark brown, with light fringes on flight feathers. Orange breast and flank, conspicuous white abdomen. Under tail cover whitish, with pale streaks. Female and immature birds have pale eyebrow, whitish throat with narrow streaks.

1 1st winter male 第一年冬天雄鳥；Po Toi 蒲台；Nov-08, 08 年 11 月；Kou Chon Hong 高俊雄
2 adult male, race *chrysolaus* 雄成鳥, *chrysolaus* 亞種；Shek Kong 石崗；Jan-07, 07 年 1 月；Helen Chan 陳燕芳
3 adult male, race *chrysolaus* 雄成鳥, *chrysolaus* 亞種；Shek Kong 石崗；Jan-07, 07 年 1 月；Law Kam Man 羅錦文
4 adult female 雌成鳥；Pui O 貝澳；Mar-08, 08 年 3 月；Ken Fung 馮漢城
5 adult female 雌成鳥；Pui O 貝澳；Mar-08, 08 年 3 月；Michelle and Peter Wong 江敏兒、黃理沛
6 adult male, race *orii* 雄成鳥；*orii* 亞種；Shek Kong 石崗；Feb-07, 07 年 2 月；Cherry Wong 黃卓研

春季過境遷徙鳥 Spring Passage Migrant			夏候鳥 Summer Visitor			秋季過境遷徙鳥 Autumn Passage Migrant			冬候鳥 Winter Visitor		
常見月份 1	2	3	4	5	6	7	8	9	10	11	12
留鳥 Resident				迷鳥 Vagrant				偶見鳥 Occasional Visitor			

2

3

4

5

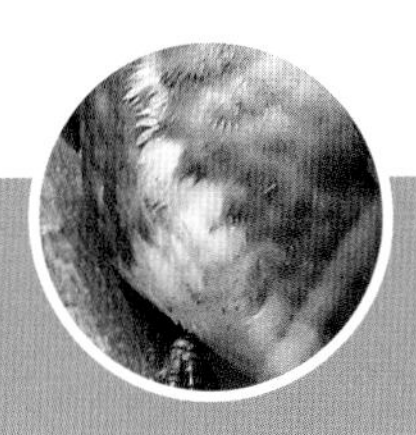

6

紅尾鶇

㊥hóng wěi dōng
㊣鶇：音東

體長length：23-25cm

Naumann's Thrush | *Turdus naumanni*

上體棕褐色，具橙棕色眉紋及頰紋，覆羽、腰部、尾羽橙紅色。下體偏白，胸部及脇部有橙紅色斑點。嘴黑色，下嘴基部黃色，腳灰褐色。

Brown upperparts. Brownish-orange supercilia and moustachial stripes. Reddish-orange wing coverts, rump and tail. Whitish underparts. Reddish-orange spots on breast and flanks. Black bill with yellow at the base of lower bill. Greyish-brown legs.

1 adult female 雌成鳥；Mount Davis 摩星嶺；Mar-16, 16 年 3 月；Aaron Lo 羅瑞華

春季過境遷徙鳥 Spring Passage Migrant			夏候鳥 Summer Visitor			秋季過境遷徙鳥 Autumn Passage Migrant			冬候鳥 Winter Visitor		
1	2	3	4	5	6	7	8	9	10	11	12

常見月份

留鳥 Resident	迷鳥 Vagrant	偶見鳥 Occasional Visitor

紫寬嘴鶇

㊟ zǐ kuān zuǐ dōng
㊟ 鶇：音東

體長length：25-28cm

Purple Cochoa | *Cochoa purpurea*

似鶇的大型林鳥，有顯眼的紫藍色冠紋，大覆羽淺紫藍色，翼端黑色，尾羽淺紫色而尾端黑。雄鳥褐紫色，雌鳥似雄鳥，上體紅褐色，下體橙褐色。虹膜深褐色，嘴和腳黑色。所有記錄被判斷為逸鳥。

Thrush-sized woodland bird. Distinctive purple stripe from forehead to crown, light purple tail with black tip. Male bird with purple brown body, female bird has rufous brown upperparts and brownish-orange underparts. Iris dark brown, bill and legs black. All records are considered as birds escaped from captivity.

1 Po Toi 蒲台：11-Oct-07, 07 年 10 月11日：Tam Yiu Leung 譚耀良
2 Po Toi 蒲台：11-Oct-07, 07 年 10 月11日：Tam Yiu Leung 譚耀良
3 Po Toi 蒲台：11-Oct-07, 07 年 10 月11日：Tam Yiu Leung 譚耀良

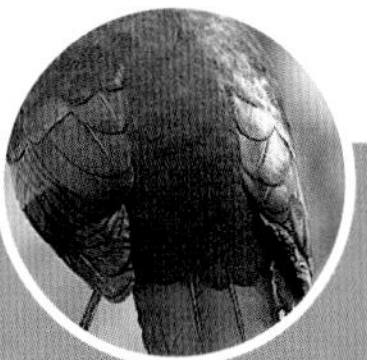

春季過境遷徙鳥 Spring Passage Migrant	夏候鳥 Summer Visitor	秋季過境遷徙鳥 Autumn Passage Migrant	冬候鳥 Winter Visitor

1 2 3 4 5 6 7 8 9 10 11 12 常見月份

留鳥 Resident	迷鳥 Vagrant	偶見鳥 Occasional Visitor

斑鶇

㊢bān dōng
[粵]鶇：音東

體長 length：23-25cm

Dusky Thrush | *Turdus eunomus*

大型鶇，背部褐色，有褐色耳羽及顎紋，而眉紋、頰下紋及喉部皆白色，對比鮮明。香港有兩個亞種，*eunomus* 亞種較多，有深褐色胸帶和紅褐色翼斑。

Large-sized thrush with brown upperparts. Brown ear coverts and malar stripes contrast with whitish supercilium, submoustachial stripes and throat. Two races occur in Hong Kong: subspecies *eunomus* has dark brown bands across the breast and reddish wing panel.

1 adult male 雄成鳥；Pui O 貝澳；Dec-08, 08 年 12 月；Herman Ip 葉紀江
2 adult male, race *eunomus*, 雄成鳥, *eunomus* 亞種；Pui O 貝澳；Dec-08, 08 年 12 月；Herman Ip 葉紀江
3 1st winter female 第一年冬天雌鳥；Shek Kong 石崗；Jan-07, 07 年 1 月；Cherry Wong 黃卓研
4 adult male 雄成鳥；Pui O 貝澳；Dec-08, 08 年 12 月；Herman Ip 葉紀江
5 1st winter female, race *eunomus*, 第一年冬天雌鳥, *eunomus* 亞種；Shek Kong 石崗；Feb-07, 07 年 2 月；Owen Chiang 深藍
6 1st winter female, race *eunomus*, 第一年冬天雌鳥, *eunomus* 亞種；Shek Kong 石崗；Jan-07, 07 年 1 月；Allen Chan 陳志雄
7 1st winter female, race *eunomus*, 第一年冬天雌鳥, *eunomus* 亞種；Shek Kong 石崗；Jan-07, 07 年 1 月；Cherry Wong 黃卓研
8 adult male 雄成鳥；Pui O 貝澳；Dec-08, 08 年 12 月；Herman Ip 葉紀江

	春季過境遷徙鳥 Spring Passage Migrant			夏候鳥 Summer Visitor			秋季過境遷徙鳥 Autumn Passage Migrant			冬候鳥 Winter Visitor		
常見月份	1	2	3	4	5	6	7	8	9	10	11	12
	留鳥 Resident				迷鳥 Vagrant				偶見鳥 Occasional Visitor			

3
4
5
6
7

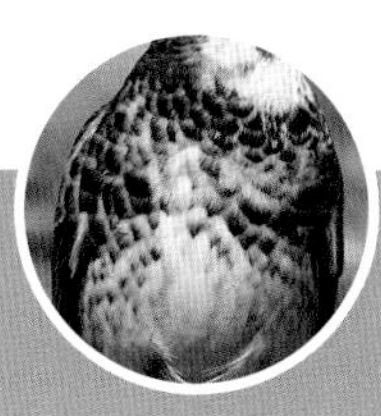

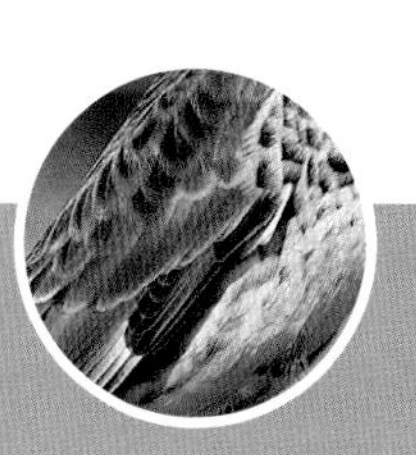

8

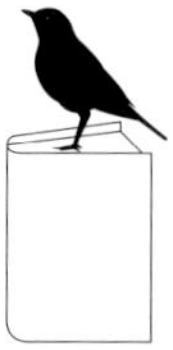

寶興歌鶇

[普]bǎo xìng gē dōng
[粵]鶇：音東

體長length：23cm

Chinese Thrush | *Turdus mupinensis*

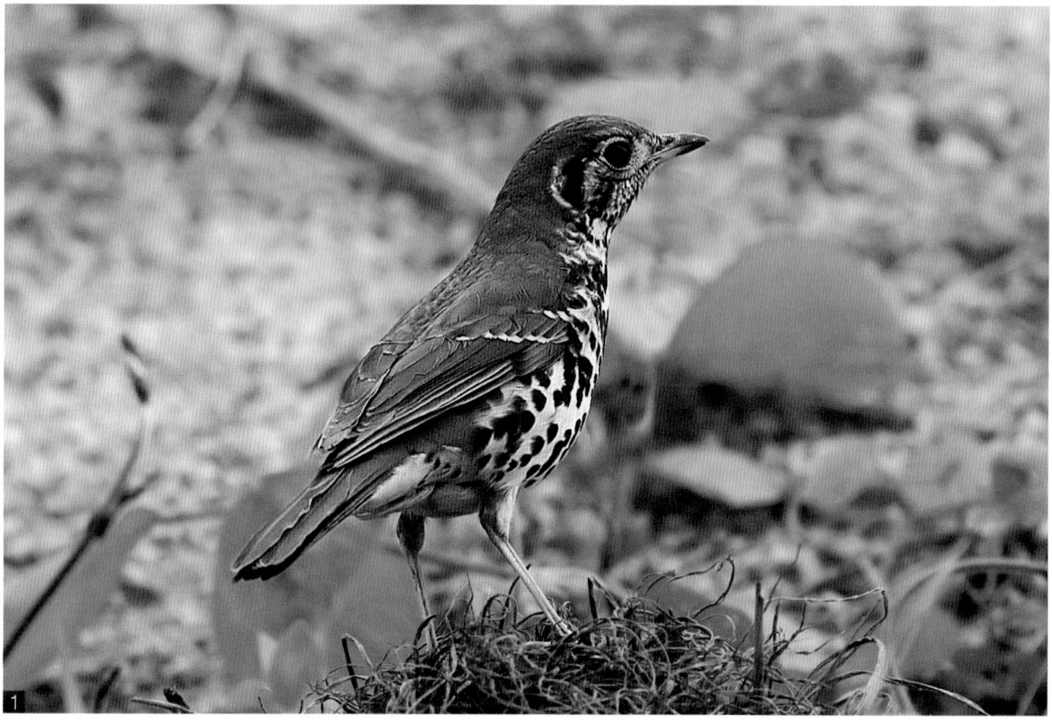

1

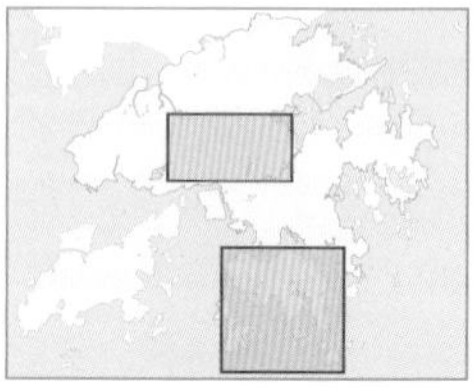

🌳

頭部褐色，耳羽顏色較淡，後側具明顯黑色斑塊，腳粉紅色。上體至尾部褐色，翼部有兩道白色翼斑。下體白色，喉部有深色縱紋，胸部至腹部有明顯而粗大的深色斑點。

Brownish head, ear coverts paler with prominent dark patch near the nape, pink legs. Upperparts to tail brown, with two wing bars. Underparts whitish, with dark streaks at throat, and heavy spotting from breast to belly.

[1] Po Toi 蒲台；Mar-06, 06 年 3 月；Allen Chan 陳志雄
[2] Po Toi 蒲台；Feb-06, 06 年 2 月；Michelle and Peter Wong 江敏兒、黃理沛
[3] Po Toi 蒲台；Mar-06, 06 年 3 月；Doris Chu 朱詠兒
[4] Po Toi 蒲台；24-Feb-06, 06 年 2 月 24 日；Michelle and Peter Wong 江敏兒、黃理沛
[5] Po Toi 蒲台；Mar-06, 06 年 3 月；Michelle and Peter Wong 江敏兒、黃理沛
[6] Po Toi 蒲台；Mar-06, 06 年 3 月；Allen Chan 陳志雄
[7] Po Toi 蒲台；2-Mar-06, 06 年 3 月 2 日；Geoff Welch

	春季過境遷徙鳥 Spring Passage Migrant			夏候鳥 Summer Visitor			秋季過境遷徙鳥 Autumn Passage Migrant			冬候鳥 Winter Visitor		
常見月份	1	2	3	4	5	6	7	8	9	10	11	12
	留鳥 Resident				迷鳥 Vagrant				偶見鳥 Occasional Visitor			

2
3
4
5
6

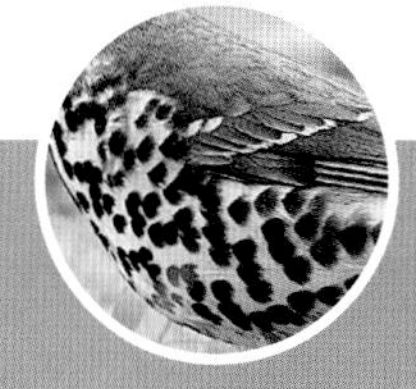

7

鵲鴝

㊂ què qú
㊄ 鵲鴝：音爵渠

體長 length：19-21cm

Oriental Magpie Robin | *Copsychus saularis*

其他名稱 Other names：Magpie Robin

俗名「豬屎渣」。黑白兩色：嘴和腳黑色，頭、背至尾上部黑色，腹、翼紋和尾緣白色，雌鳥和雄鳥相似，但頭及上體的黑色由灰色代替，幼鳥顏色更淺。叫聲響亮悅耳，變化多端，又時常發出「查」的噴氣聲。

Black-and-white robin. Black bill and legs. The head, mantle and center of tail are black, while the underparts, wing bar and outer edge of tail are white. Female resembles male but black on head and mantle is replaced by dark grey. Juveniles are streaked. Loud and melodious calls, sometimes also a long drawn-out hissing.

1 adult male 雄成鳥：Feb-03, 03年2月：Marcus Ho 何萬邦
2 adult male 雄成鳥：Chinese University of HK 香港中文大學：Feb-07, 07 年 2 月：Benson Lau 劉滙文
3 juvenile male 雄幼鳥：Kowloon Park 九龍公園：Sep-07, 07 年 9 月：Bill Man 文權溢
4 adult female 雌成鳥：Hong Kong Park 香港公園：Oct-06, 06 年 10 月：Matthew and TH Kwan 關朗曦、關子凱
5 adult female 雌成鳥：HK Zoological and Botanical Gardens 香港動植物公園：Nov-04, 04 年 11 月：Henry Lui 呂德恒
6 adult female 雌成鳥：Fo Tan 火炭：Nov-07, 07 年 11 月：Christine and Samuel Ma 馬志榮、蔡美蓮
7 juvenile 幼鳥：Ma On Shan 馬鞍山：Oct-08, 08 年 10 月：Ken Fung 馮漢城
8 adult male 雄成鳥：Feb-03, 03年2月：Marcus Ho 何萬邦
9 adult male 雄成鳥：HK Zoological and Botanical Gardens 香港動植物公園：Oct-03, 03年10月：Henry Lui 呂德恒

春季過境遷徙鳥 Spring Passage Migrant			夏候鳥 Summer Visitor			秋季過境遷徙鳥 Autumn Passage Migrant			冬候鳥 Winter Visitor		
1	2	3	4	5	6	7	8	9	10	11	12
留鳥 Resident				迷鳥 Vagrant				偶見鳥 Occasional Visitor			

常見月份

2
4
3
5
6
7
8
9

烏鶲

㊥ wū wēng

體長 length：13-14cm

Dark-sided Flycatcher | *Muscicapa sibirica*

其他名稱 Other names：Sooty Flycatcher

1

大致深灰褐色，翅膀長及尾部的三分之二。眼圈寬度不均，眼後處較粗。淺色頰下紋與深色顎紋成對比。喉部中央白色，胸口兩側有模糊的褐色斑紋，伸延到胸前相合，形成顯眼的半道白頸圈。腹及臀部均為白色，翼上有淺色邊。喜在樹冠之下活動。

A dark greyish-brown Flycatcher. Long wings reach up to two-thirds of tail. Distinctive uneven eyering, thicker behind the eye. Pale submoustachial stripe contrasts with dark malar stripe. Centre of throat white. Brown, poorly-defined marks on side of breast meet in centre, forming a prominent white half-collar. Belly and vent are white. Pale fringes on wing panels. Favours area under canopy.

1 juvenile 幼鳥；Mai Po 米埔；Sep-06, 06 年 9 月；Martin Hale 夏敖天
2 juvenile 幼鳥；Kadoorie Farm 嘉道理農場；Oct-05, 05 年 10 月；Michelle and Peter Wong 江敏兒、黃理沛
3 juvenile 幼鳥；Po Toi 蒲台；Sep-06, 06 年 9 月；Michelle and Peter Wong 江敏兒、黃理沛
4 1st winter 第一年冬天；Ho Chung 豪涌；May-04, 04 年 5 月；Michelle and Peter Wong 江敏兒、黃理沛
5 1st winter 第一年冬天；Mai Po 米埔；Sep-04, 04 年 9 月；Henry Lui 呂德恒
6 juvenile 幼鳥；Tai Po Kau 大埔滘；Oct-08, 08 年 10 月；Andy Kwok 郭匯昌
7 1st winter 第一年冬天；Po Toi 蒲台；Oct-07, 07 年 10 月；Michelle and Peter Wong 江敏兒、黃理沛
8 adult 成鳥；Tai Po Kau 大埔滘；Sep-06, 06 年 9 月；Martin Hale 夏敖天

春季過境遷徙鳥 Spring Passage Migrant	夏候鳥 Summer Visitor	秋季過境遷徙鳥 Autumn Passage Migrant	冬候鳥 Winter Visitor

常見月份 1 2 3 4 5 6 7 8 9 10 11 12

留鳥 Resident	迷鳥 Vagrant	偶見鳥 Occasional Visitor

5
2
6
3
4
7

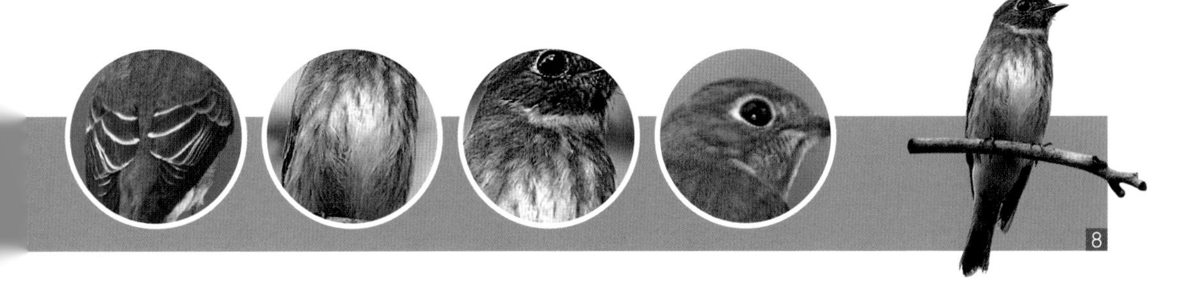
8

北灰鶲

[普] běi huī wēng

體長 length：12-14cm

Asian Brown Flycatcher | *Muscicapa dauurica*

其他名稱 Other names：闊咀鶲, Brown Flycatcher

1

大致淺灰褐色。眼先淺色，與白色眼圈連結，眼圈寬度比烏鶲平均，下嘴基黃色，顎紋明顯，部分個體有淺灰色胸帶，下體白色。喜在樹冠之下活動。

A pale grey-brown flycatcher. Pale lore links to whitish eyering. Unlike the similar Dark-sided Flycatcher, eyerings are more even in thickness. Yellow base on lower bill. Malar stripes prominent. Pale grey breast band on some individuals. Underparts white. Favours area under canopy.

1 1st winter 第一年冬天：Tai Po Kau 大埔滘：Sep-06, 06 年 9 月：Martin Hale 夏敖天
2 Mai Po 米埔：Dec-05, 05 年 12 月：Henry Lui 呂德恒
3 Po Toi 蒲台：Sep-08, 08 年 9 月：Andy Kwok 郭匯昌
4 adult 成鳥：Po Toi 蒲台：Apr-07, 07 年 4 月：Winnie Wong and Sammy Sam 森美與雲妮
5 1st winter 第一年冬天：Po Toi 蒲台：Oct-08, 08 年 10 月：Michelle and Peter Wong 江敏兒・黃理沛
6 1st winter 第一年冬天：Po Toi 蒲台：Oct-08, 08 年 10 月：Michelle and Peter Wong 江敏兒・黃理沛
7 Po Toi 蒲台：Sep-07, 07 年 9 月：Wallace Tse 謝鑑超
8 1st winter 第一年冬天：Tai Po Kau 大埔滘：Feb-07, 07 年 2 月：James Lam 林文華

春季過境遷徙鳥 Spring Passage Migrant	夏候鳥 Summer Visitor	秋季過境遷徙鳥 Autumn Passage Migrant	冬候鳥 Winter Visitor

常見月份	1	2	3	4	5	6	7	8	9	10	11	12

留鳥 Resident	迷鳥 Vagrant	偶見鳥 Occasional Visitor

2
3
4
5
6
7
8

褐胸鶲

[普] hè xiōng wēng
[粵] 褐：音喝

體長 length：13-14cm

Brown-breasted Flycatcher | *Muscicapa muttui*

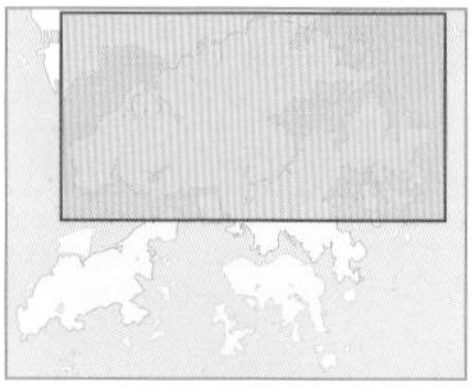

上體深褐，下體淡黃。腰、尾羽和飛羽邊緣紅褐色。白眼圈和白頰紋清晰可見，顎紋偏褐。

Brown upperparts and buffy underparts. Brownish red rump, tail and edge of flight feathers. Prominent white eyerings and moustachial stripes, and brownish malar stripe.

1 adult 成鳥；Tai Po Kau 大埔滘；Apr-09, 09 年 4 月；Michelle and Peter Wong 江敏兒、黃理沛
2 probably 1st winter 很可能是第一年冬天；Tai Po Kau 大埔滘；5-Jan-02, 02 年 1 月 5 日；Lo Kar Man 盧嘉孟
3 probably 1st winter 很可能是第一年冬天；Tai Po Kau 大埔滘；Dec-02, 02 年 12 月；Michelle and Peter Wong 江敏兒、黃理沛
4 probably 1st winter 很可能是第一年冬天；Tai Po Kau 大埔滘；Dec-02, 02 年 12 月；Michelle and Peter Wong 江敏兒、黃理沛
5 adult 成鳥；Tai Po Kau 大埔滘；Apr-09, 09 年 4 月；Michelle and Peter Wong 江敏兒、黃理沛
6 adult 成鳥；Tai Po Kau 大埔滘；Apr-09, 09 年 4 月；Michelle and Peter Wong 江敏兒、黃理沛
7 probably 1st winter 很可能是第一年冬天；Tai Po Kau 大埔滘；5-Jan-02, 02 年 1 月 5 日；Lo Kar Man 盧嘉孟

	春季過境遷徙鳥 Spring Passage Migrant			夏候鳥 Summer Visitor			秋季過境遷徙鳥 Autumn Passage Migrant			冬候鳥 Winter Visitor		
常見月份	1	2	3	4	5	6	7	8	9	10	11	12
	留鳥 Resident				迷鳥 Vagrant				偶見鳥 Occasional Visitor			

2
3
4
5
6

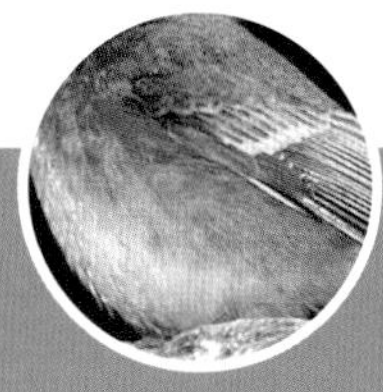

7

棕尾褐鶲

[普]zōng wěi hè wēng
[粵]褐：音喝

體長length：12-13cm

Ferruginous Flycatcher | *Muscicapa ferruginea*

其他名稱 Other names：紅褐鶲

1

2

體型細小的鶲，全身深栗褐色，頭灰色，有明顯白眼圈，喉白色。頸部有白色的半頸環，覆羽、三級飛羽、脇、腰及尾下覆羽紅褐色。喜在樹林中層活動。

Small deep chestnut-brown flycatcher with grey head, prominent white eyerings, white throat and a white half collar. Rufous brown wing coverts, tertials, flanks, vent and undertailed coverts. Usually perches on branches in mid-storey.

1 adult 成鳥：Mai Po 米埔：Apr-06, 06 年 4 月：Andy Kwok 郭匯昌
2 adult 成鳥：Po Toi 蒲台：Apr-08, 08 年 4 月：Ken Fung 馮漢城
3 1st winter 第一年冬天：Po Toi 蒲台：Apr-07, 07 年 4 月：James Lam 林文華
4 adult 成鳥：Po Toi 蒲台：Apr-08, 08 年 4 月：Ken Fung 馮漢城
5 1st winter 第一年冬天：Po Toi 蒲台：Apr-05, 05 年 4 月：Michelle and Peter Wong 江敏兒、黃理沛
6 adult 成鳥：Po Toi 蒲台：6-Apr-07, 07 年 4 月 6 日：Michelle and Peter Wong 江敏兒、黃理沛
7 adult 成鳥：Po Toi 蒲台：Apr-07, 07 年 4 月：Allen Chan 陳志雄
8 1st winter moulting into adult 第一年冬天轉換成羽：Ng Tung Chai 梧桐寨：Mar-07, 07 年 3 月：Martin Hale 夏敖天

春季過境遷徙鳥 Spring Passage Migrant	夏候鳥 Summer Visitor	秋季過境遷徙鳥 Autumn Passage Migrant	冬候鳥 Winter Visitor

常見月份 1 2 **3** **4** 5 6 7 8 **9** **10** **11** 12

留鳥 Resident	迷鳥 Vagrant	偶見鳥 Occasional Visitor

3
4
5
6
7
8

海南藍仙鶲

[普]hǎi nán lán xiān wēng

體長 length：13-14cm

Hainan Blue Flycatcher | *Cyornis hainanus*

其他名稱 Other names：海南藍鶲

1

雄鳥頭部、胸部、上背、翼、腰和尾部均為藍色，胸部的藍色伸延至腹部且漸變為白色。前冠及肩部呈亮麗輝藍色。雌鳥上體褐色，喉及胸偏黃，下體白色，尾部呈紅褐色。有時只能聞其聲而不見其影。喜在林區下層活動。

Male bird has blue head, breast, mantle, wing, rump and tail. Blue colour on breast changes gradually to white colour at belly. Shiny blue forecrown and shoulders. Female bird has brown upperparts, yellowish throat and breast band, and white underparts; tail is more rufous. Sometimes can only be heard and difficult to see. Favours low storey in forest.

1 adult male 雄成鳥：Tai Po Kau 大埔滘：Jun-07, 07 年 6 月：Allen Chan 陳志雄
2 adult male 雄成鳥：Shing Mun 城門水塘：May-07, 07 年 5 月：Jasper Lee 李君哲
3 adult male 雄成鳥：Tai Po Kau 大埔滘：Jun-07, 07 年 6 月：Lee Kai Hong 李啟康
4 juvenile 幼鳥：Tai Po Kau 大埔滘：Jul-07, 07 年 7 月：Wallace Tse 謝鑑超
5 adult male 雄成鳥：Tai Po Kau 大埔滘：Jun-07, 07 年 6 月：Wallace Tse 謝鑑超
6 adult male 雄成鳥：Shing Mun 城門水塘：May-07, 07 年 5 月：Jasper Lee 李君哲
7 adult male 雄成鳥：Tai Po Kau 大埔滘：Jun-07, 07 年 6 月：James Lam 林文華

	春季過境遷徙鳥 Spring Passage Migrant			夏候鳥 Summer Visitor			秋季過境遷徙鳥 Autumn Passage Migrant			冬候鳥 Winter Visitor		
常見月份	1	2	3	4	5	6	7	8	9	10	11	12
	留鳥 Resident				迷鳥 Vagrant				偶見鳥 Occasional Visitor			

2
3
4
5
6

7

山藍仙鶲

㊗ shān lán xiān wēng

體長 length：14-15.5cm

Hill Blue Flycatcher | *Cyornis whitei*

1

雄鳥上體深藍色而眼附近黑色，喉至脇部橙黃色；雌鳥上體褐色而眼圈淺棕色，喉至脇部淡橙色。嘴黑色，腳褐色。

Male has dark blue upperparts. Black around eyes. Yellowish-orange from throat to flanks. Female has brown upperparts and light brown eye-rings. Yellowish-orange from throat to flanks. Black bill. Brown legs.

1 adult male 雄成鳥：King's Park Garden 京士柏公園：Mar-17, 17 年 3 月：Sam Chan 陳巨輝
2 adult male 雄成鳥：Tai Po Kau 大埔滘：Feb-20, 20 年 2 月：Roman Lo. 羅文凱
3 adult male 雄成鳥：King's Park Garden 京士柏公園：Mar-17, 17 年 3 月：Kan Chi Ming 簡志明
4 adult male 雄成鳥：King's Park Garden 京士柏公園：Mar-17, 17 年 3 月：Sam Chan 陳巨輝
5 adult male 雄成鳥：King's Park Garden 京士柏公園：Mar-17, 17 年 3 月：Sam Chan 陳巨輝

春季過境遷徙鳥 Spring Passage Migrant			夏候鳥 Summer Visitor			秋季過境遷徙鳥 Autumn Passage Migrant			冬候鳥 Winter Visitor		
常見月份 1	2	3	4	5	6	7	8	9	10	11	12
留鳥 Resident				迷鳥 Vagrant				偶見鳥 Occasional Visitor			

2
3
4
5

灰紋鶲

㊟ huī wén wēng

體長 length：12.5-14cm

Grey-streaked Flycatcher | *Muscicapa griseisticta*

其他名稱 Other names：斑胸鶲、灰斑鶲

1

2

3

4

5

6

大致灰褐色，翅膀長及尾部的一半。淺色眼先，白喉上有鮮明顎紋。胸部有明顯的褐色條紋，和白色腹部成強烈對比。翼上有白邊。喜停在沒有遮閉的樹枝上。

A greyish-brown Flycatcher. Long wings reach up to half of tail. Pale lores. Strong malar stripes on white throat. Prominent brown streaks on breast contrast with white belly. White edges on wing panels. Perches on exposed branches.

1 Po Toi 蒲台：May-07, 07 年 5 月：James Lam 林文華
2 Po Toi 蒲台：May-07, 07 年 5 月：Sammy Sam Winnie Wong and 森美與雲妮
3 Po Toi 蒲台：Oct-08, 08 年 10 月：Michelle and Peter Wong 江敏兒、黃理沛
4 Po Toi 蒲台：May-07, 07 年 5 月：Jemi and John Holmes 孔思義、黃亞萍
5 Po Toi 蒲台：Sep-08, 08 年 9 月：Andy Kwok 郭匯昌
6 Po Toi 蒲台：Oct-08, 08 年 10 月：Michelle and Peter Wong 江敏兒、黃理沛

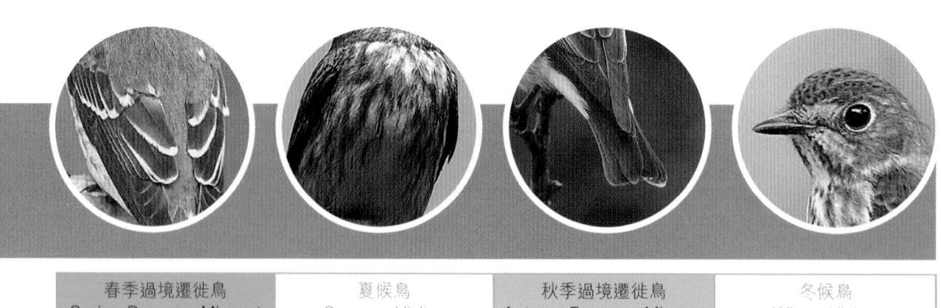

春季過境遷徙鳥 Spring Passage Migrant	夏候鳥 Summer Visitor	秋季過境遷徙鳥 Autumn Passage Migrant	冬候鳥 Winter Visitor

常見月份 1 2 3 4 5 6 7 8 9 10 11 12

留鳥 Resident	迷鳥 Vagrant	偶見鳥 Occasional Visitor

中華仙鶲

㊟ zōng huá xiān wēng

體長 length：14-15cm

Chinese Blue Flycatcher | *Cyornis glaucicomans*

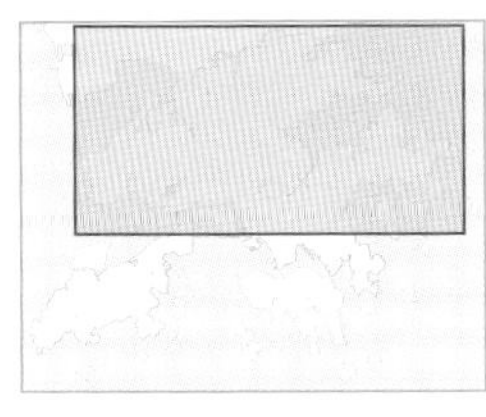

雄鳥上體藍色，前額、眉紋及肩部帶金屬閃亮，頰部藍色較深，胸部橙紅色；雌鳥上體灰褐色而眼圈淺棕色，喉白而胸部淡橙色。嘴黑色，腳深褐色。

Male has blue upperparts with metallic sheen on forehead, supercilia and shoulders. Darker blue cheeks. Reddish orange breast. Female has greyish-brown upperparts and light brown eye-rings. White throat and pale orange breast. Black bill. Dark brown legs.

1 adult male 雄成鳥：Shek Kong Catchwaters 石崗引水道：Cheng Nok Ming 鄭諾銘
2 adult male 雄成鳥：Mai Po 米埔：Dec-14, 14 年 12 月：Lee Yat Ming 李逸明
3 adult male 雄成鳥：Mai Po 米埔：Dec-14, 14 年 12 月：Kinni Ho 何建業
4 adult male 雄成鳥：Mai Po 米埔：Dec-14, 14 年 12 月：Lee Yat Ming 李逸明

春季過境遷徙鳥 Spring Passage Migrant	夏候鳥 Summer Visitor	秋季過境遷徙鳥 Autumn Passage Migrant	冬候鳥 Winter Visitor

1 2 3 4 5 6 7 8 9 10 11 12 常見月份

留鳥 Resident	迷鳥 Vagrant	偶見鳥 Occasional Visitor

白喉林鶲

㊣ bái hóu lín wēng

體長 length：15cm

Brown-chested Jungle Flycatcher | *Cyornis brunneatus*

上體為濃厚的灰褐色，喉白色，頸近白色而略帶淡斑紋。下體淡色，尾部紅褐色。喉部白色，黑色眼大，有眼後較粗的白色眼圈。嘴長而粗，上嘴深色下嘴偏黃。

Rich greyish brown upperparts white throat, white neck with danker marks. Pale underparts. Reddish brown tail. White throat. Large black eyes and white uneven eye-rings which are thicker behind the eye. Long and thick bill. Upperbill dark, lower bill yellowish.

[1] Mai Po 米埔；Sept-05, 05 年 9 月；Tam Yiu Leung 譚耀良
[2] Mai Po 米埔；Sept-05, 05 年 9 月；Tam Yiu Leung 譚耀良
[3] Tai Po Kau 大埔滘；Aug-04, 04 年 8 月；Michelle and Peter Wong 江敏兒、黃理沛
[4] adult 成鳥；Tai Po Kau 大埔滘；Aug-08, 08 年 8 月；Ken Fung 馮漢城

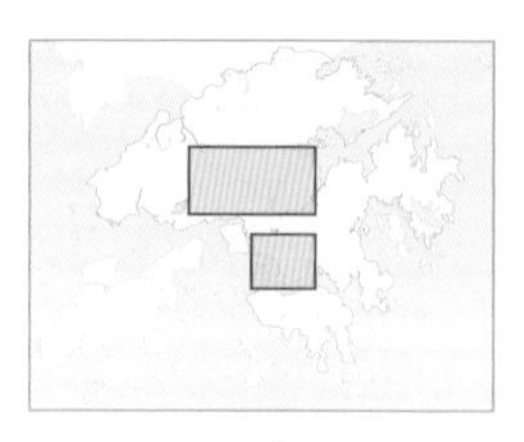

春季過境遷徙鳥 Spring Passage Migrant	夏候鳥 Summer Visitor	秋季過境遷徙鳥 Autumn Passage Migrant	冬候鳥 Winter Visitor

常見月份 1 2 3 4 5 6 7 **8** **9** **10** **11** 12

留鳥 Resident	迷鳥 Vagrant	偶見鳥 Occasional Visitor

小仙鶲

㊟ xiǎo xiān wēng

體長 length：11-14cm

Small Niltava | *Niltava macgrigoriae*

小型鶲。雄鳥上體深藍色，面頰黑色，前額、頸側的羽毛有閃亮的藍色。喉部黑色，胸部深藍色，腹部淺藍色，尾下覆羽白色。雌鳥上體褐色，下體褐色較淺，頸側有亮藍色斑。

Small-sized flycatcher. Male bird has dark blue upperparts. Cheeks are blackish. Feathers on forehead, sides of neck are shiny blue. Throat is black, breast is dark blue, belly is light blue and vent is white. Female has brown upperparts and paler underparts. It also has a prominent shiny blue patch at the side of the neck.

1 adult male 雄成鳥：Feb-04, 04 年 2 月：Michelle and Peter Wong 江敏兒、黃理沛
2 adult male 雄成鳥：Tai Po Kau 大埔滘：Feb-08, 08 年 2 月：Wallace Tse 謝鑑超
3 adult male 雄成鳥：Chung Mei 涌尾：Feb-04, 04 年 2 月：Michelle and Peter Wong 江敏兒、黃理沛
4 adult male 雄成鳥：Tai Po Kau 大埔滘：Feb-08, 08 年 2 月：Wallace Tse 謝鑑超
5 adult male 雄成鳥：Tai Po Kau 大埔滘：Feb-08, 08 年 2 月：Wallace Tse 謝鑑超
6 adult male 雄成鳥：Chung Mei 涌尾：Feb-04, 04 年 2 月：Michelle and Peter Wong 江敏兒、黃理沛

春季過境遷徙鳥 Spring Passage Migrant	夏候鳥 Summer Visitor	秋季過境遷徙鳥 Autumn Passage Migrant	冬候鳥 Winter Visitor

1 2 3 4 5 6 7 8 9 10 11 12 常見月份

留鳥 Resident	迷鳥 Vagrant	偶見鳥 Occasional Visitor

棕腹大仙鶲

[普]zōng fù dà xiān wēng

體長 length：18cm

Fujian Niltava | *Niltava davidi*

1

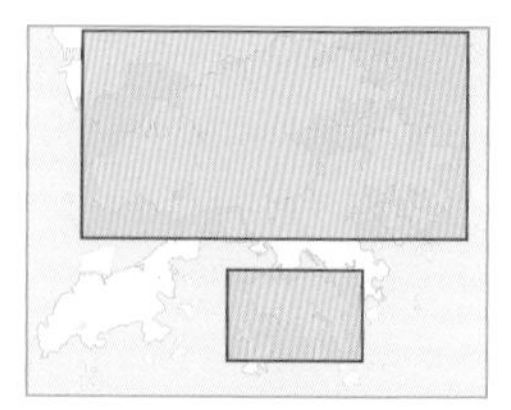

大型鶲。雄鳥上體及尾上深藍色，臉偏黑，前額、頸側、肩及腰的羽毛有閃亮的藍色。胸及腹部鮮橙褐色，伸延至臀部且漸變為白色。雌鳥上體褐色，下體褐色較淺，尾部紅褐色較濃。頸上有一道偏白彎月紋，頸兩側各有一小塊藍色，非常明顯，是雌鳥的特徵。喜在林區陰暗的下層活動。

Large-sized flycatcher. Male bird has dark blue upperparts and uppertail. Face is blackish. Feathers on forehead, sides of neck, shoulders and rump are shiny blue. Breast and belly are bright orange rufous, fading gradually to white towards vent. Female bird has brown upperparts and paler on underparts but tail is more rufous. A whitish crescent across the neck and a small blue patch on both sides of the neck are distinctive. Prefers shaded area of lower storeys.

1 adult male 雄成鳥；Ha Fa Shan 下花山；Feb-08, 08 年 2 月；Eling Lee 李佩玲
2 adult male 雄成鳥；HK Trail 港島徑；Jan-07, 07 年 1 月；Maison Fung 馮振威
3 adult male 雄成鳥；Shatin Pass 沙田坳；Nov-04, 04 年 11 月；Michelle and Peter Wong 江敏兒、黃理沛
4 adult female 雌成鳥；Tai Po Kau 大埔滘；1 Jan-13, 13年1月1日；Beetle Cheng 鄭諾銘
5 adult male 雄成鳥；Ha Fa Shan 下花山；Feb-08, 08 年 2 月；Jasper Lee 李君哲
6 adult male 雄成鳥；Ha Fa Shan 下花山；Feb-08, 08 年 2 月；Jasper Lee 李君哲
7 adult male 雄成鳥；Ha Fa Shan 下花山；Feb-08, 08 年 2 月；Jasper Lee 李君哲
8 adult male 雄成鳥；Ha Fa Shan 下花山；Feb-08, 08 年 2 月；Jasper Lee 李君哲
9 adult male 雄成鳥；Ha Fa Shan 下花山；Feb-08, 08 年 2 月；Jasper Lee 李君哲

春季過境遷徙鳥 Spring Passage Migrant			夏候鳥 Summer Visitor			秋季過境遷徙鳥 Autumn Passage Migrant			冬候鳥 Winter Visitor		
1	2	3	4	5	6	7	8	9	10	11	12

常見月份

留鳥 Resident	迷鳥 Vagrant	偶見鳥 Occasional Visitor

白腹姬鶲

[普] bái fù jī wēng

體長 length：16-17cm

Blue-and-white Flycatcher | *Cyanoptila cyanomelana*

其他名稱 Other names：白腹鶲

1

2

大型鶲。雄鳥上體藍色，下體白色，臉、喉及胸部均為黑色。雌鳥頭、背及尾上為褐色，下體白色，飛羽紅褐色。第一年度冬的雄鳥，頭及背部褐色，腰和尾部藍色。喜在樹冠活動。

Large flycatcher. Male bird has blue upperparts and white underparts. Black face, throat and breast. Female bird has brown head, mantle and upper tail. Underparts white and rufous on flight feathers. First winter male has brown head and mantle, and blue rump and tail. Favours canopy.

1. adult male, race *cumatalis* 雄成鳥, *cumatalis* 亞種：Po Toi 蒲台：Apr-07, 07 年 4 月：Winnie Wong and Sammy Sam 森美與雲妮
2. adult male, race *cyanomelana* 雄成鳥, *cyanomelana* 亞種：Cheung Chau 長洲：Oct-07, 07 年 10 月：Matthew and TH Kwan 關朗曦、關子凱
3. female 雌鳥：Mai Po 米埔：Oct-08, 08 年 10 月：Cheng Nok Ming 鄭諾銘
4. 1st winter male 第一年冬天雄鳥：7-Oct-07, 07 年 10 月 7 日：Michelle and Peter Wong 江敏兒、黃理沛
5. 1st winter male 第一年冬天雄鳥：Po Toi 蒲台：Oct-08, 08 年 10 月：Wallace Tse 謝鑑超
6. 1st winter male 第一年冬天雄鳥：Po Toi 蒲台：Oct-07, 07 年 10 月：Michelle and Peter Wong 江敏兒、黃理沛
7. 1st winter male 第一年冬天雄鳥：7-Oct-07, 07 年 10 月 7 日：Michelle and Peter Wong 江敏兒、黃理沛
8. adult male, race *cyanomelana* 雄成鳥, *cyanomelana* 亞種：Po Toi 蒲台：Jun-08, 08 年 6 月：Chan Kai Wai 陳佳瑋

春季過境遷徙鳥 Spring Passage Migrant	夏候鳥 Summer Visitor	秋季過境遷徙鳥 Autumn Passage Migrant	冬候鳥 Winter Visitor

常見月份	1	2	3	4	5	6	7	8	9	10	11	12

留鳥 Resident	迷鳥 Vagrant	偶見鳥 Occasional Visitor

銅藍鶲

㊟ tóng lán wēng

體長 length：15-17cm

Verditer Flycatcher | *Eumyias thalassinus*

全身呈閃亮藍綠色，尾下覆羽末端白色。雄鳥有黑色眼先，雌鳥顏色較暗淡，眼先較纖細和不明顯。喜停於明顯易見的地方。

Glossy bluish-green body. White tips on undertail coverts. Male has black lore. Female is duller and has smaller black lore which is not obvious. Perches prominently.

1 breeding male 繁殖羽雄鳥：Andy Cheung 張玉良
2 non-breeding adult male 非繁殖羽雄成鳥：Lam Tsuen 林村：11-Jan-09, 09 年 1 月 11 日：Jacky Chan 陳家華
3 non-breeding adult female 非繁殖羽雌成鳥：Wun Yiu 碗窰：Jan-03, 03 年 1 月：Henry Lui 呂德恒
4 non-breeding adult female 非繁殖羽雌成鳥：Wun Yiu 碗窰：Jan-03, 03 年 1 月：Henry Lui 呂德恒
5 non-breeding adult male 非繁殖羽雄成鳥：Tai Po 大埔：Feb-08, 08 年 2 月：James Lam 林文華
6 breeding female 繁殖羽雌鳥：Lions Nature Education Center 獅子會自然教育中心：Dec-08, 08 年 12 月：Michelle and Peter Wong 江敏兒、黃理沛
7 breeding male 繁殖羽雄鳥：Shing Mun 城門水塘：Feb-07, 07 年 2 月：Law Kam Man 羅錦文
8 breeding male 繁殖羽雄鳥：Tai Po 大埔：Feb-08, 08 年 2 月：Andy Kwok 郭匯昌

春季過境遷徙鳥 Spring Passage Migrant	夏候鳥 Summer Visitor	秋季過境遷徙鳥 Autumn Passage Migrant	冬候鳥 Winter Visitor

常見月份 1 2 3 4 5 6 7 8 9 10 11 12

留鳥 Resident	迷鳥 Vagrant	偶見鳥 Occasional Visitor

3
6
4
5
7

8

白喉短翅鶇

[普] bái hóu duǎn chì dōng
[粵] 鶇：音東

體長length：11-13cm

Lesser Shortwing | *Brachypteryx leucophrys*

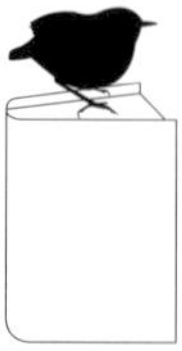

1

2

3

🌳

上體深褐色，上背、尾上覆羽和翼較為紅褐色。喉灰白色，眉紋短而不明顯，眼圈淡褐色。胸部羽毛沾深褐色，腹部中央羽色較白。脇部灰褐色，尾下覆羽淡褐色。腳長而壯碩，翼和尾短小。活躍於下層叢林。喜歡在較高的山區繁殖，在較低處越冬。

Upperparts dark brown, but mantle, uppertail coverts, wings are more rufous. Throat greyish white, short and indistinct eyebrow and buffish brown eye-rings. Breast is mottled with brown changing to white at the centre of belly. Flanks are greyish brown and undertail coverts are brownish white. It has long and strong legs, short wings and a very short tail. Favours forest floor. Breeds on higher ground and descends to lowland forests in winter.

[1] Tai Po Kau 大埔滘：Jan-04, 04 年 1 月：Michelle and Peter Wong 江敏兒、黃理沛
[2] Lung Fu Shan 龍虎山：Oct-07, 07 年 10 月：Tony Hung 洪敦熹
[3] Tai Po Kau 大埔滘：Jan-04, 04 年 1 月：Michelle and Peter Wong 江敏兒、黃理沛
[4] Tai Po Kau 大埔滘：Jan-04, 04 年 1 月：Michelle and Peter Wong 江敏兒、黃理沛

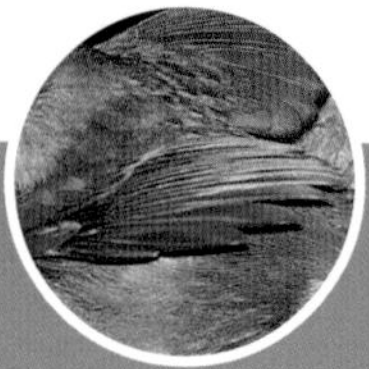

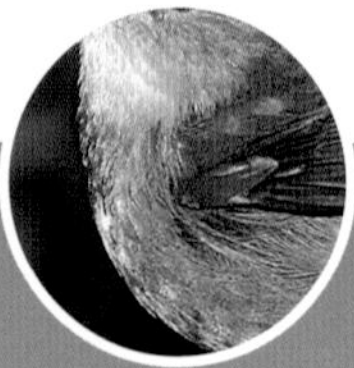

4

	春季過境遷徙鳥 Spring Passage Migrant			夏候鳥 Summer Visitor			秋季過境遷徙鳥 Autumn Passage Migrant			冬候鳥 Winter Visitor		
常見月份	1	2	3	4	5	6	7	8	9	10	11	12
	留鳥 Resident				迷鳥 Vagrant				偶見鳥 Occasional Visitor			

藍歌鴝

㊟lán gē qú
㊟鴝：音渠

體長length：13-14cm

Siberian Blue Robin | *Larvivora cyane*

嘴黑色，腳長及淡色，尾藍色，下體偏白。雄鳥上體深藍色，臉部至胸部兩側黑色。雌鳥頭、背、翼及脇部褐色，胸部有褐色鱗狀紋。

Black bill. Long and pale legs. Blue tail and whitish underparts. Male has deep blue upperparts and black from face to the sides of breast. Female has brown head, back, wings and flanks, and brown scales on breast.

1 juvenile female 雌幼鳥：Tsuen Wan 荃灣：Owen Chiang 深藍
2 juvenile female 雌幼鳥：Tsuen Wan 荃灣：Owen Chiang 深藍

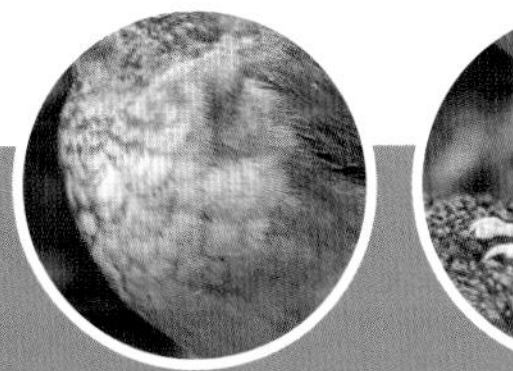

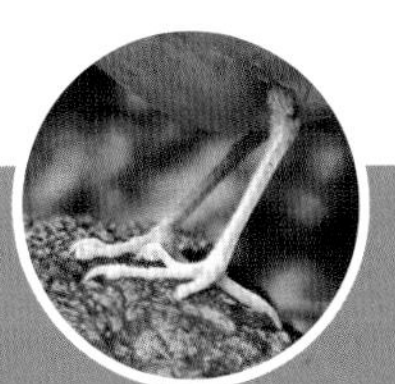

春季過境遷徙鳥 Spring Passage Migrant	夏候鳥 Summer Visitor	秋季過境遷徙鳥 Autumn Passage Migrant	冬候鳥 Winter Visitor

1 2 3 **4** 5 6 7 8 **9** **10** 11 12 常見月份

留鳥 Resident	迷鳥 Vagrant	偶見鳥 Occasional Visitor

紅尾歌鴝

[普]hóng wěi gē qú
[粵]鴝：音渠

體長length：13-14cm

Rufous-tailed Robin | *Larvivora sibilans*

1

全身褐色。嘴深色，腳偏淡。上體橄欖褐色，耳羽有縱紋，尾栗色。下體白色，胸及脇部有橄欖褐色的半月形斑紋。常顫動尾部，叫聲是輕輕的「格格」聲。

Generally brown. Dark bill. Pale legs. Olive-brown upperparts with streaks on ear-coverts, chestnut tail. Underparts white, with olive-brown crescent-shaped scales on the breast and flanks. Constantly flicks tail. Call a quiet rattle.

1 Kwai Chung 葵涌；Mar-07, 07 年 3 月；Sammy Sam and Winnie Wong 森美與雲妮
2 Kwai Chung 葵涌；Mar-07, 07 年 3 月；Martin Hale 夏敖天
3 Lamma Island 南丫島；Mar-05, 05 年 3 月；Cherry Wong 黃卓研
4 Siu Lek Yuen 小瀝源；Feb-08, 08 年 2 月；Ken Fung 馮漢城
5 10-Mar-07, 07 年 3 月 10 日；Michelle and Peter Wong 江敏兒、黃理沛
6 Kwai Chung 葵涌；Mar-07, 07 年 3 月；Allen Chan 陳志雄
7 Kwai Chung 葵涌；Mar-07, 07 年 3 月；Raymond Cheng 鄭兆文
8 Kwai Chung 葵涌；Mar-07, 07 年 3 月；Wallace Tse 謝鑑超

	春季過境遷徙鳥 Spring Passage Migrant			夏候鳥 Summer Visitor			秋季過境遷徙鳥 Autumn Passage Migrant			冬候鳥 Winter Visitor		
常見月份	1	2	3	4	5	6	7	8	9	10	11	12
	留鳥 Resident				迷鳥 Vagrant				偶見鳥 Occasional Visitor			

2
3
4
5
6
7

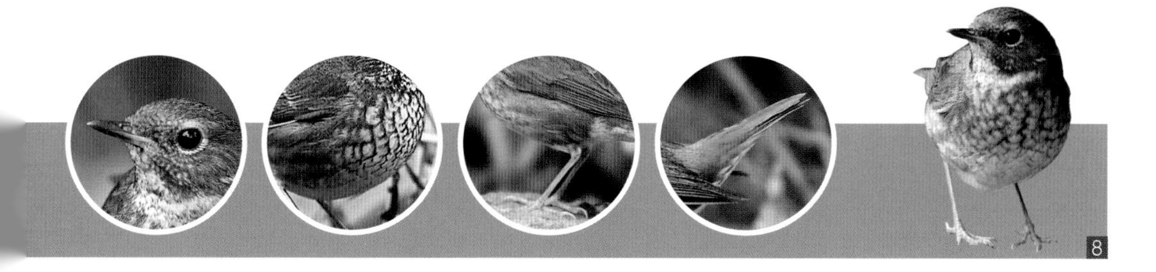
8

日本歌鴝

[普] rì běn gē qú
[粵] 鴝：音渠

體長 length：14-15cm

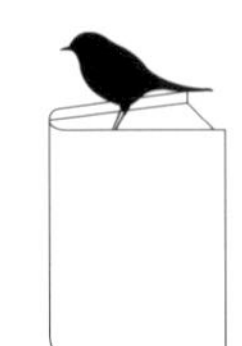

Japanese Robin | *Larvivora akahige*

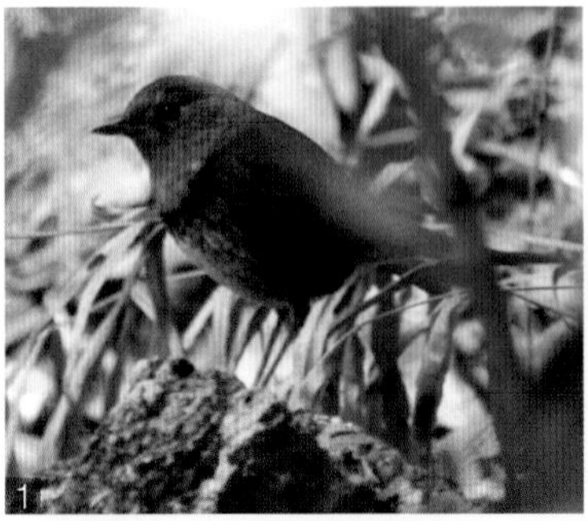
1

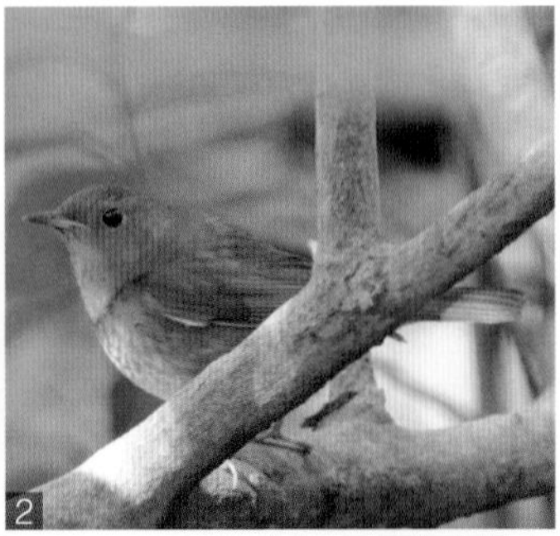
2

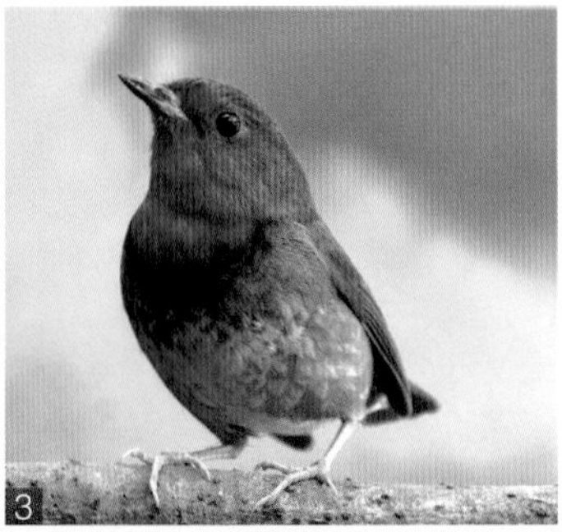
3

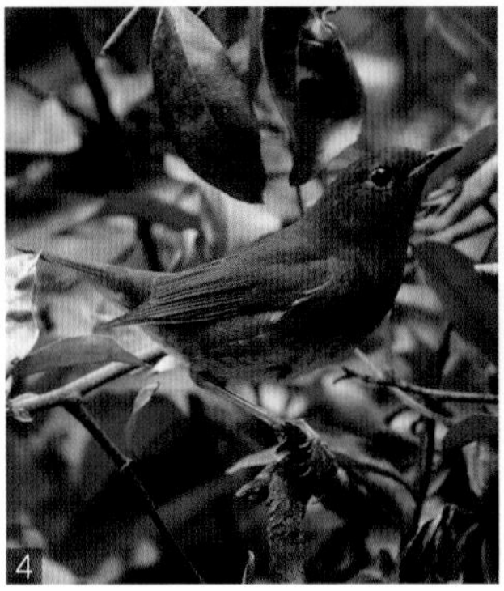
4

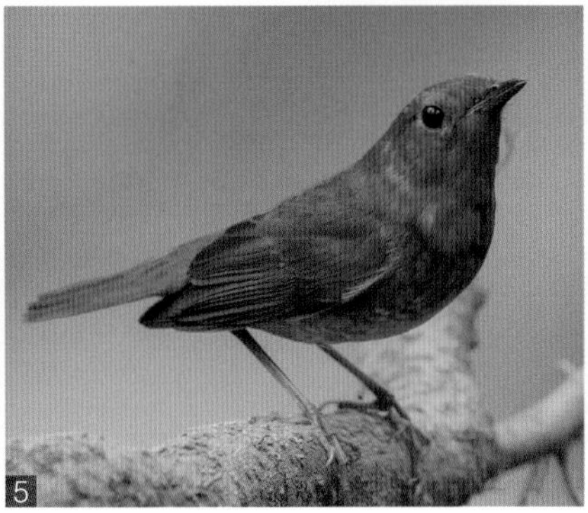
5

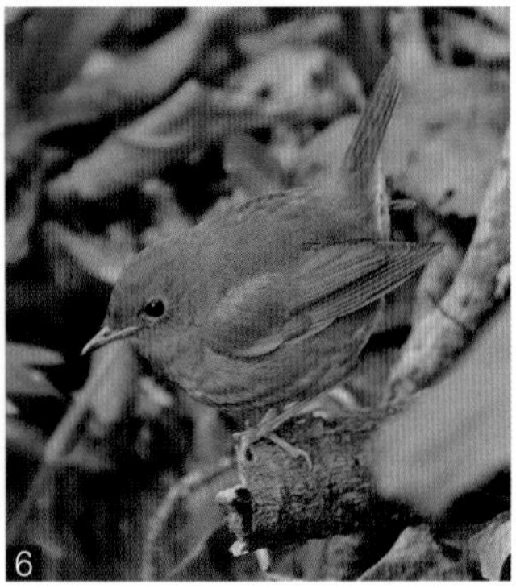
6

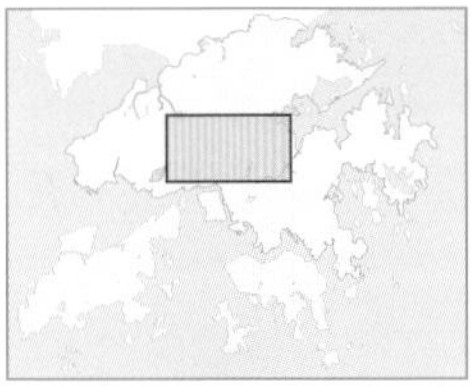

🌳

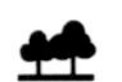

嘴黑色，腳淡色，上體褐色，下體偏白，常豎起尾部。雄鳥臉和胸紅褐色，和淡色腹部之間有一條黑色幼橫紋。

Black bill and pale legs. Brown upperparts and whitish underparts. Male with reddish brown face and breast and a thin black band between breast and belly.

[1] adult male 雄成鳥；Po Toi 蒲台；20-Nov-07, 07 年 11 月 20 日；Geoff Welch
[2] adult male 雄成鳥；Tai Lam 大欖；Jan-15, 15 年 1 月；John Clough
[3] adult male 雄成鳥；Tai Lam 大欖；Dec-19, 19 年 12 月；Roman Lo 羅文凱
[4] adult male 雄成鳥；Tai Lam 大欖；Nov-19, 19 年 11 月；Kwok Tsz Ki 郭子祈
[5] adult male 雄成鳥；Tai Lam 大欖；Nov-20, 20 年 11月；Matthew Kwan 關朗曦
[6] adult female 雌成鳥；KFBG 嘉道理農場；Feb 18, 18 年 2 月；Henry Lui 呂德恒
[7] male 雄鳥；Ng Tung Chai 梧桐寨；8-Mar-09, 09 年 3 月 8 日；Michelle and Peter Wong 江敏兒、黃理沛

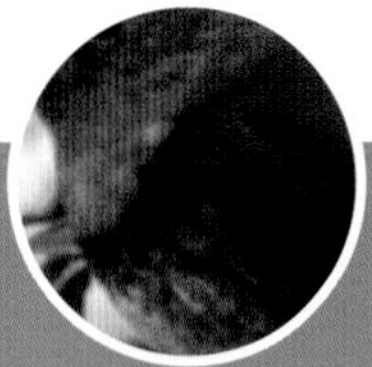

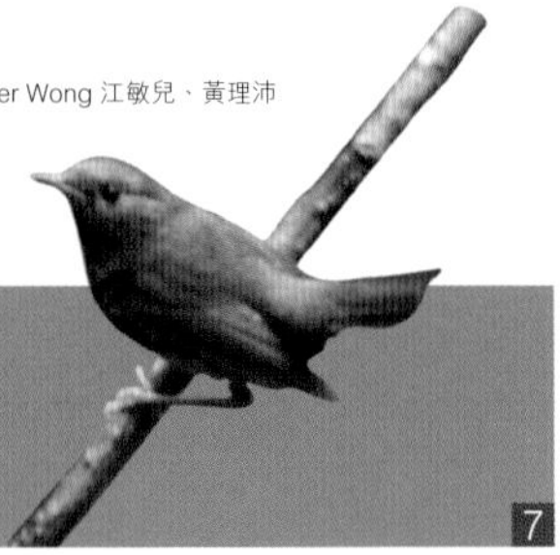
7

	春季過境遷徙鳥 Spring Passage Migrant			夏候鳥 Summer Visitor			秋季過境遷徙鳥 Autumn Passage Migrant			冬候鳥 Winter Visitor		
常見月份	1	2	3	4	5	6	7	8	9	10	11	12
	留鳥 Resident				迷鳥 Vagrant				偶見鳥 Occasional Visitor			

白尾藍地鴝

㊔ bái wěi lán dì qú
㊣ 鴝：音渠

體長 length：17-19cm

White-tailed Robin | *Myiomela leucura*

其他名稱 Other names：白尾斑地鴝

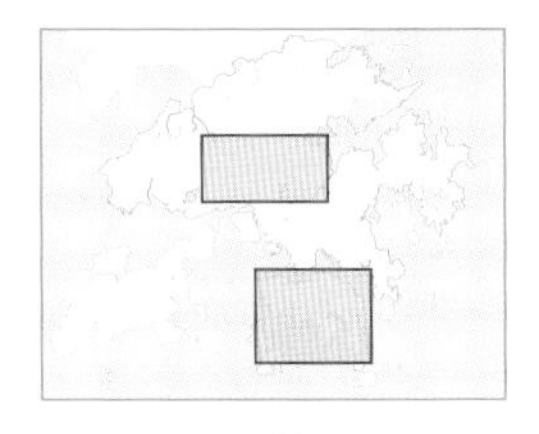

全身深藍色，嘴和腳黑色。前額和小覆羽有亮藍斑塊。尾基部兩側有白斑。雌鳥和雄鳥相似，但全身褐色，沒有藍斑。狀況不明，冬季時在適合的生境有個別記錄。

Overall dark brown. Black bill and legs. Forehead and shoulder have glossy blue patch. Prominent white patches on base of tail. Female resembles male but overall brown and without glossy blue patches. Status uncertain. A few records in suitable habitat during winter.

1 Tai Po Kau 大埔滘：Michelle and Peter Wong 江敏兒、黃理沛

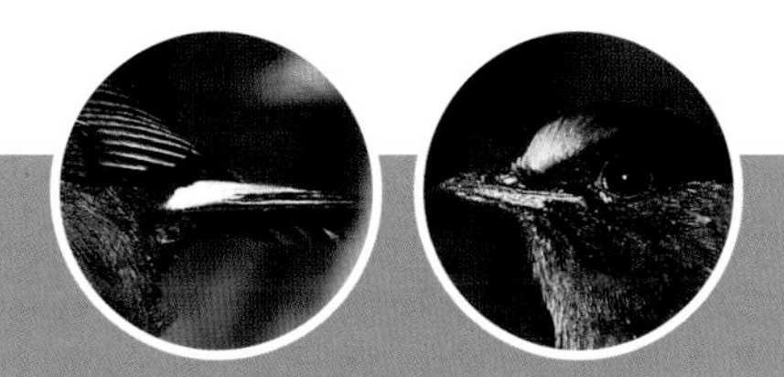

春季過境遷徙鳥 Spring Passage Migrant			夏候鳥 Summer Visitor			秋季過境遷徙鳥 Autumn Passage Migrant			冬候鳥 Winter Visitor			
1	2	3	4	5	6	7	8	9	10	11	12	常見月份
留鳥 Resident				迷鳥 Vagrant				偶見鳥 Occasional Visitor				

藍喉歌鴝

[普]lán hóu gē qú
[粵]鴝：音渠

體長length：13-15cm

Bluethroat | *Luscinia svecica*

其他名稱Other names：藍點頦

上體深褐色，有白色眉紋。繁殖期雄鳥喉部和胸很特別，有栗、藍、黑、白四色組成的圖案；雌鳥喉部白色，頰下有黑紋伸延至胸及脇部。下體白色，有時沾淡黃，飛行時外側尾羽基部明顯栗色。亞種 *svecica* 的藍喉上有紅色斑點。叫聲為重複的「cheech」聲。

Dark brown upperparts with prominent white eyebrow. Breeding male has distinctive chestnut, blue, black and white pattern on the throat and breast. Female has white throat and black submoustachial stripes which extend to breast and flanks. The white underparts may be buffy yellow. The outer tail feathers have obvious chestnut base in flight. Red spot on blue throat for the race *svecica*. Call a repeated "cheech".

1 adult male 雄成鳥；Long Valley 塱原；Dec-06, 06 年 12 月；Andy Kwok 郭匯昌
2 adult female 雌成鳥；Long Valley 塱原；Dec-06, 06 年 12 月；Michelle and Peter Wong 江敏兒、黃理沛
3 adult male 雄成鳥；Long Valley 塱原；Dec-06, 06 年 12 月；Michelle and Peter Wong 江敏兒、黃理沛
4 adult male 雄成鳥；Long Valley 塱原；18-Dec-05, 05 年 12 月 18 日；Lo Kar Man 盧嘉孟
5 adult male 雄成鳥；Long Valley 塱原；Jan-08, 08 年 1 月；Pang Chun Chiu 彭俊超
6 adult female 雌成鳥；Long Valley 塱原；Dec-06, 06 年 12 月；Cherry Wong 黃卓研
7 non-breeding adult male moulting into breeding plumage 非繁殖羽雄成鳥轉換繁殖羽；Long Valley 塱原；Feb-08, 08 年 2 月；Cheng Nok Ming 鄭諾銘
8 adult male 雄成鳥；Long Valley 塱原；13-Jan-07, 07 年 1 月 13 日；Doris Chu 朱詠兒

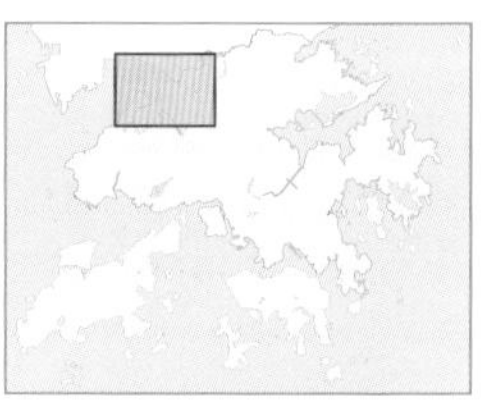

春季過境遷徙鳥 Spring Passage Migrant				夏候鳥 Summer Visitor			秋季過境遷徙鳥 Autumn Passage Migrant			冬候鳥 Winter Visitor		
常見月份	1	2	3	4	5	6	7	8	9	10	11	12
留鳥 Resident				迷鳥 Vagrant				偶見鳥 Occasional Visitor				

2
3
4
5
6
7

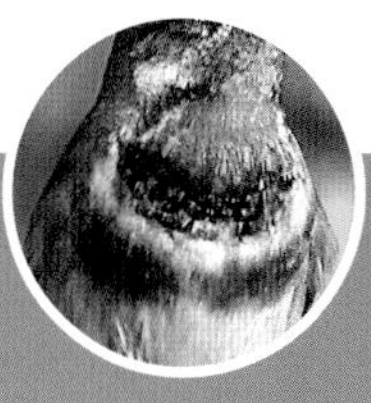
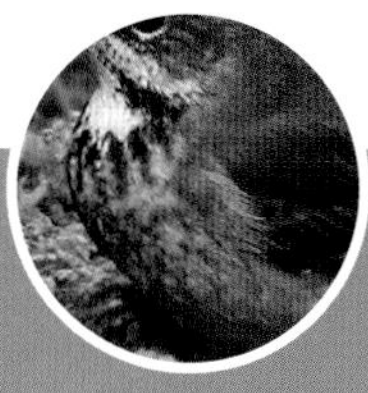

8

紅喉歌鴝

[普]hóng hóu gē qú
[粵]鴝：音渠

體長length：14-16cm

Siberian Rubythroat | *Calliope calliope*

其他名稱Other names：紅點頦

1

全身大致褐色，體型壯碩。嘴深色，腳偏淡，上體褐色，有明顯白眉紋和頰紋。本種的特徵是成年雄鳥喉部紅色，雌鳥喉部粉紅色。下體色較淺，第一年度冬鳥像巨型的褐柳鶯。叫聲像憂怨的哨聲。

Robust brown robin. Dark bill. Pale legs. Brown upperparts with prominent white eyebrow and moustachial stripe. Male has a distinctive red throat, which is pinkish on females. Paler underparts. First winter bird resembles robust Dusky Warblers. Call a plaintive whistle.

1 1st winter male 第一年冬天雄鳥；Mui Wo 梅窩；Jan-07, 07 年 1 月；Allen Chan 陳志雄
2 adult male 雄成鳥；Shek Kong 石崗；Feb-08, 08 年 2 月；Yue Pak Wai 余柏維
3 1st winter female 第一年冬天雌鳥；Pui O 貝澳；Feb-08, 08 年 2 月；Andy Kwok 郭匯昌
4 male 雄鳥；Tsuen Wan 荃灣；Owen Chiang 深藍
5 adult male 雄成鳥；Long Valley 塱原；Dec-06, 06 年 12 月；Andy Kwok 郭匯昌
6 adult male 雄成鳥；Long Valley 塱原；Dec-06, 06 年 12 月；Michelle and Peter Wong 江敏兒、黃理沛
7 1st winter male 第一年冬天雄鳥；Shek Kong 石崗；Feb-08, 08 年 2 月；Yue Pak Wai 余柏維
8 adult male 雄成鳥；Long Valley 塱原；Dec-06, 06 年 12 月；Michelle and Peter Wong 江敏兒、黃理沛

	春季過境遷徙鳥 Spring Passage Migrant			夏候鳥 Summer Visitor			秋季過境遷徙鳥 Autumn Passage Migrant			冬候鳥 Winter Visitor		
常見月份	1	2	3	4	5	6	7	8	9	10	11	12
	留鳥 Resident				迷鳥 Vagrant				偶見鳥 Occasional Visitor			

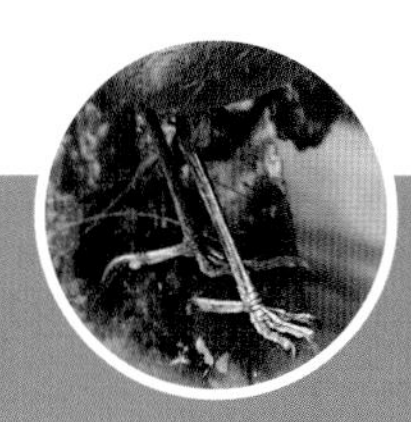

8

紅脇藍尾鴝

㊟hóng xié lán wěi qú
㊟鴝：音渠

體長 length：13-15cm

Red-flanked Bluetail | *Tarsiger cyanurus*

雌鳥和雄鳥共同的特徵是喉部白色、脇橙黃色，尾藍色，嘴及腳黑色。雄鳥上體鮮藍色有淡白眉，雌鳥及未成年幼鳥則為橄欖褐色；下體白色胸及脇沾黃褐色。叫聲有沙啞的「wheest」和輕輕的「chack chack」聲。

White throat, rufous orange flanks and blue tails are characteristic of both sexes. Black bill and legs. Adult male has bright blue upperparts with white eyebrow, while female and immature are olive-brown. Underparts are white with buffy breast and flank. Call a coarse whistle and a quiet "chack chack".

1 adult male 雄成鳥：Shek Kong 石崗：Mar-06, 06 年 3 月：Andy Kwok 郭匯昌
2 adult female 雌成鳥：Pui O 貝澳：Mar-08, 08 年 3 月：Isaac Chan 陳家強
3 adult female 雌成鳥：Mai Po 米埔：Dec-05, 05 年 12 月：Henry Lui 呂德恒
4 female / 1st winter male 雌鳥 / 第一年冬天雄鳥：Po Toi 蒲台：Feb-08, 08 年 2 月：Chan Kai Wai 陳佳瑋
5 adult male 雄成鳥：Pui O 貝澳：Feb-08, 08 年 2 月：Danny Ho 何國海
6 female / 1st winter male 雌鳥 / 第一年冬天雄鳥：Lantau Island 大嶼山：Feb-08, 08 年 2 月：Eling Lee 李佩玲
7 adult male 雄成鳥：Shek Kong 石崗：Feb-06, 06 年 2 月：Lee Kai Hong 李啟康
8 adult male 雄成鳥：Pui O 貝澳：Feb-08, 08 年 2 月：Ken Fung 馮漢城

	春季過境遷徙鳥 Spring Passage Migrant			夏候鳥 Summer Visitor			秋季過境遷徙鳥 Autumn Passage Migrant			冬候鳥 Winter Visitor		
常見月份	1	2	3	4	5	6	7	8	9	10	11	12
	留鳥 Resident				迷鳥 Vagrant				偶見鳥 Occasional Visitor			

2
3
6
4
5
7

8

灰背燕尾

㊟huī bèi yàn wěi

體長length：22-25cm

Slaty-backed Forktail | *Enicurus schistaceus*

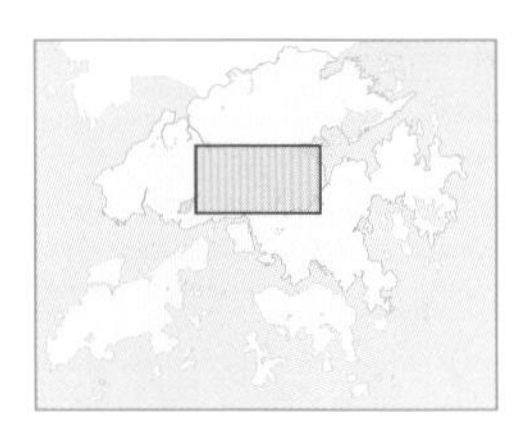

嘴及臉黑色，頭頂及上體灰色，下體及腰白色，長尾黑白分明像剪刀，腳淡色。幼鳥的頭和背深褐色，胸部有鱗狀斑。靜立時尾巴不停搖擺，在河溪環境活動。

Black bill and face. Slaty-grey crown and mantle. White underparts and rump. Black and white long scissor-like tail. Pale legs. Immature bird has olive-brown on head and mantle with scaly patterns on breast. Continually twitches tail at rest. Favours rivers and stream.

1 adult 成鳥：Tai Po Kau 大埔滘：Oct-04, 04 年 10 月：Henry Lui 呂德恆
2 adult 成鳥：Tai Po Kau 大埔滘：Oct-04, 04 年 10 月：Henry Lui 呂德恆
3 adult 成鳥：Tai Po Kau 大埔滘：Oct-04, 04 年 10 月：Cherry Wong 黃卓研
4 adult 成鳥：Tai Po Kau 大埔滘：Oct-04, 04 年 10 月：Henry Lui 呂德恆
5 adult 成鳥：Tai Po Kau 大埔滘：27-Nov-04, 04 年 11 月 27 日：Michelle and Peter Wong 江敏兒，黃理沛
6 adult 成鳥：Tai Po Kau 大埔滘：Dec-04, 04 年 12 月：Pippen Ho 何志剛

春季過境遷徙鳥 Spring Passage Migrant	夏候鳥 Summer Visitor	秋季過境遷徙鳥 Autumn Passage Migrant	冬候鳥 Winter Visitor

常見月份 1 2 3 4 5 6 7 8 9 10 11 12

留鳥 Resident	迷鳥 Vagrant	偶見鳥 Occasional Visitor

2

3

4

5

6

紫嘯鶇

[普]zǐ xiào dōng
[粵]鶇：音東

體長length：29-35cm

Blue Whistling Thrush | *Myophonus caeruleus*

嘴及腳黑色。全身看似黑色，但如果光線良好，獨特的深紫色和淺色斑點便明顯可見。叫聲是清脆嘹亮的長嘯聲，遠處可聞，常不停開合尾羽。

Black bill and legs. Body appears black, but deep violet plumage with lighter spangles is unmistakable under good light conditions. Call a loud and long whistle, which could be heard from long distance. Often fans tail.

1 adult 成鳥：Po Toi 蒲台：Feb-08, 08 年 2 月：Cherry Wong 黃卓研
2 juvenile 幼鳥：Chinese University of HK 香港中文大學：May-07, 07 年 5 月：Eling Lee 李佩玲
3 adult 成鳥：Discovery Bay 愉景灣：Apr-07, 07 年 4 月：Sonia and Kenneth Fung 馮啟文、蕭敏晶
4 adult 成鳥：Cheung Chau 長洲：Dec-04, 04 年 12 月：Henry Lui 呂德恒
5 adult 成鳥：Chinese University of HK 香港中文大學：Feb-07, 07 年 2 月：Banson Lau 劉滙文
6 adult 成鳥：Kwai Chung 葵涌：Jan-07, 07 年 1 月：James Lam 林文華
7 adult 成鳥：Siu Lek Yuen 小瀝源：Mar-08, 08 年 3 月：Ken Fung 馮漢城

	春季過境遷徙鳥 Spring Passage Migrant			夏候鳥 Summer Visitor			秋季過境遷徙鳥 Autumn Passage Migrant			冬候鳥 Winter Visitor		
常見月份	1	2	3	4	5	6	7	8	9	10	11	12
	留鳥 Resident				迷鳥 Vagrant				偶見鳥 Occasional Visitor			

2
3
4
5
6
7

白眉姬鶲

普 bái méi jī wēng

體長 length：13-13.5cm

Yellow-rumped Flycatcher | *Ficedula zanthopygia*

其他名稱 Other names：白眉鶲

1

雄鳥上體黑色，眉紋白色。喉、胸、腰和腹部黃色，尾下覆羽白色，翼上有明顯白斑。雌鳥和幼鳥上體橄欖褐色，腰淡黃色，有白色翼帶，喉及胸部皮黃色，腹部淡黃褐色，尾下覆羽白色。喜在開闊樹林下層活動。雄性在香港較為罕見，大多數的觀察記錄為雌鳥或幼鳥。

Male has black upperparts with white supercilium, bright yellow throat, breast, rump and belly, white vent and prominent white wing patch. Female and juvenile birds have olive-brown upperparts, pale yellow rump, white wing patches, buffish throat and breast, pale yellowish brown belly, and white undertail coverts. Favours low storey in open wooded area. In Hong Kong, male is very rare and most of the records were females or juveniles.

1 1st summer male 第一年夏天雄鳥；Po Toi 蒲台；Apr-08, 08 年 4 月；Michelle and Peter Wong 江敏兒、黃理沛
2 1st winter male 第一年冬天雄鳥；Mai Po 米埔；Sep-06, 06 年 9 月；Martin Hale 夏敖天
3 1st summer male 第一年夏天雄鳥；Po Toi 蒲台；Apr-08, 08 年 4 月；Michelle and Peter Wong 江敏兒、黃理沛
4 1st winter female 第一年冬天雌鳥；Mai Po 米埔；30-Sep-04, 04 年 9 月 30 日；Michelle and Peter Wong 江敏兒、黃理沛
5 1st summer male 第一年夏天雄鳥；Po Toi 蒲台；Apr-08, 08 年 4 月；Michelle and Peter Wong 江敏兒、黃理沛

春季過境遷徙鳥 Spring Passage Migrant	夏候鳥 Summer Visitor	秋季過境遷徙鳥 Autumn Passage Migrant	冬候鳥 Winter Visitor

常見月份 1 2 3 **4** 5 6 7 **8** **9** **10** **11** 12

留鳥 Resident	迷鳥 Vagrant	偶見鳥 Occasional Visitor

2
3
4
5

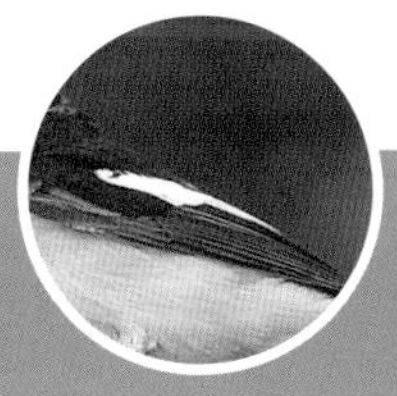

黃眉姬鶲

[普] huáng méi jī wēng

體長 length：13-13.5cm

Narcissus Flycatcher | *Ficedula narcissina*

其他名稱 Other names：黃眉鶲

1

雄鳥上體黑色，眉紋、喉、胸及腹鮮黃色，臀部白色，對比鮮明。翼上有明顯白斑，個別喉部染有橙色。雌鳥上體橄欖褐色，沒有黃腰或白色翼帶。喉及胸淡黃色，下體污白色，尾上覆羽及尾上紅褐色。喜在開闊樹林下層活動。

Black upperparts on male contrast with bright yellow supercilium, throat, breast and belly, and white vent. Prominent white wing patch. Some individuals have orange tinge on throat. Female upperparts are olive-brown, no yellow rump or white wing patch. Throat and breast are buffish. Underparts are dirty white. Uppertail coverts and upper tail are rufescent. Favours low storey in open wooded area.

1 adult male 雄成鳥：Po Toi 蒲台：Oct-08, 08 年 10 月：Michelle and Peter Wong 江敏兒、黃理沛
2 adult male 雄成鳥：Po Toi 蒲台：Oct-08, 08 年 10 月：Andy Kwok 郭匯昌
3 adult male 雄成鳥：Po Toi 蒲台：Oct-08, 08 年 10 月：Michelle and Peter Wong 江敏兒、黃理沛
4 adult male 雄成鳥：17-Apr-05, 05 年 4 月 17 日：Michelle and Peter Wong 江敏兒、黃理沛
5 adult male 雄成鳥：Po Toi 蒲台：Oct-08, 08 年 10 月：Lee Kai Hong 李啟康
6 adult female 雌成鳥：Po Toi 蒲台：Nov-07, 07 年 11 月：Michelle and Peter Wong 江敏兒、黃理沛
7 adult female 雌成鳥：Po Toi 蒲台：Nov-07, 07 年 11 月：Michelle and Peter Wong 江敏兒、黃理沛
8 1st summer male 第一年夏天雄鳥：Mai Po 米埔：Apr-07, 07 年 4 月：Cherry Wong 黃卓研

春季過境遷徙鳥 Spring Passage Migrant	夏候鳥 Summer Visitor	秋季過境遷徙鳥 Autumn Passage Migrant	冬候鳥 Winter Visitor

常見月份 1 2 3 4 5 6 7 8 9 10 11 12

留鳥 Resident	迷鳥 Vagrant	偶見鳥 Occasional Visitor

綠背姬鶲 [普]lǜ bèi jī wēng

體長 length：13-13.5cm

Green-backed Flycatcher | *Ficedula elisae*

1

2

3

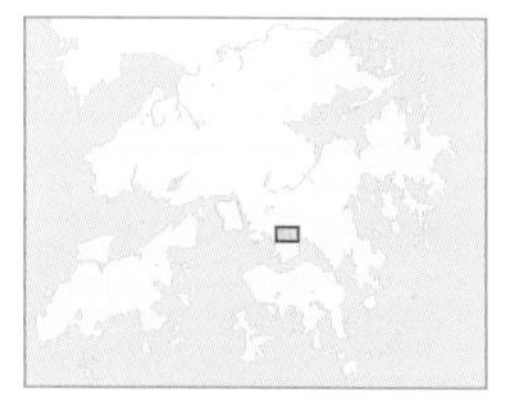

上體灰綠色，眼圈皮黃色，眼前有黃色短紋。翼和尾黑色，覆羽和飛羽羽緣有白邊。下體和腰部黃色，嘴和腳黑色。有機會與黃眉姬鶲的雌性亞種(*elisae*)混淆。

Greyish green upperparts. Yellowish eyering and a short yellow stripe in front of eye. Black wing and tail. Wing coverts and flight feathers fringed white. Yellow underparts and rump. Black bill and legs. A possibility of getting confused with the female subspecies（*elisae*）of the Narcissus Flycatcher.

1 Po Toi 蒲台：Mar-09, 09 年 3 月：Allen Chan 陳志雄
2 female 雌鳥：Po Toi 蒲台：3-Apr-05, 05 年 4 月 3 日：Michelle and Peter Wong 江敏兒、黃理沛
3 female 雌鳥：Po Toi 蒲台：3-Apr-05, 05 年 4 月 3 日：Michelle and Peter Wong 江敏兒、黃理沛
4 female/immature 雌鳥/未成年鳥：Tai Po Kau 大埔滘：27-Nov-05, 05 年 11 月 27 日：Michelle and Peter Wong 江敏兒、黃理沛
5 female/immature 雌鳥/未成年鳥：Tai Po Kau 大埔滘：27-Nov-05, 05 年 11 月 27 日：Michelle and Peter Wong 江敏兒、黃理沛
6 female 雌鳥：Po Toi 蒲台：3-Apr-05, 05 年 4 月 3 日：Michelle and Peter Wong 江敏兒、黃理沛
7 Po Toi 蒲台：Mar-09, 09 年 3 月：Allen Chan 陳志雄
8 Po Toi 蒲台：Mar-09, 09 年 3 月：Allen Chan 陳志雄
9 female/immature 雌鳥/未成年鳥：Tai Po Kau 大埔滘：27-Nov-05, 05 年 11 月 27 日：Michelle and Peter Wong 江敏兒、黃理沛

春季過境遷徙鳥 Spring Passage Migrant	夏候鳥 Summer Visitor	秋季過境遷徙鳥 Autumn Passage Migrant	冬候鳥 Winter Visitor

常見月份 1 2 3 4 5 6 7 8 9 10 11 12

留鳥 Resident	迷鳥 Vagrant	偶見鳥 Occasional Visitor

鴝姬鶲

㊟ qú jī wēng
㊟ 鴝：音渠

體長 length：12.5-13.5cm

Mugimaki Flycatcher | *Ficedula mugimaki*

其他名稱 Other names：鴝鶲, Robin Flycatcher

雄鳥似白眉鶲及黃眉鶲，可從其黑色上體及眼後白色短眉紋判別。喉及胸部橙色，有明顯白色翼帶，腹部、臀部及尾羽基部均為白色。雌鳥上體褐色，眉紋較淡，喉及上胸有不同程度的橙色，腹部及臀部白色，但沒有雄鳥尾羽基部的白色。個別鳥的覆羽及三級飛羽上有淺色邊緣。喜於林區的中至下層活動。

Male is similar to Yellow-rumped and Narcissus Flycatchers but differs by black upperparts with short white eyebrow behind the eye. Throat and breast are orange in colour. Prominent white wing patch. Belly, vent and the base of outertail feathers are white. Female has brown upperparts and fainter supercilium. Orange on throat and upper breast variable. Belly and vent are white but no white on the base of outertail feathers. Some individuals have pale fringes on wing coverts and tertials. Favours mid to low storey.

1 adult male 雄成鳥；Siu Lek Yuen 小瀝源；Jan-09, 09 年 1 月；Michelle and Peter Wong 江敏兒、黃理沛
2 female 雌鳥；Po Toi 蒲台；Oct-08, 08 年 10 月；Michelle and Peter Wong 江敏兒、黃理沛
3 1st winter male 第一年冬天雄鳥；Mui Siu Hang 梅樹坑；Jan-07, 07 年 1 月；Andy Kwok 郭匯昌
4 adult male 雄成鳥；Siu Lek Yuen 小瀝源；Jan-09, 09 年 1 月；Ken Fung 馮漢城
5 1st winter male 第一年冬天雄鳥；Mui She Hang 梅樹坑；Jan-07, 07 年 1 月；Andy Kwok 郭匯昌
6 1st winter male 第一年冬天雄鳥；Po Toi 蒲台；Oct-08, 08 年 10 月；James Lam 林文華
7 female 雌鳥；Tai Po 大埔；Jan-07, 07 年 1 月；Cherry Wong 黃卓研
8 adult male 雄成鳥；Siu Lek Yuen 小瀝源；Jan-09, 09 年 1 月；Michelle and Peter Wong 江敏兒、黃理沛

春季過境遷徙鳥 Spring Passage Migrant	夏候鳥 Summer Visitor	秋季過境遷徙鳥 Autumn Passage Migrant	冬候鳥 Winter Visitor

常見月份	1	2	3	4	5	6	7	8	9	10	11	12

留鳥 Resident	迷鳥 Vagrant	偶見鳥 Occasional Visitor

2
3
4
5
6
7

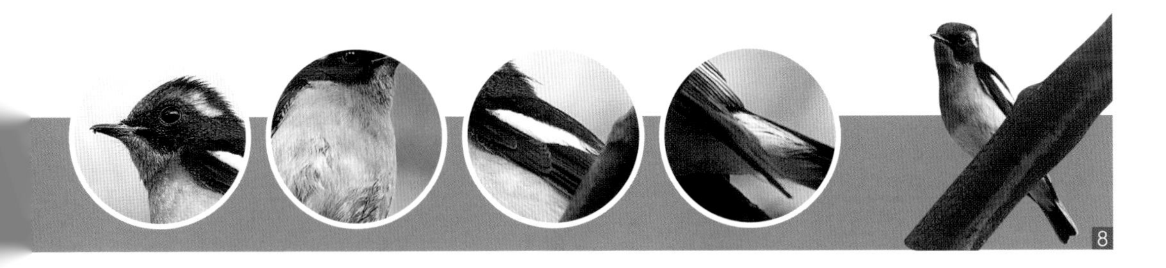
8

銹胸藍姬鶲

㊟xiù xiōng lán jī wēng

體長 length：13-13.5cm

Slaty-backed Flycatcher | *Ficedula hodgsonii*

雄鳥外型與棕腹大仙鶲相似但體型明顯較小，上體、前額、耳羽、頸側、翼上覆羽和腰藍色，但沒有閃亮斑塊，胸橙色上延至喉部，尾羽黑色，外側尾羽基部有白斑。雌鳥與鴝姬鶲雌鳥相似，翼較短，喉、胸及下體灰褐色。

Male distinguished from Fujian Niltava by smaller size. Blue upperparts, forehead, ear coverts, sides of neck, shoulders and rump are blue, but without shiny feathers. Orange throat and breast. Tail fethers black in colour, with white patch at the base of outer tail feathers. Female distinguished from Mugimaki flycatcher by shorter wings, greyish brown throat, breast and underparts.

1 female 雌鳥：Tsuen Wan 荃灣：Feb-08, 08 年 2 月：Michelle and Peter Wong 江敏兒、黃理沛
2 female 雌鳥：Tsuen Wan 荃灣：Feb-08, 08 年 2 月：Michelle and Peter Wong 江敏兒、黃理沛
3 female 雌鳥：Tso Kung Tam 曹公潭：Feb-08, 08 年 2 月：Ken Fung 馮漢城
4 female 雌鳥：Tsuen Wan 荃灣：Feb-08, 08 年 2 月：James Lam 林文華
5 female 雌鳥：Tso Kung Tam 曹公潭：Feb-08, 08 年 2 月：Ken Fung 馮漢城
6 female 雌鳥：Andy Cheung 張玉良
7 female 雌鳥：Andy Cheung 張玉良
8 female 雌鳥：Tsuen Wan 荃灣：Feb-08, 08 年 2 月：Michelle and Peter Wong 江敏兒、黃理沛

春季過境遷徙鳥 Spring Passage Migrant	夏候鳥 Summer Visitor	秋季過境遷徙鳥 Autumn Passage Migrant	冬候鳥 Winter Visitor

常見月份 1 2 3 4 5 6 7 8 9 10 11 12

留鳥 Resident	迷鳥 Vagrant	偶見鳥 Occasional Visitor

2
3
4
5
6
7
8

橙胸姬鶲

㊟ chéng xiōng jī wēng

體長 length：13-14.5cm

Rufous-gorgeted Flycatcher | *Ficedula strophiata*

其他名稱 Other names：橙胸鶲

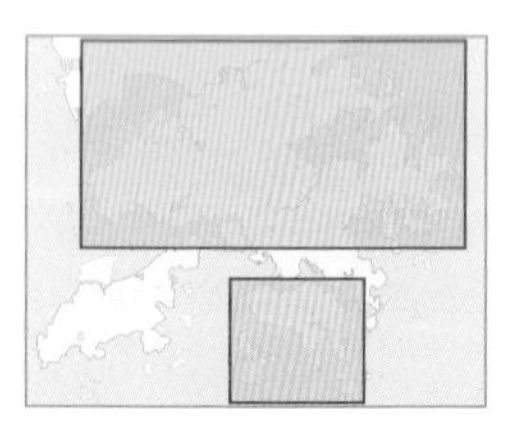

上體灰褐色，前額偏白，尾羽黑色而兩旁有白斑，翼橄欖色。下體灰色，胸部橙色。嘴和腳黑色。

Greyish brown upperparts. Whitish forehead. Black tail with white patch in both sides. Wings are olive. Grey underparts and orange throat-patch. Black bill and legs.

1 adult male 雄成鳥：Pok Fu Lam 薄扶林：Feb-10, 10 年 2 月：Michelle and Peter Wong 江敏兒、黃理沛
2 adult female 雌成鳥：Tai Po Kau 大埔滘：Feb-08, 08 年 2 月：Andy Kwok 郭匯昌
3 adult female 雌成鳥：Tai Po Kau 大埔滘：Feb-08, 08 年 2 月：Andy Kwok 郭匯昌
4 adult female 雌成鳥：Tai Po Kau 大埔滘：Feb-08, 08 年 2 月：Andy Kwok 郭匯昌
5 adult female 雌成鳥：Po Toi 蒲台：Nov-09, 09 年 11 月：Michelle and Peter Wong 江敏兒、黃理沛
6 adult female 雌成鳥：Po Toi 蒲台：Nov-09, 09 年 11 月：Michelle and Peter Wong 江敏兒、黃理沛
7 adult male 雄成鳥：Pok Fu Lam 薄扶林：Feb-10, 10 年 2 月：Michelle and Peter Wong 江敏兒、黃理沛

春季過境遷徙鳥 Spring Passage Migrant	夏候鳥 Summer Visitor	秋季過境遷徙鳥 Autumn Passage Migrant	冬候鳥 Winter Visitor

常見月份 1 2 3 4 5 6 7 8 9 10 11 12

留鳥 Resident	迷鳥 Vagrant	偶見鳥 Occasional Visitor

2
3
4
5
6
7

紅胸姬鶲

㊟ hóng xiōng jī wēng

體長 length：11.5cm

Red-breasted Flycatcher | *Ficedula parva*

其他名稱 Other names：紅胸鶲

1

上體灰褐色，喉偏白，嘴基淡色，雄鳥額上有白色幼紋，頸部有偏紅褐色頸紋，翼橄欖褐色，下體淡色，尾羽深褐色而兩旁有白斑。嘴和腳黑色。

Greyish-brown upperparts. Whitish throat. The base of bill is paler. Adult male has narrow white strip at forehead, and rufous-brown strip at neck. Olive-brown wings. Pale underparts. Dark brown tail with white patch in both sides. Black bill and legs.

1 1st winter male 第一年冬天雄鳥；Po Toi 蒲台；Nov-07, 07 年 11 月；James Lam 林文華
2 1st summer female 第一年夏天雌鳥；Po Toi 蒲台；Apr-08, 08 年 4 月；Cherry Wong 黃卓研
3 1st winter female 第一年冬天雌鳥；23-Dec-07, 07 年 12 月 23 日；Michelle and Peter Wong 江敏兒、黃理沛
4 1st winter female 第一年冬天雌鳥；Shek Kong 石崗；Jan-09, 09 年 1 月；Michelle and Peter Wong 江敏兒、黃理沛
5 1st winter female 第一年冬天雌鳥；25-Dec-07, 07 年 12 月 25 日；Michelle and Peter Wong 江敏兒、黃理沛
6 adult female 雌成鳥；Po Toi 蒲台；Apr-05, 05 年 4 月；Michelle and Peter Wong 江敏兒、黃理沛
7 1st winter female 第一年冬天雌鳥；Shek Kong 石崗；Jan-09, 09 年 1 月；Michelle and Peter Wong 江敏兒、黃理沛

	春季過境遷徙鳥 Spring Passage Migrant				夏候鳥 Summer Visitor		秋季過境遷徙鳥 Autumn Passage Migrant			冬候鳥 Winter Visitor		
常見月份	1	2	3	4	5	6	7	8	9	10	11	12

留鳥 Resident	迷鳥 Vagrant	偶見鳥 Occasional Visitor

2
3
4
5
6

7

紅喉姬鶲

㊟ hóng hóu jī wēng

體長 length：11.5cm

Red-throated Flycatcher | *Ficedula albicilla*

其他名稱 Other names：紅喉鶲

體型細小的鶲，大致灰褐色，嘴黑色，下體白色。尾黑色但基部外側羽毛白色，尾部經常上下擺動。喉部灰褐色，繁殖時中央部分會變為橙紅色，兩側灰色。

A small greyish-brown flycatcher. Blackish bill. Underparts whitish. Black tail with white base on outer tail feathers. Often flicks tail. Throat brownish-grey when not breeding, but turns to orange-red with grey on sides in breeding plumage.

1 non-breeding adult male 非繁殖羽雄成鳥；Kam Tin 錦田；Mar-07, 07 年 3 月；Martin Hale 夏敖天
2 1st winter 第一年冬天；Shek Kong 石崗；Jan-09, 09 年 1 月；Michelle and Peter Wong 江敏兒、黃理沛
3 adult female 雌成鳥；Pui O 貝澳；Feb-08, 08 年 2 月；Chan Kai Wai 陳佳瑋
4 1st winter 第一年冬天；Shek Kong 石崗；Jan-09, 09 年 1 月；Cherry Wong 黃卓研
5 adult female 雌成鳥；Jacky Chan 陳家華
6 1st winter 第一年冬天；Jacky Chan 陳家華
7 1st winter 第一年冬天；Mai Po 米埔；27-Dec-03, 03 年 12 月 27 日；Michelle and Peter Wong 江敏兒、黃理沛

春季過境遷徙鳥 Spring Passage Migrant			夏候鳥 Summer Visitor			秋季過境遷徙鳥 Autumn Passage Migrant			冬候鳥 Winter Visitor		
常見月份 1	2	3	4	5	6	7	8	9	10	11	12
留鳥 Resident				迷鳥 Vagrant				偶見鳥 Occasional Visitor			

2
3
4
5
6
7

白眉藍姬鶲

㊟ bái méi lán jī wēng

體長 length：11.5-12cm

Ultramarine Flycatcher | *Ficedula superciliaris*

其他名稱 Other names：白眉藍鶲

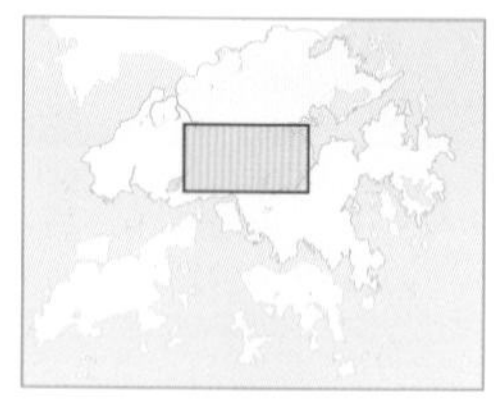

頭、上體具有藍色光澤。頸側和胸側有黑斑。飛羽和尾羽黑色。下體白色，尾基有白色小斑。雌鳥上體灰褐，下體沾淡黃。首年度冬鳥全身淡色，上體、腰及尾有輝藍色光澤，翼上覆羽淡色，有明顯白色翼斑。

Head and upperparts has shiny blue colour. Both sides of neck and breast have black colour patches. Underparts white with small white patches at the base of tail. Female bird has greyish brown upperparts, and buff yellow underparts. First witner bird has plain face. Upperparts overall pale grey. Rump and sides of the tail has shining cobalt blue. Pale blue wing coverts with prominent white wing bar.

1 male 雄鳥：Kadoorie Farm 嘉道理農場：Apr 07, 07 年 4 月：Michelle and Peter Wong 江敏兒、黃理沛
2 1st winter male 第一年冬天雄鳥：Kadoorie Farm 嘉道理農場：Dec-06, 06 年 12 月：Maison Fung 馮振威
3 1st winter male 第一年冬天雄鳥：Kadoorie Farm 嘉道理農場：Dec-06, 06 年 12 月：Maison Fung 馮振威

春季過境遷徙鳥 Spring Passage Migrant			夏候鳥 Summer Visitor			秋季過境遷徙鳥 Autumn Passage Migrant			冬候鳥 Winter Visitor		
常見月份 1	2	3	4	5	6	7	8	9	10	11	12

留鳥 Resident	迷鳥 Vagrant	偶見鳥 Occasional Visitor

黑喉紅尾鴝

[普] hēi hóu hóng wěi qú
[粵] 鴝：音渠

體長 length：14-19cm

Hodgson's Redstart | *Phoenicurus hodgsoni*

雄鳥上體大致灰色。面部到胸部黑色，下體至尾部紅褐色。雄鳥與北紅尾鴝很相似，可以其較不突出的頭部、較灰的體羽及較小的翼上白斑區分。雌鳥翼上沒有白斑。與赭紅尾鴝的分別是雄性黑喉紅尾鴝整體較灰，翼上的白斑更是主要特徵。

Male is generally grey upperparts, black from face to breast, rufous underparts and tail. Male is similar to Daurian Redstart, but with a less distinct crown, greyer mantle, and smaller white patch on the wing. Female has no white on the wing. To separate from Black Redstart, male Hodgson's Redstart is greyer overall, and the white spot on the wing is the best field mark.

[1] female / 1st winter male or female 雌鳥 / 第一年冬天雄鳥或雌鳥：Po Toi 蒲台：Geoff Welch

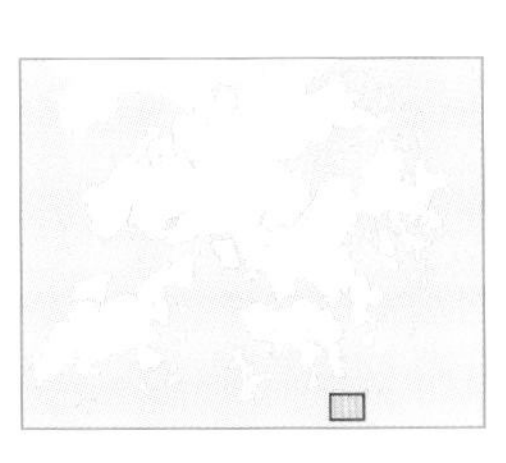

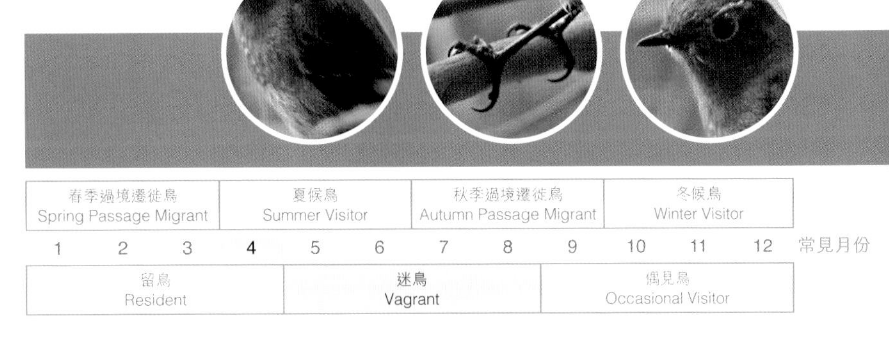

春季過境遷徙鳥 Spring Passage Migrant			夏候鳥 Summer Visitor			秋季過境遷徙鳥 Autumn Passage Migrant			冬候鳥 Winter Visitor			
1	2	3	**4**	5	6	7	8	9	10	11	12	常見月份
留鳥 Resident				**迷鳥 Vagrant**				偶見鳥 Occasional Visitor				

赭紅尾鴝

[普] zhě hóng wěi qú
[粵] 赭：音者；鴝：音渠

體長 length：14-15cm

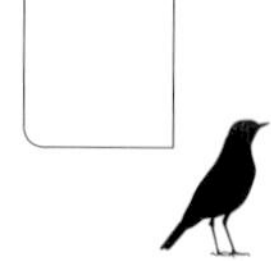

Black Redstart | *Phoenicurus ochruros*

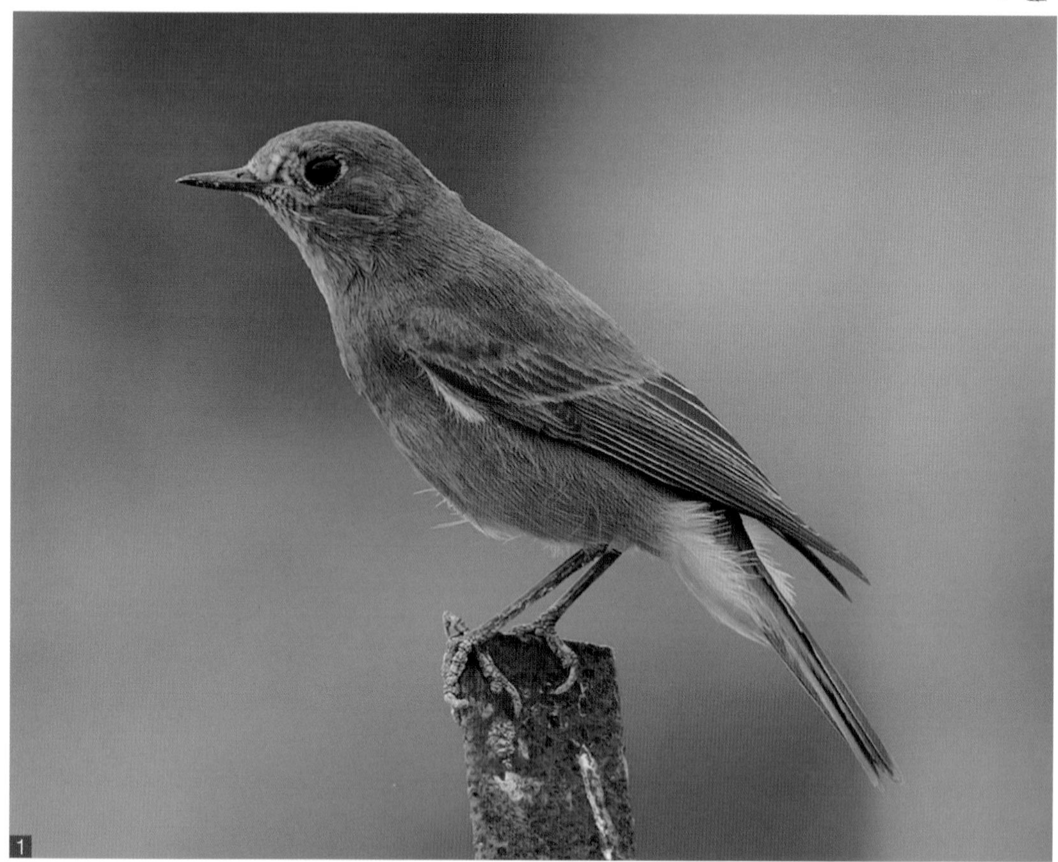

1

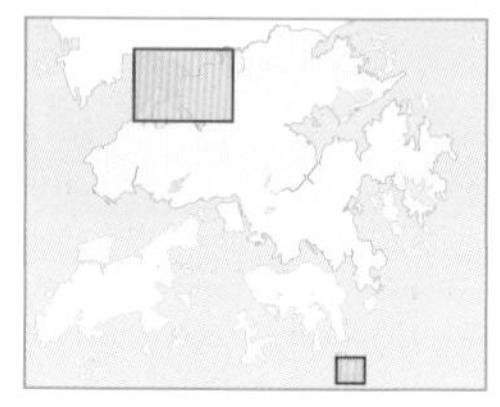

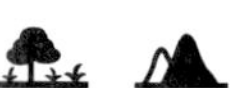

貌似常見的北紅尾鴝但沒有白色翼斑。雄鳥上體及喉至上胸大致黑色，下胸至尾下橙色。雌鳥大致橄欖褐色，下體較淺色，外側尾羽橙色。嘴和腳黑色。

Resembles common Daurian Redstart but lacks white wing patches. Male has mainly black upperparts and black from throat to upper breast. Orange from lower breast to undertail. Female has mainly olive brown body with paler underparts. Orange outer tail feathers. Black bill and legs.

1 adult female 雌成鳥；Long Valley 塱原；Nov-17, 17 年 11 月；Allen Chan 陳志雄
2 adult female 雌成鳥；Long Valley 塱原；Nov-17, 17 年 11 月；Kan Chi Ming 簡志明
3 adult female 雌成鳥；Long Valley 塱原；Nov-17, 17 年 11 月；Kan Chi Ming 簡志明
4 adult female 雌成鳥；Long Valley 塱原；Nov-17, 17 年 11 月；Kan Chi Ming 簡志明
5 adult female 雌成鳥；Long Valley 塱原；Dec-17, 17 年 12 月；Ng Sze On 吳思安

	春季過境遷徙鳥 Spring Passage Migrant			夏候鳥 Summer Visitor			秋季過境遷徙鳥 Autumn Passage Migrant			冬候鳥 Winter Visitor		
常見月份	1	2	3	4	5	6	7	8	9	10	11	12
	留鳥 Resident				迷鳥 Vagrant				偶見鳥 Occasional Visitor			

2
3
4
5

北紅尾鴝

[普]bĕi hóng wĕi qú
[粵]鴝：音渠

體長 length：15cm

Daurian Redstart | *Phoenicurus auroreus*

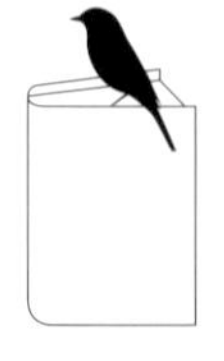

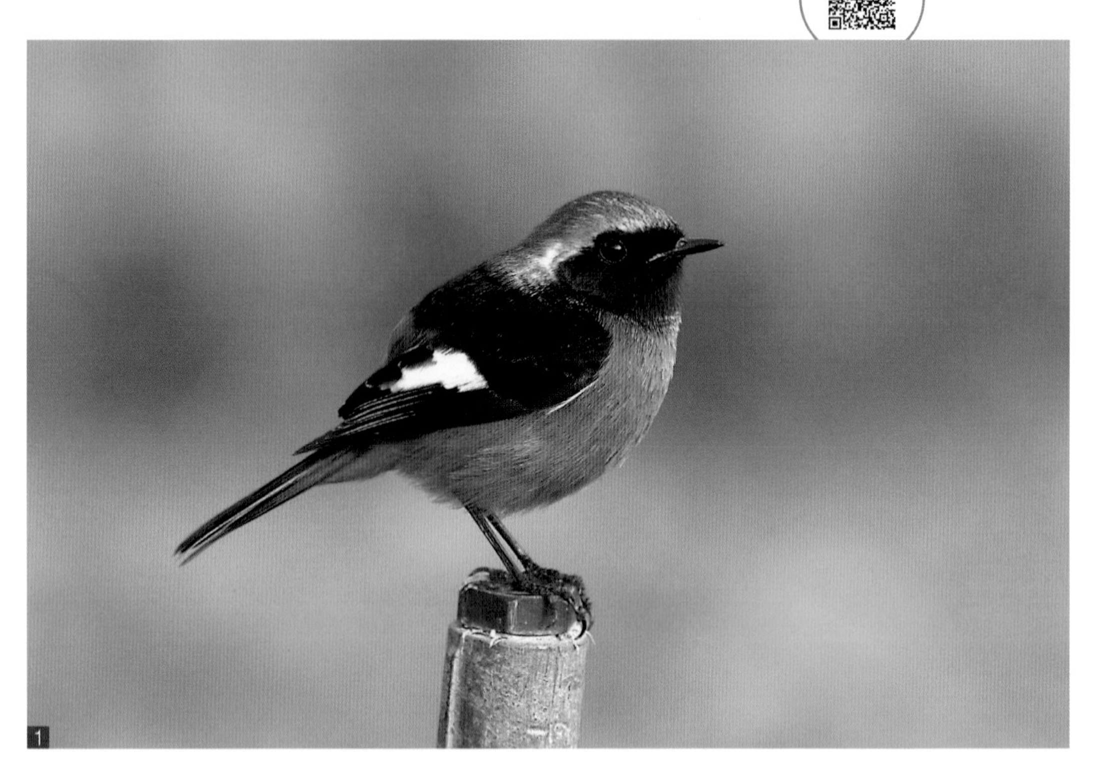

1

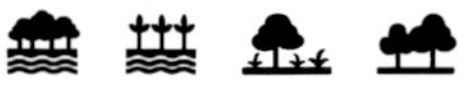

雌鳥和雄鳥共同的特徵是腰及尾羽紅褐色，有明顯白色翼斑。嘴和腳黑色。雄鳥下體為特別的橙栗色，上體及面部黑色，頭及後枕銀灰色。雌鳥主要為橄欖褐色。常顫動尾巴。

Rufous rump and tail, prominent white wing patches are characteristics of both sexes. Black bill and legs. Male has distinctive rufous orange underparts, black upperparts and face, silvery-grey crown and nape. Female is olive-brown instead. Shivers tail constantly.

1 adult male 雄成鳥：Sai Kung 西貢：Jan-09, 09 年 1 月：Aka Ho
2 adult female 雌成鳥：Mai Po 米埔：Dec-06, 06 年 12 月：Matthew and TH Kwan 關朗曦、關子凱
3 adult female 雌成鳥：Lion Rock Counlry Park 獅子山郊野公園：Jan-07, 07 年 1 月：Martin Hale 夏敖天
4 adult male 雄成鳥：Mai Po 米埔：9-Jan-07, 07 年 1 月 9 日：Christina Chan 陳燕明
5 adult male 雄成鳥：Po Toi 蒲台：3-Mar-06, 06 年 3 月 3 日：Doris Chu 朱詠兒
6 adult male 雄成鳥：Mar-06, 06 年 3 月：Lee Kai Hong 李啟康
7 adult male 雄成鳥：Danny Ho 何國海
8 adult female 雌成鳥：Po Toi 蒲台：Feb-08, 08 年 2 月：Cherry Wong 黃卓研

	春季過境遷徙鳥 Spring Passage Migrant			夏候鳥 Summer Visitor			秋季過境遷徙鳥 Autumn Passage Migrant			冬候鳥 Winter Visitor		
常見月份	1	2	3	4	5	6	7	8	9	10	11	12
	留鳥 Resident				迷鳥 Vagrant				偶見鳥 Occasional Visitor			

紅尾水鴝

[普]hóng wěi shuǐ qú
[粵]鴝：音渠

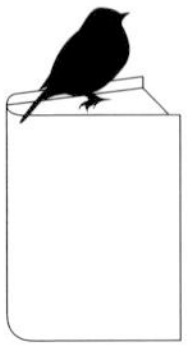

體長length：12-13cm

Plumbeous Water Redstart | *Phoenicurus fuliginosus*

其他名稱Other names：鉛色水鴝

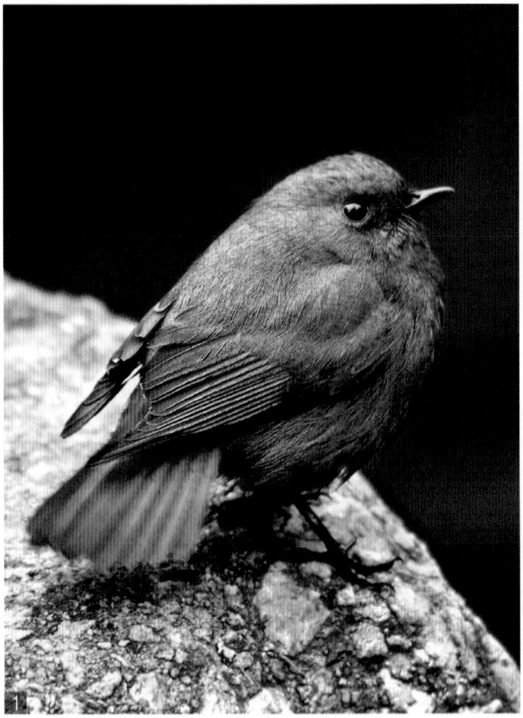
1

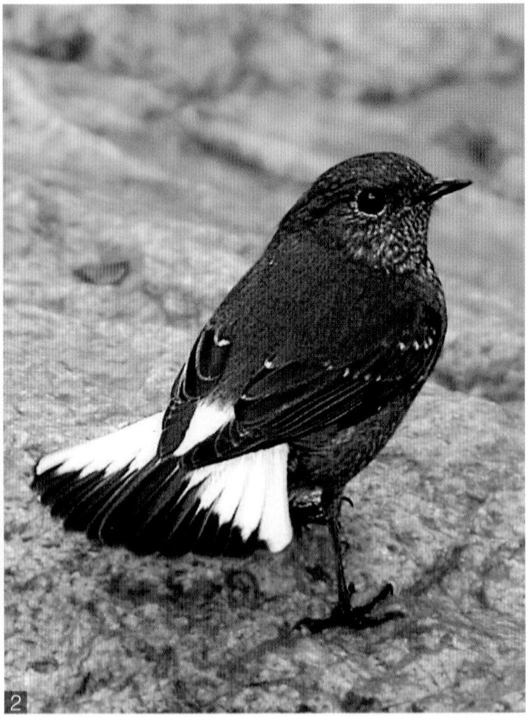
2

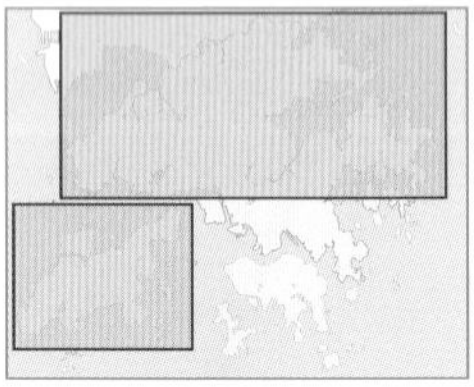

🌊

嘴黑色，腳深色。雄鳥為獨特的石板藍色，腰、臀及尾栗色。雌鳥及未成年鳥上體灰褐色，下體灰色，有淡色的鱗狀紋，腰、臀及外側尾羽基部皆為白色。喜愛停在溪澗中的石上，不斷開合尾羽。

Black bill and dark legs. Male bird has unmistakable with slaty-blue body and chestnut rump, vent and tail. Female and immature birds have brownish grey upperparts, grey underparts with grey scaly pattern. Rump, vent and outer tail base are white. Usually perches on rocks in mid-stream, fanning its conspicuous tail.

[1] adult male 雄成鳥；Kowloon Bay 九龍灣；Dec-06, 06 年 12 月；Andy Kwok 郭匯昌
[2] adult female 雌成鳥；Siu Lek Yuen 小瀝源；Feb-03, 03 年 2 月；Henry Lui 呂德恒
[3] adult female 雌成鳥；Tai Po 大埔；Jan-07, 07 年 1 月；Michelle and Peter Wong 江敏兒、黃理沛
[4] adult male 雄成鳥；Ngau Tau Kok 牛頭角；Owen Chiang 深藍
[5] adult female 雌成鳥；Tai Om 大庵；Feb-07, 07 年 2 月；Danny Ho 何國海
[6] adult male 雄成鳥；Chung Mei 涌尾；Feb-05, 05 年 2 月；Michelle and Peter Wong 江敏兒、黃理沛
[7] adult female 雌成鳥；Andy Cheung 張玉良
[8] adult male 雄成鳥；Kowloon Bay 九龍灣；Dec-06, 06 年 12 月；Pippen Ho 何志剛

春季過境遷徙鳥 Spring Passage Migrant	夏候鳥 Summer Visitor	秋季過境遷徙鳥 Autumn Passage Migrant	冬候鳥 Winter Visitor

常見月份	1	2	3	4	5	6	7	8	9	10	11	12

留鳥 Resident	迷鳥 Vagrant	偶見鳥 Occasional Visitor

3
4
5
6
7
8

藍磯鶇

㊟ lán jī dōng
㊟ 磯鶇：音基東

體長 length：20-23cm

Blue Rock Thrush | *Monticola solitarius*

嘴及腳黑色。*pandoo*亞種雄鳥全身閃亮藍色，與*philippensis*亞種的上體羽色相同，但腹部深栗色，兩種都有淡黑及近白色的鱗狀斑紋。雌鳥上體由深褐至灰藍都有，下體密佈黑色鱗狀斑紋。常直立於岩石上或屋頂。

Black bill and legs. Males range from all glossy blue (race *pandoo*) to glossy blue with a deep chestnut belly (race *philippensis*), with whitish and black scales on body. Females have dark brown to greyish blue upperparts with paler scaly underparts. Usually stands upright on exposed rocks or buildings.

1 adult male, race *philippensis* 雄成鳥：*philippensis* 亞種：Tuen Mun 屯門：Dec-08, 08 年 12 月：Kitty Koo 古愛婉
2 female 雌鳥：Aberdeen 香港仔：Feb-07, 07 年 2 月：Andy Kwok 郭匯昌
3 adult male, race *philippensis* 雄成鳥, *philippensis* 亞種：Tuen Mun 屯門：Jan-09, 09 年 1 月：Danny Ho 何國海
4 juvenile male, race *pandoo* 雄幼鳥, *pandoo* 亞種：Plover Cove 船灣淡水湖：Jan-07, 07 年 1 月：Martin Hale 夏敖天
5 adult male, race *philippensis* 雄成鳥, *philippensis* 亞種：Tuen Mun 屯門：Jan-09, 09 年 1 月：Andy Kwok 郭匯昌
6 juvenile male, race *philippensis* 雄幼鳥, *philippensis* 亞種：Po Toi 蒲台：Sep-06, 06 年 9 月：Allen Chan 陳志雄
7 female 雌鳥：Tung Chung 東涌：Dec-08, 08 年 12 月：Lau Chu Kwong 劉柱光

	春季過境遷徙鳥 Spring Passage Migrant			夏候鳥 Summer Visitor			秋季過境遷徙鳥 Autumn Passage Migrant			冬候鳥 Winter Visitor		
常見月份	1	2	3	4	5	6	7	8	9	10	11	12
	留鳥 Resident				迷鳥 Vagrant				偶見鳥 Occasional Visitor			

3
4
5
6

7

栗腹磯鶇

[普]lì fù jī dōng
[粵]磯鶇：音基東

體長length：21-23cm

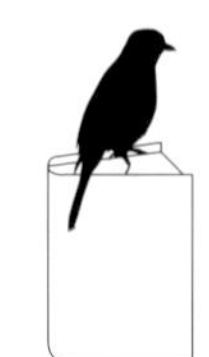

Chestnut-bellied Rock Thrush | *Monticola rufiventris*

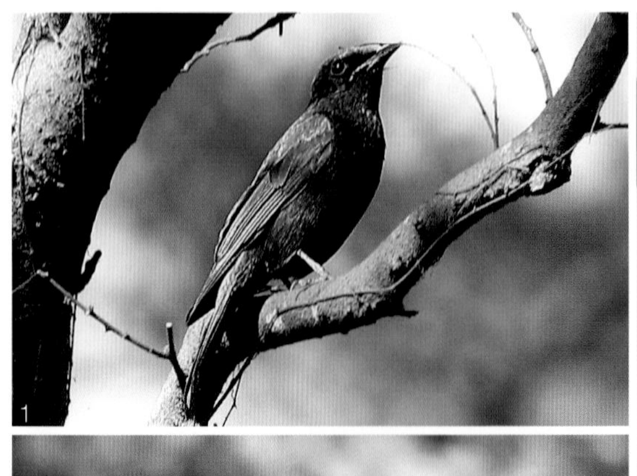

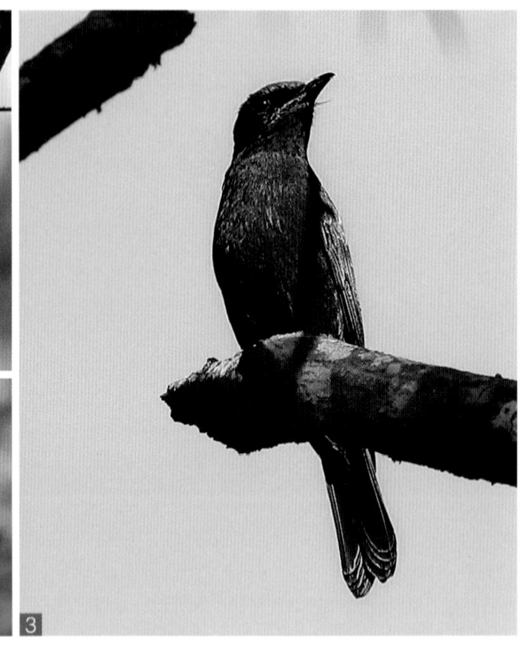

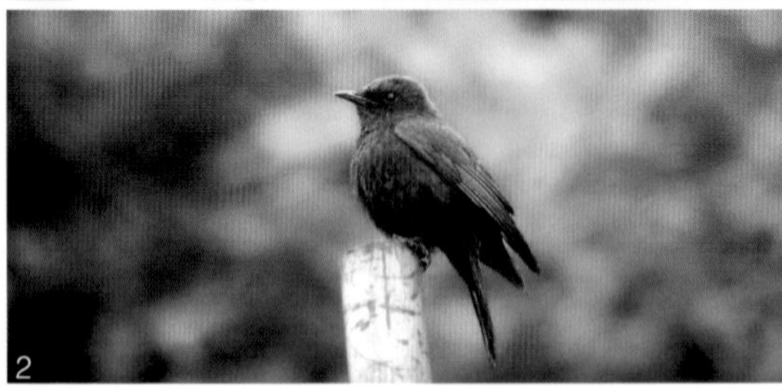

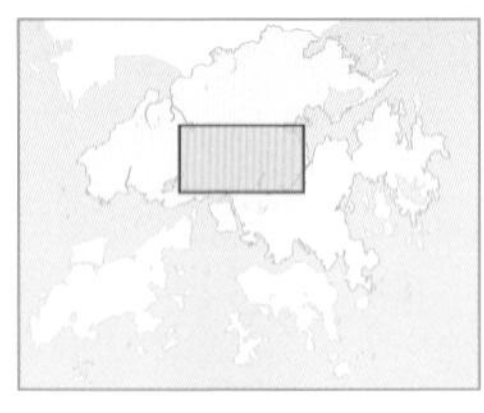

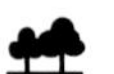

嘴和腳深色。雄鳥前額鮮藍色，和深黑的臉部成對比，腹部栗色延長至喉部，可與紅腹的藍磯鶇亞種*philippensis*區分。雌鳥褐色，頸部有皮黃色月牙形斑，下體滿佈深褐色的扇貝形斑紋。未成年鳥褐黃色，有黑色鱗紋。常站在樹頂，緩慢上下擺動尾巴。

Dark bill and legs. Male distinguished from Blue Rock Thrush race *philippensis* by contrasting black face, bright blue forehead and chestnut underparts up to the throat. Female has brown upperparts with buffy crescent on sides of neck and brown scaly underparts. Immatures are rusty buff with bold black scales. Perches on tree tops and slowly flicks its tail.

1 adult male 雄成鳥；Kadoorie Farm 嘉道理農場；Mar-05, 05 年 3 月；Michelle and Peter Wong 江敏兒、黃理沛
2 adult male 雄成鳥；Kwokjai 張振國
3 adult male 雄成鳥；Kadoorie Farm 嘉道理農場；Mar-05, 05 年 3 月；Michelle and Peter Wong 江敏兒、黃理沛
4 adult female 雌成鳥；Kadoorie Farm 嘉道理農場；Feb-07, 07 年 2 月；Danny Ho 何國海

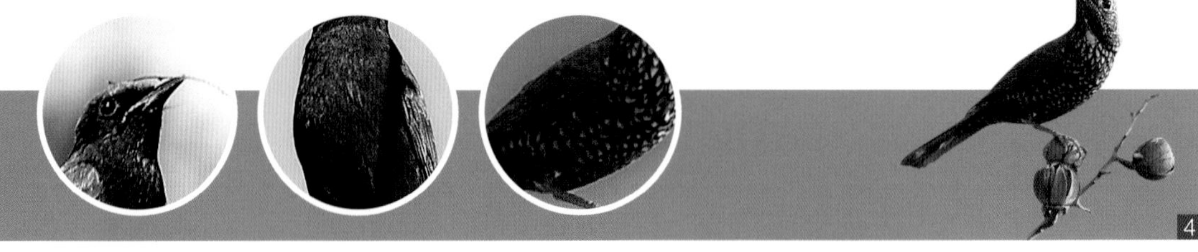

春季過境遷徙鳥 Spring Passage Migrant	夏候鳥 Summer Visitor	秋季過境遷徙鳥 Autumn Passage Migrant	冬候鳥 Winter Visitor

常見月份	1	2	3	4	5	6	7	8	9	10	11	12

留鳥 Resident	迷鳥 Vagrant	偶見鳥 Occasional Visitor

白喉磯鶇

[普] bái hóu jī dōng
[粵] 磯鶇：音基東

體長 length：16-19cm

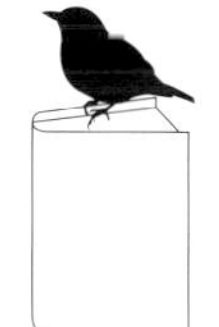

White-throated Rock Thrush | *Monticola gularis*

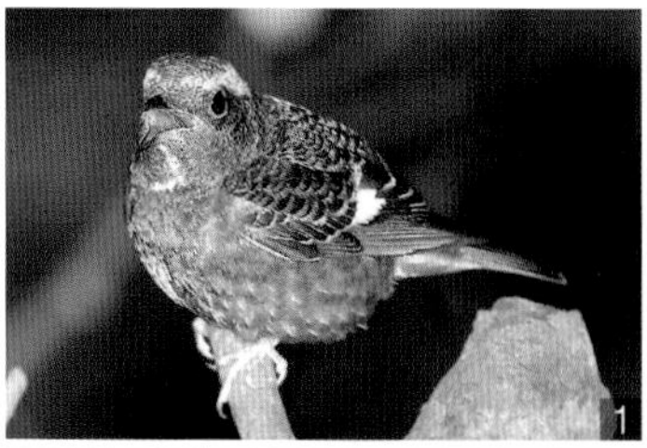
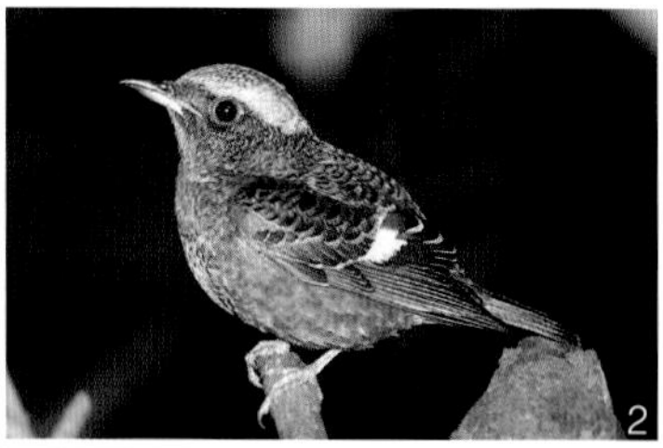
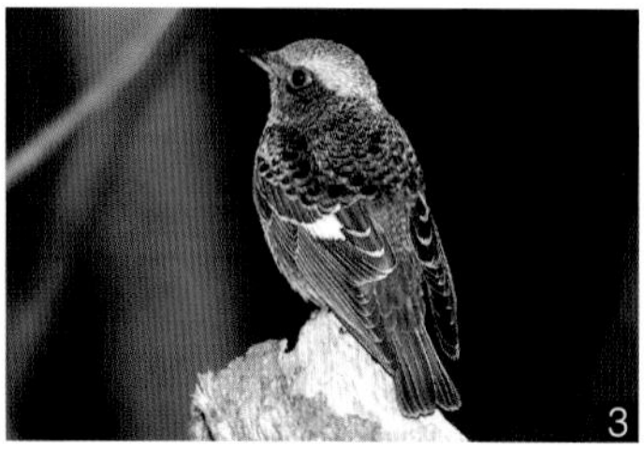
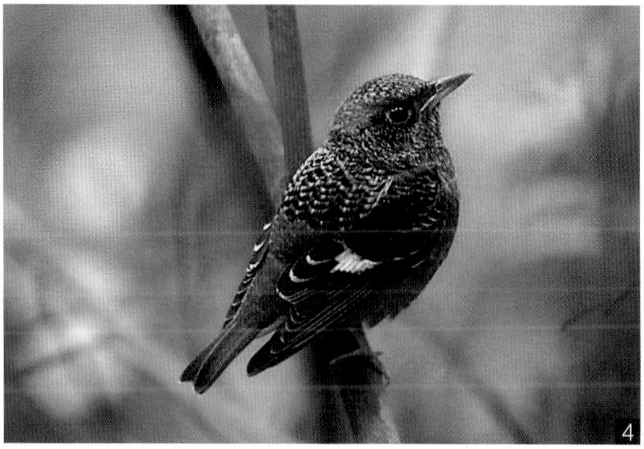
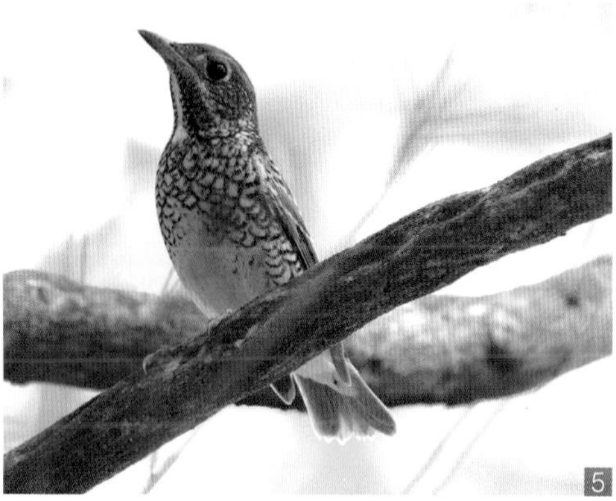

有顯眼白喉，嘴短、尾短，上體深褐色，具有黃褐色羽緣。雄鳥頭頂至後枕石板藍色，耳羽黑色，眼先和面頰橙色；翼上覆羽具有藍斑，次級飛羽基部有明顯白點；胸及腰部橙色，腹部和脇部顏色較淡。雌鳥頭部及耳羽深褐色，眼先淡褐，下體白色，面頰、胸部、脇部及腰部具有明顯月形斑紋。

Distinctive white throat, short bill, short tail. Upperparts dark brown, with yellowish feather fringe. Male bird has dark blue crown to hind nape, dark ear coverts, orange lores and cheek. It has distinctive blue patch on lesser covert, and white spot on the base of secondaries. Orange breast and rump, with paler color at belly and flank. Female has darkish brown crown and ear coverts, and pale lores. Underparts whitish, with brownish crescent pattern at cheek, breast, flank and rump.

[1] adult non-breeding / 1st winter male 非繁殖羽成鳥 / 第一年冬天雄鳥：Tai Po Kau 大埔滘：Jan-05, 05 年 1 月：Michelle and Peter Wong 江敏兒、黃理沛
[2] adult non-breeding / 1st winter male 非繁殖羽成鳥 / 第一年冬天雄鳥：Tai Po Kau 大埔滘：Jan-05, 05 年 1 月：Michelle and Peter Wong 江敏兒、黃理沛
[3] adult non-breeding / 1st winter male 非繁殖羽成鳥 / 第一年冬天雄鳥：Tai Po Kau 大埔滘：Jan-05, 05 年 1 月：Michelle and Peter Wong 江敏兒、黃理沛
[4] adult non-breeding / 1st winter male 非繁殖羽成鳥 / 第一年冬天雄鳥：Tai Po Kau 大埔滘：8-Jan-05, 05 年 1 月 8 日：Doris Chu 朱詠兒
[5] adult female 雌成鳥：Ho Man Tin 何文田：Oct-19, 19 年 10 月：Roman Lo 羅文凱

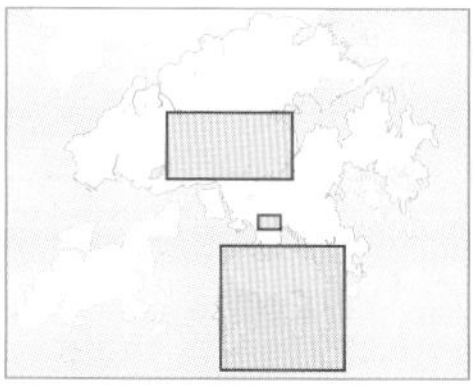
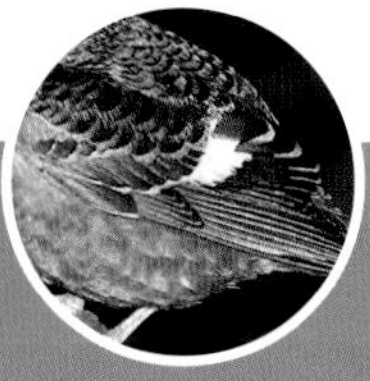
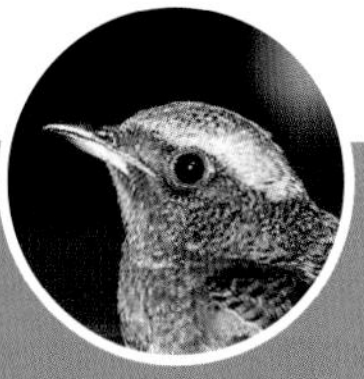

春季過境遷徙鳥 Spring Passage Migrant	夏候鳥 Summer Visitor	秋季過境遷徙鳥 Autumn Passage Migrant	冬候鳥 Winter Visitor

1 2 3 4 5 6 7 8 9 **10** **11** 12 常見月份

留鳥 Resident	迷鳥 Vagrant	偶見鳥 Occasional Visitor

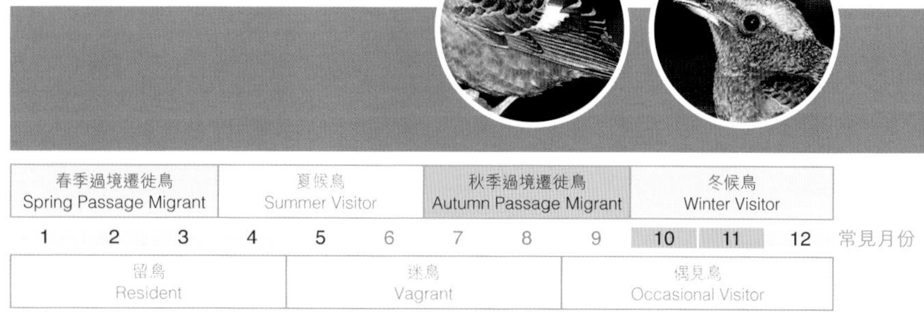

黑喉石䳭

㊟ hēi hóu shí jí
㊟ 䳭：音即

體長 length：12.5cm

Stejneger's Stonechat | *Saxicola stejnegeri*

其他名稱 Other names：Common Stonechat

1

2

嘴和腳深色。雄鳥頭部及飛羽明顯黑色，上體深褐，頸、翼及尾上覆羽有白斑，下體淡褐色。雌鳥及幼鳥頭及上體黑色為淡棕色取代。飛行時翼斑白色較明顯。叫聲像石頭互碰的「即—即—」。常佇立於灌叢頂、圍欄及電線高處，非常顯眼。

Dark bill and legs. Male has distinctive black head and dark brown upperparts, with white patches on neck, wings and upper tail coverts. Underparts are buffy brown. On female and immature black on head and upperparts is replaced by pale brown. White wing patches only seen in flight. Call is an occasional "tack". Perches prominently on shrubs, fences or wires.

1 breeding male 繁殖羽雄鳥：Long Valley 塱原：Nov-08, 08 年 11 月：James Lam 林文華
2 adult female 雌成鳥：Long Valley 塱原：Nov-04, 04 年 11 月：Jemi and John Holmes 孔思義、黃亞萍
3 non-breeding adult male 非繁殖羽雄成鳥：Long Valley 塱原：Oct-04, 04 年 10 月：Henry Lui 呂德恆
4 breeding male 繁殖羽雄鳥：HK Wetland Park 香港濕地公園：Jan-07, 07 年 1 月：Law Kam Man 羅錦文
5 non-breeding adult male 非繁殖羽雄成鳥：Long Valley 塱原：Nov-08, 08 年 11 月：Ken Fung 馮漢城
6 adult female 雌成鳥：Long Valley 塱原：4-Nov-08, 08 年 11 月 4 日：Sung Yik Hei 宋亦希
7 non-breeding adult male 非繁殖羽雄成鳥：Long Valley 塱原：Dec-06, 06 年 12 月：Cherry Wong 黃卓研
8 non-breeding adult male 非繁殖羽雄成鳥：Long Valley 塱原：Nov-08, 08 年 11 月：Danny Ho 何國海

春季過境遷徙鳥 Spring Passage Migrant	夏候鳥 Summer Visitor	秋季過境遷徙鳥 Autumn Passage Migrant	冬候鳥 Winter Visitor

常見月份	1	2	3	4	5	6	7	8	9	10	11	12

留鳥 Resident	迷鳥 Vagrant	偶見鳥 Occasional Visitor

3
4
5
6
7

8

灰林鵖

㊟ huī lín jí
㊟ 鵖：音即

體長length：14-15cm

Grey Bush Chat | *Saxicola ferreus*

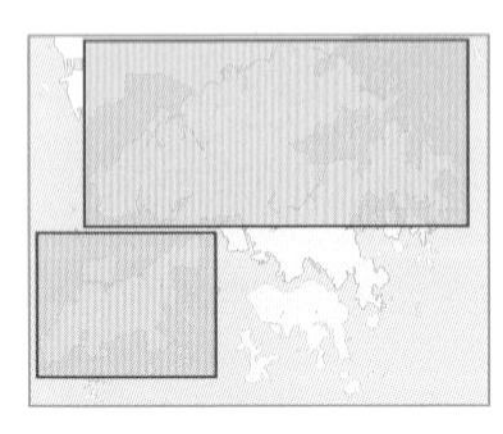

嘴短黑色，頭部深灰色，面頰黑色，有顯眼白色眉紋。雄鳥上體、翼和尾部深灰色，有淡色斑紋，喉至下體白色，胸部沾灰，腳黑色。雌鳥全身偏褐，腰部及尾羽兩側紅褐色。

Bill short and black. Head dark grey with black cheeks and prominent white supercilium. Upperparts, wings and tail dark grey, with pale narrow stripes. Throat to underparts white, breast tinted grey. Black legs. Female is overall brownish, with reddish brown rump and outertail feathers.

1 1st winter male 第一年冬天雄鳥 Mar-07, 07 年 3 月；Lee Kai Hong 李啟康
2 adult female 雌成鳥；Mai Po 米埔；Francis Chu 崔汝棠
3 adult female 雌成鳥；Mai Po 米埔；Francis Chu 崔汝棠
4 1st winter male 第一年冬天雄鳥；Po Toi 蒲台；Dec-07, 07 年 12 月；Michelle and Peter Wong 江敏兒、黃理沛
5 adult male 雄成鳥；Mui Wo 梅窩；Feb-07, 07 年 2 月；Maison Fung 馮振威
6 adult male and female 雄成鳥和雌成鳥；Tai Mo Shan 大帽山；Oct-06, 06 年 10 月；Henry Lui 呂德恒
7 adult female 雌成鳥；Tai Mo Shan 大帽山；Oct-06, 06 年 10 月；Henry Lui 呂德恒
8 1st winter male 第一年冬天雄鳥；Kwai Chung 葵涌；Mar-07, 07 年 3 月；Pippen Ho 何志剛

春季過境遷徙鳥 Spring Passage Migrant			夏候鳥 Summer Visitor			秋季過境遷徙鳥 Autumn Passage Migrant			冬候鳥 Winter Visitor		
1	2	3	4	5	6	7	8	9	10	11	12

常見月份

留鳥 Resident	迷鳥 Vagrant	偶見鳥 Occasional Visitor

2
3
4
5
6
7

8

棕腹仙鶲

㊟ zōng fù xiān wēng

體長 length：15-18cm

Rufous-bellied Niltava | *Niltava sundara*

大型鶲。雄鳥及雌鳥外型與棕腹大仙鶲很相似。雄鳥上體藍色，面黑色，頭頂、頸側、肩、腰及尾的羽毛均為閃亮的藍色。胸、腹及臀部為鮮橙褐色。與棕腹大仙鶲雄鳥的區別在胸部均勻的鮮橙褐色延伸至臀部，頭部的閃亮藍色由前額延伸至後枕。雌鳥上體與下體均是褐色，尾下覆羽褐色較淺，腰及尾紅褐色，頸前有一道偏白彎月紋，頸兩側各有一小塊藍色斑。與棕腹大仙鶲雌鳥的區別在胸部均勻的褐色延伸至下體。腳灰色。喜在林區陰暗下層活動。所有記錄被判斷為逸鳥。

Large-sized flycatcher. Resembles a Fujian Niltava both male and female. Male bird has blue upperpart and black face. Feathers on head, sides of neck, shoulders, rump and tail are shiny blue. Breast to vent are bright orange rufous. Male distinguished from male Fujian Niltava by shiny blue from the forehead to the nape and uniform bright orange rufous from breast to underpart even to the vent. Female bird mainly brown. Undertail coverts are pale brown. Rump and tail are rufous. A whitish crescent across the neck and a small blue patch on both sides of the neck. Female distinguished from female Fujian Niltava by uniform brown from breast to underpart. Grey legs. Prefers shaded area of lower storey. All records are considered as birds escaped from captivity.

1 Tai Po Kau 大埔滘：16-Feb-08, 08 年 2 月 16 日：Yue Pak Wai 余柏維
2 Tai Po Kau 大埔滘：16-Feb-08, 08 年 2 月 16 日：Yue Pak Wai 余柏維

註：雖然缺乏更多圖片作參考，但按圖片判斷極有可能是棕腹仙鶲。
Note: Although lack of more photos for reference, it is highly probable a Rufous-bellied Niltava accoding to the photo.

春季過境遷徙鳥 Spring Passage Migrant			夏候鳥 Summer Visitor			秋季過境遷徙鳥 Autumn Passage Migrant			冬候鳥 Winter Visitor		
1	2	3	4	5	6	7	8	9	10	11	12
留鳥 Resident				迷鳥 Vagrant				偶見鳥 Occasional Visitor			

常見月份

白斑黑石䳭

㊟bái bān hēi shí jí
㊟䳭：音即

體長length：13-14cm

Pied Bush Chat | *Saxicola caprata*

1

2

全身煙黑色，兩邊翅膀都有一條顯眼的白紋，腰，下腹及翼斑白色，是常見的籠鳥。所有記錄被判斷為逸鳥。

Smoky black body with a white narrow band on each side of the wing. White rump, vent and wing patch. A common captive bird. All records are considered as birds escaped from captivity.

1 male 雄鳥：Kam Tin 錦田：29-Dec-02, 02 年 12 月 29 日：Lo Kar Man 盧嘉孟
2 male 雄鳥：Shek Kip Mei 石硤尾：Jun-03, 03 年 6 月：Cherry Wong 黃卓研
3 male 雄鳥：Shek Kip Mei 石硤尾：Jun-03, 03 年 6 月：Cherry Wong 黃卓研

3

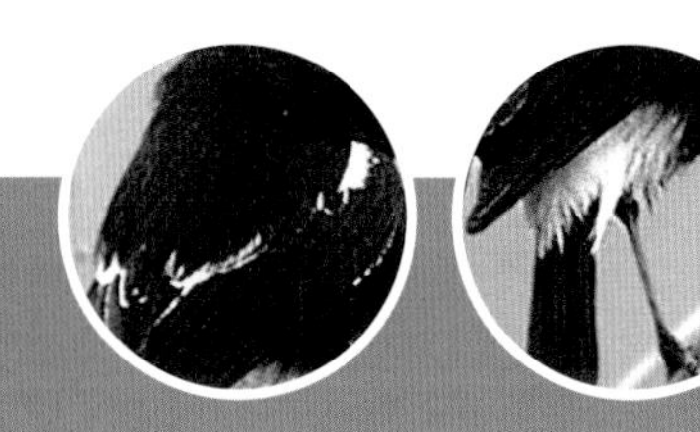

春季過境遷徙鳥 Spring Passage Migrant	夏候鳥 Summer Visitor	秋季過境遷徙鳥 Autumn Passage Migrant	冬候鳥 Winter Visitor

1 2 3 4 5 6 7 8 9 10 11 12 常見月份

留鳥 Resident	迷鳥 Vagrant	偶見鳥 Occasional Visitor

橙腹葉鵯

㊗chéng fù yè bēi
㊣鵯：音卑

體長length：17-19.8cm

Orange-bellied Leafbird | *Chloropsis hardwickii*

1

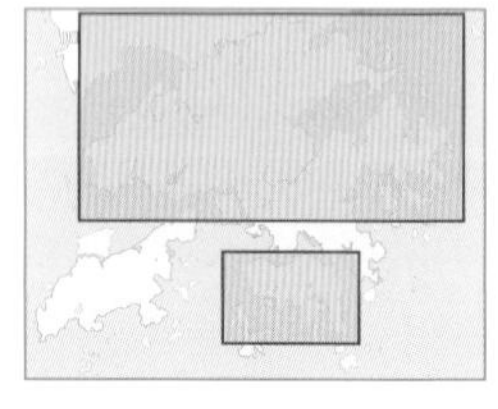

全身大致綠色。雄鳥面頰、喉、嘴和腳黑色，有顯眼的紫銀色頰下紋，翼上有亮藍色肩斑，尾部深藍色，下體橙色。雌鳥和幼鳥全身綠色。叫聲為重複的「chip-eee」，也經常模仿其他林鳥的叫聲。

Mainly bright green. Male has black cheeks, wings are throat, bill and legs, and with prominent silvery lilac moustachial stripe. Wings are blue in colour, tail dark blue. Underparts orange. Female all green apart from faint lilac moustacial and juveniles completely green. Call a repeated "chip-eee". Often imitates other woodland bird songs.

1 male 雄鳥：Tai Po Kau 大埔滘：Feb-07, 07 年 2 月：Allen Chan 陳志雄
2 male 雄鳥：Tai Po Kau 大埔滘：Feb-07, 07 年 2 月：Man Kuen Yat Bill 文權溢
3 male 雄鳥：Tai Po Kau 大埔滘：Mar-08, 08 年 3 月：Eling Lee 李佩玲
4 juvenile 幼鳥：Tai Po Kau 大埔滘：Feb-04, 04 年 2 月：Henry Lui 呂德恒
5 male 雄鳥：Tai Mo Shan 大帽山：Jan-05, 05 年 1 月：Jemi and John Holmes 孔思義、黃亞萍
6 male 雄鳥：Tai Po Kau 大埔滘：Mar-08, 08 年 3 月：Chan Kin Chung Gary 陳建中
7 male 雄鳥：Tai Po Kau 大埔滘：Mar-08, 08 年 3 月：Chan Kin Chung Gary 陳建中

	春季過境遷徙鳥 Spring Passage Migrant			夏候鳥 Summer Visitor			秋季過境遷徙鳥 Autumn Passage Migrant			冬候鳥 Winter Visitor		
常見月份	1	2	3	4	5	6	7	8	9	10	11	12
	留鳥 Resident				迷鳥 Vagrant				偶見鳥 Occasional Visitor			

2
3
4
5
6

7

紅胸啄花鳥

㊟hóng xiōng zhuó huā niǎo

體長 length：7-9cm

Fire-breasted Flowerpecker | *Dicaeum ignipectus*

體型細小。雄鳥臉部和嘴部深色，頭、上體至腰深藍綠色，尾黑色。喉至下體淡黃褐色，胸前有一小片紅斑，紅斑之下有細長黑紋。雌鳥上體深褐色，腹部淡黃褐色，脇部兩旁較深色。叫聲為沙啞而從容的「得－得－」聲。

Small-sized. Male has darkish cheeks and bill. Head, upperparts to rump greenish blue, tail black. Throat to lowerpart pale yellowish brown, with a small red patch on the breast, and darkish stripe below it. Female has yellowish brown upperparts, underparts are buffy, with pale brownish flanks. Call is an unhurried "duk-duk".

1 adult male 雄成鳥；Tai Po Kau 大埔滘；Mar-03, 03 年 3 月；Henry Lui 呂德恒
2 adult male 雄成鳥；Lung Fu Shan 龍虎山；Feb-08, 08 年 2 月；Cheng Nok Ming 鄭諾銘
3 1st summer male 第一年夏天雄鳥；Tuen Mun 屯門；Dec-06, 06 年 12 月；Law Kam Man 羅錦文
4 adult male 雄成鳥；Tai Po Kau 大埔滘；Mar-03, 03 年 3 月；Cherry Wong 黃卓研
5 female / juvenile 雌鳥 / 幼鳥；Tai Po Kau 大埔滘；Jan-04, 04 年 1 月；Cherry Wong 黃卓研
6 1st summer male 第一年夏天雄鳥；Tai Po Kau 大埔滘；Apr-06, 06 年 4 月；Henry Lui 呂德恒
7 female / juvenile 雌鳥 / 幼鳥；Wonderland Villas 華景山莊；Dec-06, 06 年 12 月；Matthew and TH Kwan 關朗曦、關子凱
8 adult male 雄成鳥；Tai Po Kau 大埔滘；Mar-03, 03 年 3 月；Henry Lui 呂德恒

	春季過境遷徙鳥 Spring Passage Migrant			夏候鳥 Summer Visitor			秋季過境遷徙鳥 Autumn Passage Migrant			冬候鳥 Winter Visitor		
常見月份	1	2	3	4	5	6	7	8	9	10	11	12
	留鳥 Resident				迷鳥 Vagrant				偶見鳥 Occasional Visitor			

2
3
4
5
6
7

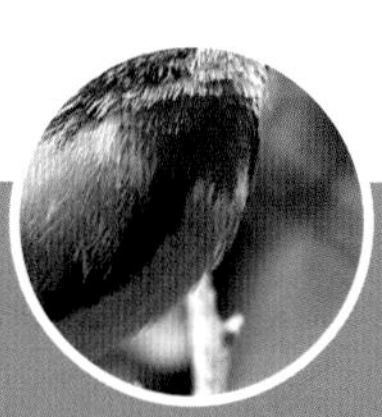

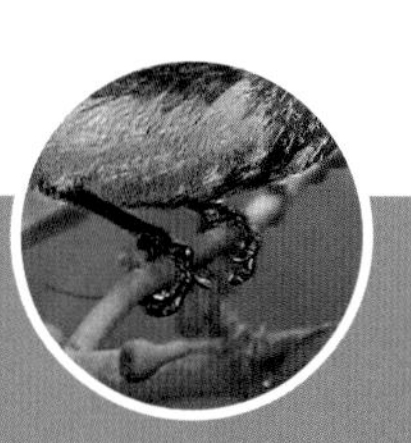

8

朱背啄花鳥

[普] zhū bèi zhuó huā niǎo

體長 length：7-9cm

Scarlet-backed Flowerpecker | *Dicaeum cruentatum*

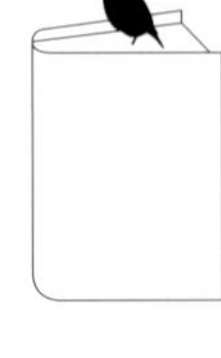

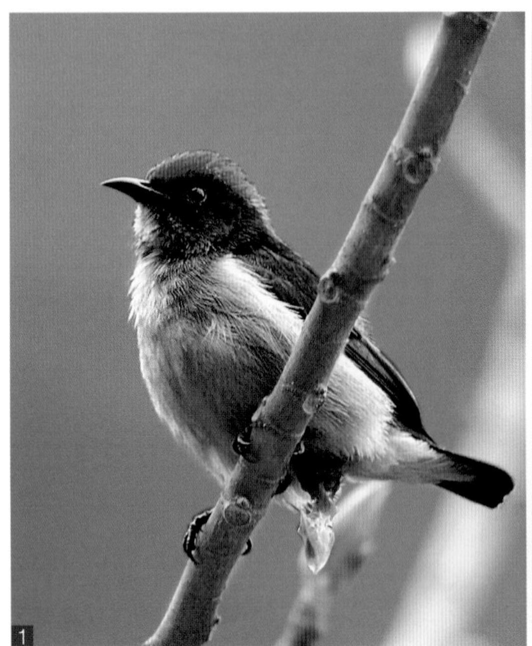

1

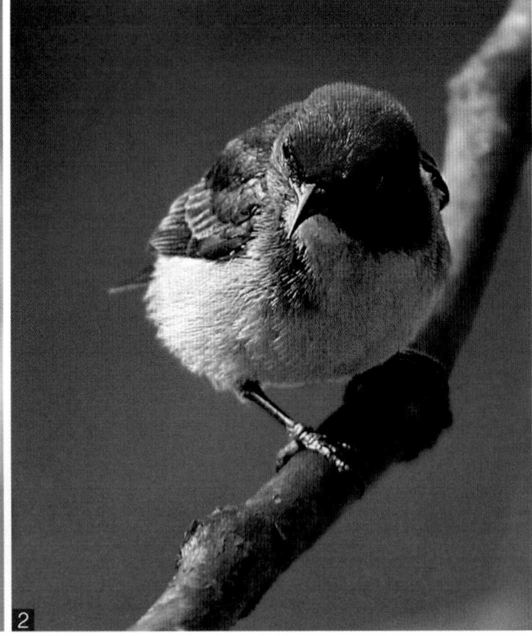

2

體型細小。雄鳥的前額、後枕、上身以至腰部朱紅色，臉頰深灰色，翼深藍色而帶有金屬光澤。雌鳥上身褐綠色，下體顏色較淡，腰部朱紅色，而幼鳥則沒有朱紅色。飛行或在樹上跳動時，會發出輕快的「的一的一」聲。

Small-sized. Male has unmistakable bright crimson crown, extending along upperparts to rump. Dark grey cheeks, wings are darkish glossy blue. Female has pale brownish green upperparts with crimson rump, underparts are paler. Juvenile has no crimson colour. Call a series of "dik-dik", especially in flight.

1 adult male 雄成鳥；Tai Po Kau 大埔滘；Feb-06, 06 年 2 月；Andy Kwok 郭匯昌
2 1st summer male 第一年夏天雄鳥；Shing Mun 城門水塘；Jan-07, 07 年 1 月；Jasper Lee 李君哲
3 adult male 雄成鳥；Tai Po Kau 大埔滘；Jan-09, 09 年 1 月；Lee Yat Ming 李逸明
4 1st summer male 第一年夏天雄鳥；Shing Mun 城門水塘；Jan-07, 07 年 1 月；Jasper Lee 李君哲
5 adult female 雌成鳥；Shing Mun 城門水塘；Jan-07, 07 年 1 月；Jasper Lee 李君哲
6 adult male 雄成鳥；Ng Tung Chai 梧桐寨；Aug-07, 07 年 8 月；Danny Ho 何國海
7 adult female 雌成鳥；Shek Kong 石崗；Jan-09, 09 年 1 月；Pippen Ho 何志剛
8 adult male 雄成鳥；Tai Po Kau 大埔滘；Feb-06, 06 年 2 月；Andy Kwok 郭匯昌

	春季過境遷徙鳥 Spring Passage Migrant			夏候鳥 Summer Visitor			秋季過境遷徙鳥 Autumn Passage Migrant			冬候鳥 Winter Visitor		
常見月份	1	2	3	4	5	6	7	8	9	10	11	12
	留鳥 Resident				迷鳥 Vagrant				偶見鳥 Occasional Visitor			

3
4
5
6
7

8

純色啄花鳥

[普] chún sè zhuó huā niǎo

體長 length：7.5-9cm

Plain Flowerpecker | *Dicaeum minullum*

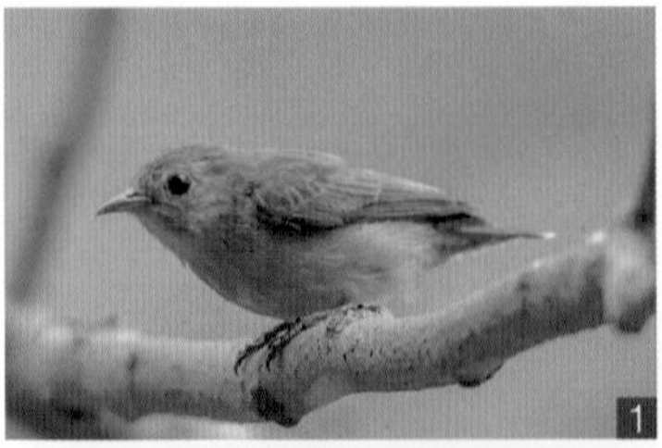
1

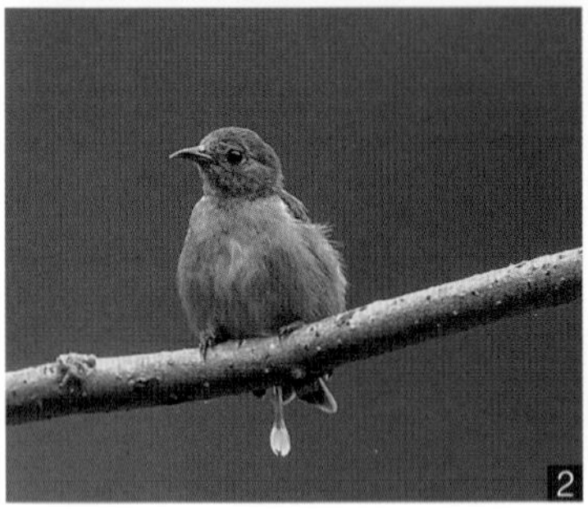
2

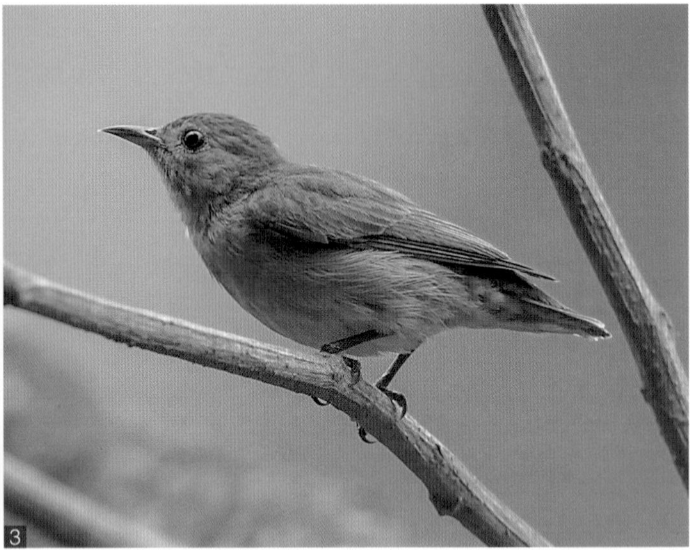
3

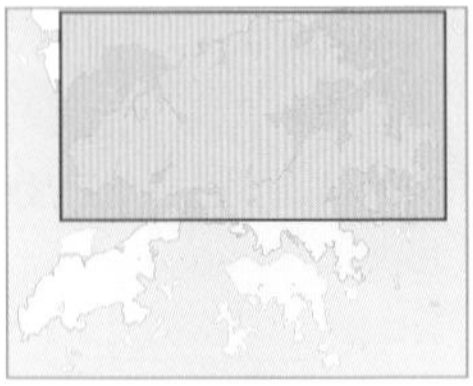

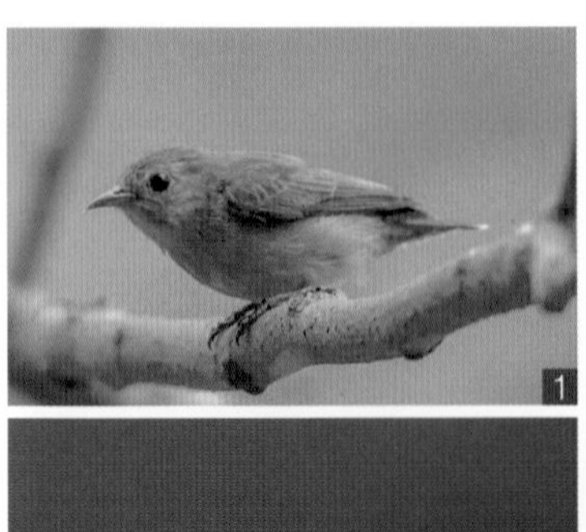

雌雄相似，上體橄欖綠色，翼角沾白，下體淺灰色，腹部中央沾奶黃色。黑色嘴細，腳深灰藍色。

Male and female have similar appearance. Olive green upperparts. Alulae tinged with white. Light grey underparts. Middle of belly tinged with milky yellow. Thin black bill. Dark greyish-blue legs.

[1] adult 成鳥：Tai Po Kau 大埔滘：Apr-19, 19 年 4 月：Wong Wong Yung 王煌容
[2] adult 成鳥：Tai Po Kau 大埔滘：Apr-19, 19 年 4 月：Wong Wong Yung 王煌容
[3] adult 成鳥：Tai Po Kau 大埔滘：Apr-19, 19 年 4 月：Wong Wong Yung 王煌容
[4] adult 成鳥：Tai Po Kau 大埔滘：Apr-19, 19 年 4 月：Wong Wong Yung 王煌容

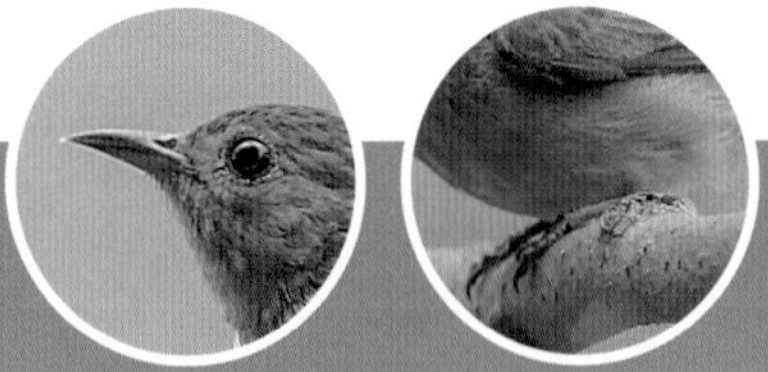

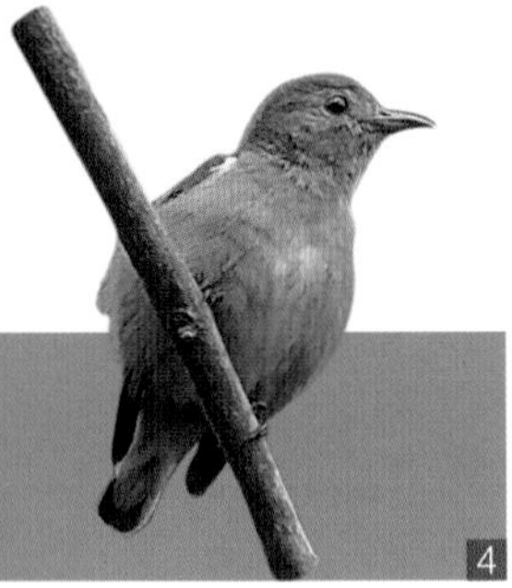
4

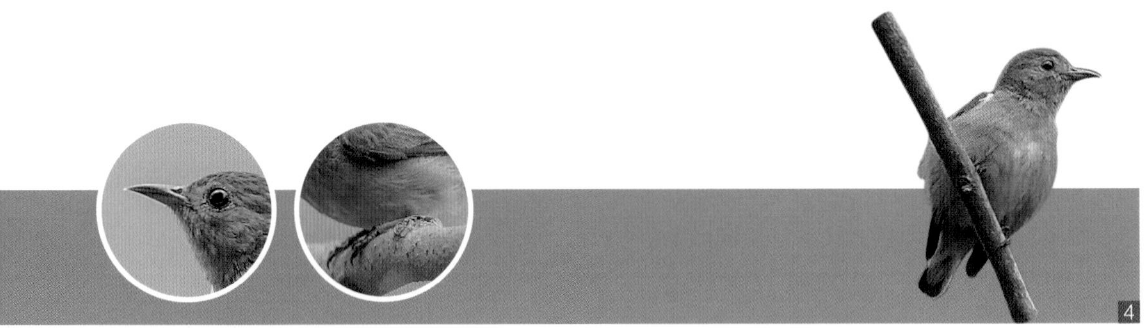

春季過境遷徙鳥 Spring Passage Migrant	夏候鳥 Summer Visitor	秋季過境遷徙鳥 Autumn Passage Migrant	冬候鳥 Winter Visitor

常見月份 1 2 3 4 5 6 7 8 9 10 11 12

留鳥 Resident	迷鳥 Vagrant	偶見鳥 Occasional Visitor

黃腹花蜜鳥

㊟ huáng fù huā mì niǎo

體長 length：10-11.4cm

Olive-backed Sunbird | *Cinnyris jugularis*

上體橄欖褐色，下體黃色。雄鳥喉至胸部是具金屬光澤的黑紫色，有上栗紅下黑色的胸帶。黑色嘴幼長向下彎，腳黑色。

Olive brown upperparts with yellow underparts. Male has metallic dark purple sheen from throat to breast. Chestnut red over black breast band. Black curved bill. Black legs.

1 adult male 雄成鳥：Po Toi 蒲台：Apr-16, 16 年 4 月：Aaron Lo 羅瑞華
2 adult male 雄成鳥：Po Toi 蒲台：Apr-16, 16 年 4 月：Aaron Lo 羅瑞華

春季過境遷徙鳥 Spring Passage Migrant	夏候鳥 Summer Visitor	秋季過境遷徙鳥 Autumn Passage Migrant	冬候鳥 Winter Visitor

1 2 3 4 5 6 7 8 9 10 11 12 常見月份

留鳥 Resident	迷鳥 Vagrant	偶見鳥 Occasional Visitor

藍喉太陽鳥

[普] lán hóu tài yáng niăo

體長 length：14-15cm

Mrs. Gould's Sunbird | *Aethopyga gouldiae*

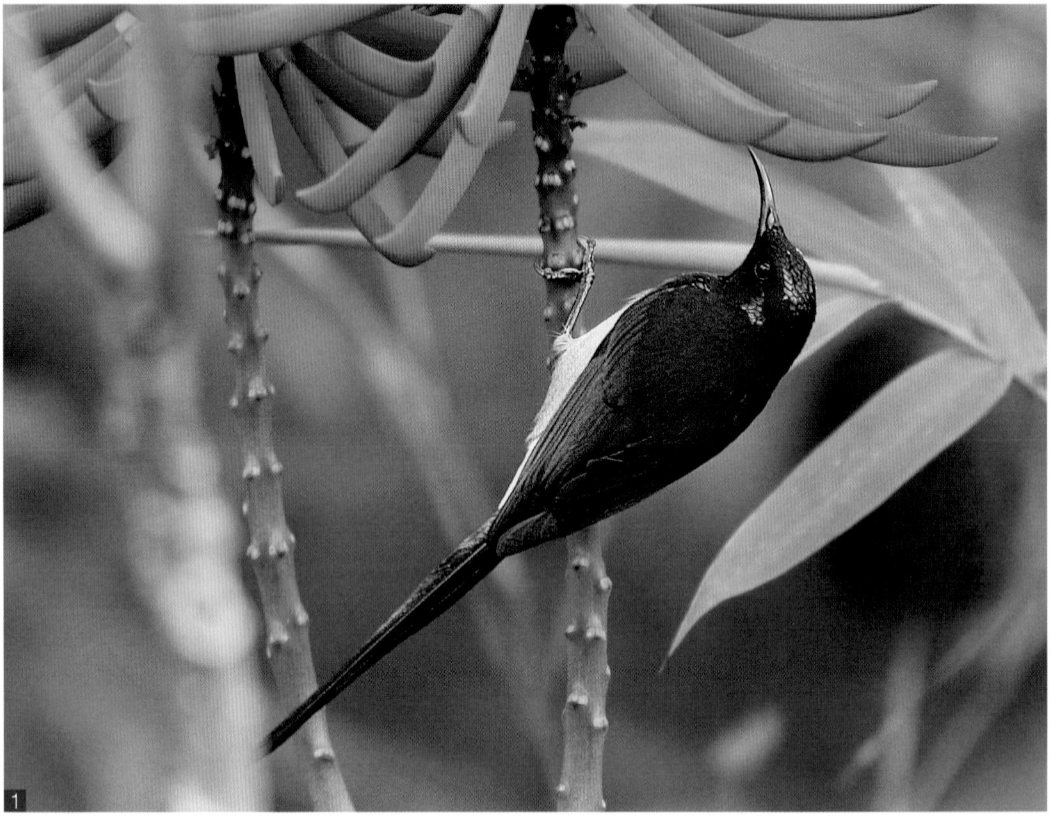

1

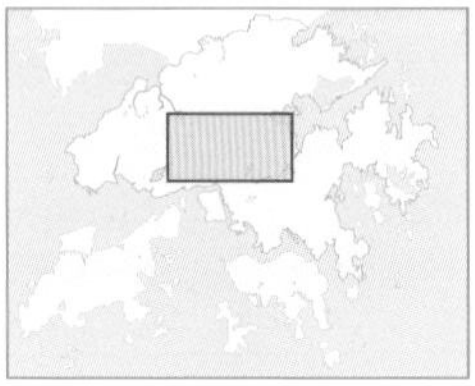

🌳

雄性上胸、頸及背部皆紅色。長長的藍色尾部是主要的特徵。翼橄欖綠色，下體綠黃色。雌鳥上體橄欖綠，下體綠黃，腰淺黃。近年偶見於大埔滘。

Male with red upper breast, neck and back. Prominent long blue tail. Olive-green wings. Underparts greenish yellow. Female bird has overall olive-green colour, pale yellow rump and underparts greenish yellow. Occasionally noted in Tai Po Kau in recent years.

[1] adult male 雄成鳥：Tai Po Kau 大埔滘：Mar-07, 07 年 3 月：Martin Hale 夏敖天
[2] 1st summer male, race *dabryii* 第一年夏天雄鳥, *dabryii* 亞種：Tai Po Kau 大埔滘：Mar-07, 07 年 3 月：KY Lee 容姨姨 / 李啟源
[3] 1st summer male, race *dabryii* 第一年夏天雄鳥, *dabryii* 亞種：Tai Po Kau 大埔滘：Mar-07, 07 年 3 月：KY Lee 容姨姨 / 李啟源
[4] 1st summer male, race *dabryii* 第一年夏天雄鳥, *dabryii* 亞種：Tai Po Kau 大埔滘：Mar-07, 07 年 3 月：Andy Kwok 郭匯昌
[5] 1st summer male, race *dabryii* 第一年夏天雄鳥, *dabryii* 亞種：Tai Po Kau 大埔滘：Mar-07, 07 年 3 月：Andy Kwok 郭匯昌
[6] adult male 雄成鳥：Tai Po Kau 大埔滘：Jan-07, 07 年 1 月：Maison Fung 馮振威
[7] 1st summer male, race *dabryii* 第一年夏天雄鳥, *dabryii* 亞種：Tai Po Kau 大埔滘：Mar-07, 07 年 3 月：KY Lee 容姨姨 / 李啟源
[8] 1st winter male, race *dabryii* 第一年冬天雄鳥, *dabryii* 亞種：Tai Po Kau 大埔滘：Feb-05, 05 年 2 月：Cherry Wong 黃卓研

春季過境遷徙鳥 Spring Passage Migrant	夏候鳥 Summer Visitor	秋季過境遷徙鳥 Autumn Passage Migrant	冬候鳥 Winter Visitor

常見月份	1	2	3	4	5	6	7	8	9	10	11	12

留鳥 Resident	迷鳥 Vagrant	偶見鳥 Occasional Visitor

2
3
4
5
6
7

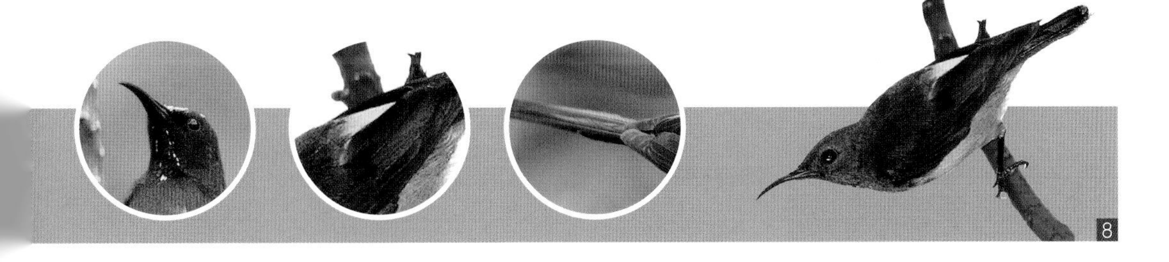

8

叉尾太陽鳥

[普] chā wěi tài yáng niǎo

體長 length：10cm

Fork-tailed Sunbird | *Aethopyga christinae*

體型細小。嘴尖細及向下彎。雄鳥湖水藍色的頭帶金屬光澤，臉部紅色，上體綠色，腰黃色，尾部有兩條細長尾羽，看起來像「叉尾」，喉部和胸部紅色，下體淡黃色。雌鳥上身綠色，腰黃色，喉和下身淡黃色，沒有叉尾。叫聲為急速而輕柔略帶金屬感的「zwink-zwink」。

Small bird with decurved bill. Male has glossy greenish blue head, red cheeks, green upperparts and yellow rump, with two elongated tail feathers forming a "forked tail". Throat and breast red, underparts pale yellow. Female has green upperparts and yellow rump. Throat to underparts buffy. No "fork" on tail. Call is a soft and frequent "zwink-zwink" and a metallic trill.

1 adult male 雄成鳥；Tai Po Kau 大埔滘；Mar-07, 07 年 3 月；Allen Chan 陳志雄
2 adult male 雄成鳥；Tai Po Kau 大埔滘；Mar-07, 07 年 3 月；Eling Lee 李佩玲
3 adult male 雄成鳥；Kadoorie Farm 嘉道理農場；20-Feb-08, 08 年 2 月 20 日；Anita Lee 李雅婷
4 adult male 雄成鳥；Tai Po Kau 大埔滘；Mar-07, 07 年 3 月；Allen Chan 陳志雄
5 adult female 雌成鳥；Tai Po Kau 大埔滘；Mar-07, 07 年 3 月；James Lam 林文華
6 adult male 雄成鳥；Tai Po Kau 大埔滘；Mar-07, 07 年 3 月；Jemi and John Holmes 孔思義、黃亞萍
7 adult female 雌成鳥；Tai Po Kau 大埔滘；Mar-07, 07 年 3月；Martin Hale 夏敖天
8 adult female 雌成鳥；Tai Po Kau 大埔滘；Feb-07, 07 年 2 月；Andy Kwok 郭匯昌

	春季過境遷徙鳥 Spring Passage Migrant			夏候鳥 Summer Visitor			秋季過境遷徙鳥 Autumn Passage Migrant			冬候鳥 Winter Visitor		
常見月份	1	2	3	4	5	6	7	8	9	10	11	12
	留鳥 Resident				迷鳥 Vagrant				偶見鳥 Occasional Visitor			

2
3
4
5
6
7

8

家麻雀

㊟ jiā má què

體長 length：16-18cm

House Sparrow | *Passer domesticus*

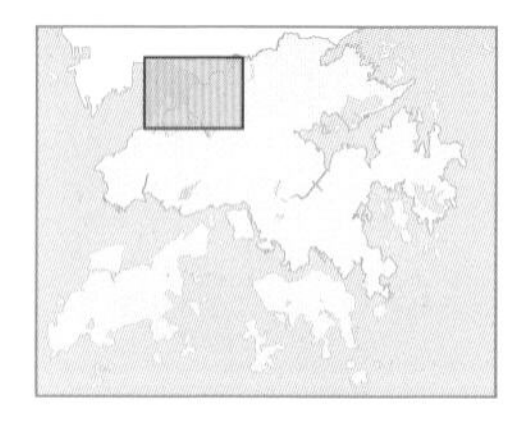

背部栗紅色具黑白縱紋，下體灰白色。雄鳥頭頂和腰部灰色，眼後栗紅色，頦部黑色，面頰白色。雌鳥顏色較淡，有淺色眉紋。嘴黑色，雌鳥嘴基黃色，腳淺褐色。

Chestnut red back with black and white streaks. Greyish white underparts. Male has grey crown and rump. Chestnut red ear-coverts. Black throat and white cheeks. Female is paler with pale supercilia. Black bill. Female has yellow at base of bill. Light brown legs.

1 adult male 雄成鳥：Long Valley 塱原：Nov-12, 12 年 11 月：Allen Chan 陳志雄
2 adult female 雌成鳥：Long Valley 塱原：Nov-12, 12 年 11 月：Allen Chan 陳志雄
3 adult female 雌成鳥：Long Valley 塱原：Nov-17, 17 年 11 月：Henry Lui 呂德恒
4 adult female 雌成鳥：Long Valley 塱原：Nov-17, 17 年 11 月：Henry Lui 呂德恒

春季過境遷徙鳥 Spring Passage Migrant			夏候鳥 Summer Visitor			秋季過境遷徙鳥 Autumn Passage Migrant			冬候鳥 Winter Visitor		
1	2	3	4	5	6	7	8	9	10	11	12

常見月份

留鳥 Resident	迷鳥 Vagrant	偶見鳥 Occasional Visitor

山麻雀

㊟ shān má què

體長 length：14-15cm

Russet Sparrow | *Passer cinnamomeus*

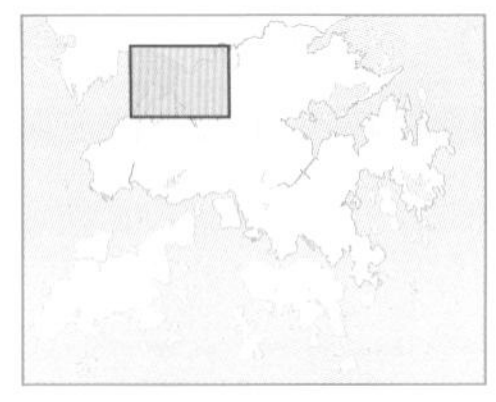

頭部栗色，面頰白色，有黑而短的過眼線，嘴和喉黑色。上背栗色，有黑色縱紋，喉至下體奶白色。雌鳥顏色較暗，頭部淡灰褐色，有白色粗眉紋和深灰色過眼線。嘴灰色，腳粉紅色。

Chestnut head, white cheeks, short dark eyestripe. Upperparts chestnut with black stripes. Throat to underparts milky white. Female is duller, with pale greyish brown head, thick white supercilium and grey eyestripe. Grey bill and pink legs.

1 adult male 雄成鳥：Tai Po 大埔：Jan-08, 08 年 1 月：Pippen Ho 何志剛
2 adult male 雄成鳥：Wun Yiu 碗窰：28-Jan-08, 08 年 1 月 28 日：Owen Chiang 深藍
3 adult male 雄成鳥：Tai Po 大埔：Jan-08, 08 年 1 月：Pippen Ho 何志剛

春季過境遷徙鳥 Spring Passage Migrant			夏候鳥 Summer Visitor			秋季過境遷徙鳥 Autumn Passage Migrant			冬候鳥 Winter Visitor			
1	2	3	4	5	6	7	8	9	10	11	12	常見月份
留鳥 Resident				迷鳥 Vagrant				偶見鳥 Occasional Visitor				

樹麻雀

㊟ shù má què

體長 length：14-15cm

Eurasian Tree Sparrow | *Passer montanus*

其他名稱 Other names：麻雀 Tree Sparrow

1

香港最常見的麻雀。頭部褐色，白色面頰上有黑斑，眼先、喉部和嘴黑色。上體褐色，有黑色縱紋，下體淡灰褐色，脇部沾褐，腳粉紅色。幼鳥面部和喉部黑色不明顯。

The most common sparrow in Hong Kong. Brown head, white cheeks marked with a black patch. Lores, throat and bill black. Upperparts brown with dark stripes. Underparts pale grey, flanks brown. Juvenile has paler cheek patches and is duller overall.

1 adult 成鳥；Kowloon Park 九龍公園；Jun-04, 04 年 6 月；Cherry Wong 黃卓研
2 adult 成鳥；Cheung Chau 長洲；Nov-04, 04 年 11 月；Henry Lui 呂德恆
3 adult 成鳥；Mai Po 米埔；Jun-07, 07 年 6 月；Cherry Wong 黃卓研
4 juvenile 幼鳥；Kowloon Park 九龍公園；Aug-07, 07 年 8 月；Man Kuen Yat Bill 文權溢
5 adult and juvenile 成鳥和幼鳥；Victoria Park 維多利亞公園；Aug-06, 06 年 8 月；Chan Kai Wai 陳佳瑋
6 adult 成鳥；Long Valley 塱原；Dec-04, 04 年 12 月；Henry Lui 呂德恆

	春季過境遷徙鳥 Spring Passage Migrant			夏候鳥 Summer Visitor			秋季過境遷徙鳥 Autumn Passage Migrant			冬候鳥 Winter Visitor		
常見月份	1	2	3	4	5	6	7	8	9	10	11	12
	留鳥 Resident				迷鳥 Vagrant				偶見鳥 Occasional Visitor			

2
3
4
5

6

白腰文鳥

㊟ bái yāo wén niǎo

體長 length：11-12cm

White-rumped Munia | *Lonchura striata*

其他名稱 Other names：White-backed Munia

1

嘴部黑色，呈圓錐形，前額、眼先和喉部黑色。頭、上體和胸部深褐色，腰部和腹部明顯白色，尾羽和飛羽黑色，常成群活動。

Large, black conical bill. Black forehead, lores and throat. Head, upperparts and breast dark brown, with prominent white rump and belly. Black tail and flight feathers. Often in small flocks.

1 adult 成鳥：Hong Kong Park 香港公園：Nov-04, 04 年 11 月：Henry Lui 呂德恆
2 adult 成鳥：Lamma Island 南丫島：Sep-05, 05 年 9 月：Pippen Ho 何志剛
3 juvenile 幼鳥：Shing Mun 城門水塘：18-Oct-04, 04 年 10 月 18 日：Wong Hok Sze 王學思
4 adult 成鳥：Kowloon Park 九龍公園：Sep-06, 06 年 9 月：Joyce Tang 鄧玉蓮
5 adult 成鳥：Kowloon Park 九龍公園：Mar-07, 07 年 3 月：Benson, Wui-Man Lau 劉滙文
6 adult 成鳥：Kowloon Park 九龍公園：Aug-03, 03 年 8 月：Henry Lui 呂德恆
7 adult 成鳥：Mai Po 米埔：Dec-07, 07 年 12 月：Cherry Wong 黃卓研

	春季過境遷徙鳥 Spring Passage Migrant			夏候鳥 Summer Visitor			秋季過境遷徙鳥 Autumn Passage Migrant			冬候鳥 Winter Visitor		
常見月份	1	2	3	4	5	6	7	8	9	10	11	12
	留鳥 Resident				迷鳥 Vagrant				偶見鳥 Occasional Visitor			

2
3
4
5
6

7

斑文鳥

㊟bān wén niǎo

體長length：12cm

Scaly-breasted Munia | *Lonchura punctulata*

其他名稱 Other names：Spotted Munia

嘴部黑色，呈圓錐形，眼先和喉部深褐色。頭、上體和胸部褐色，下體白色，胸和脇的羽毛邊緣褐色像鱗片，尾羽和飛羽褐色。常成群活動，有時與其他文鳥混在一起。叫聲為輕柔而重複的「mit-mit…」聲。

Large, black conical bill. Dark brown lores and throat. Head, upperparts and breast brown, underparts white, feathers at the breast and flanks fringed brown and appear scaly. Brown tail and flight feathers. Often in groups, sometimes mixes with other munias. The call is a soft "mit-mit…".

1 adult 成鳥：Po Toi 蒲台：Mar-07, 07 年 3 月：Pippen Ho 何志剛
2 adult 成鳥：Long Valley 塱原：Aug-04, 04 年 8 月：Henry Lui 呂德恒
3 juvenile 幼鳥：Long Valley 塱原：Dec-07, 07 年 12 月：Isaac Chan 陳家強
4 adult 成鳥：Lok Ma Chau 落馬洲：Nov-06, 06 年 11 月：Martin Hale 夏敖天
5 adult and juvenile 成鳥和幼鳥：Mai Po 米埔：Apr-08, 08 年 4 月：Michael Schmitz
6 adult 成鳥：Long Valley 塱原：12-Sep-07, 07 年 9 月 12 日：Owen Chiang 深藍

	春季過境遷徙鳥 Spring Passage Migrant			夏候鳥 Summer Visitor			秋季過境遷徙鳥 Autumn Passage Migrant			冬候鳥 Winter Visitor		
常見月份	1	2	3	4	5	6	7	8	9	10	11	12
	留鳥 Resident				迷鳥 Vagrant				偶見鳥 Occasional Visitor			

2 3 4 5

6

黃胸織雀

㊥huáng xiōng zhī què

體長length：15cm

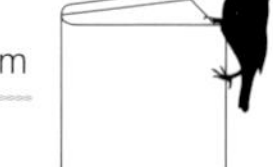

Baya Weaver | *Ploceus philippinus*

其他名稱 Other names：黃胸織布鳥

嘴黑而粗厚，全身黃褐色，上體有深色縱紋。繁殖期雄鳥面頰黑色，頭頂和後枕鮮黃色。雌鳥顏色較淡。春夏間會築起成酒樽狀的巢，掛在樹上。屬外來鳥種。米埔魚塘一帶曾有一個小種群繁殖，2005年後再無野化種群記錄。

Bill thick and black. Overall plumage brownish yellow, with dark streaks on upperparts. Breeding male has prominent black cheeks and bright yellow crown and nape. Female is paler in colour. Builds bottle-like nest on trees in summer. There was an introduced small breeding population at fishponds in Mai Po. No feral population after 2005.

1 adult male 雄成鳥；Mai Po 米埔；Jun-03, 03 年 6 月；Matthew and TH Kwan 關朗曦、關子凱
2 adult male 雄成鳥；Mai Po 米埔；14-Sep-04, 04 年 9 月 14 日；Owen Chiang 深藍
3 adult male 雄成鳥；Mai Po 米埔；May-04, 04 年 5 月；Cherry Wong 黃卓研
4 adult female 雌成鳥；Mai Po 米埔；Aug-04, 04 年 8 月；Cherry Wong 黃卓研

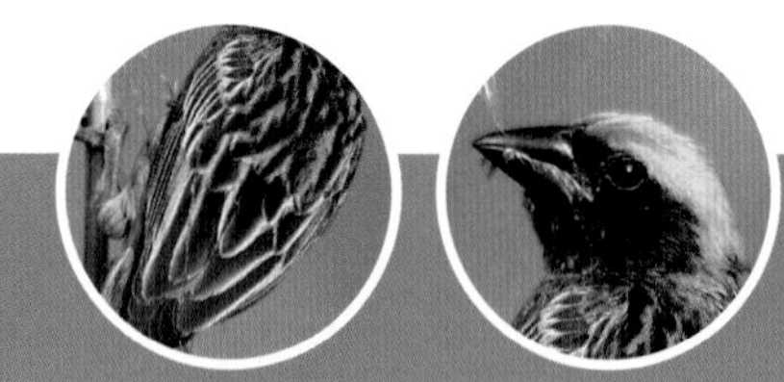

春季過境遷徙鳥 Spring Passage Migrant	夏候鳥 Summer Visitor	秋季過境遷徙鳥 Autumn Passage Migrant	冬候鳥 Winter Visitor

常見月份 1 2 3 4 5 6 7 8 9 10 11 12

留鳥 Resident	迷鳥 Vagrant	偶見鳥 Occasional Visitor

山鶺鴒

㊌ shān jí líng
㊋ 鶺鴒：音脊零

體長 length：16-18cm

Forest Wagtail | *Dendronanthus indicus*

其他名稱 Other names：林鶺鴒

上體灰褐色，有明顯的淡色眼眉。翼上有兩組黑白相間的粗橫斑，胸部有斷開的黑色「工」字紋，下體淡色。尾部常左右擺動，可以此和其他鶺鴒區別。

Upperparts brownish grey in colour. Prominent pale supercilium. Two sets of thick black-and-white bands on wings. Broken crosswise "H" mark on breast. Underparts pale. Swings tail from side to side.

1 Mai Po 米埔：Sep-06, 06 年 9 月：Martin Hale 夏敖天
2 Tai Po Kau 大埔滘：Sep-05, 05 年 9 月：Francis Chu 崔汝棠
3 Mai Po 米埔：Sep-06, 06 年 9 月：Pippen Ho 何志剛
4 Kowloon Hills Catchwater 九龍水塘引水道：Jan-04, 04 年 1 月：Henry Lui 呂德恆

春季過境遷徙鳥 Spring Passage Migrant	夏候鳥 Summer Visitor	秋季過境遷徙鳥 Autumn Passage Migrant	冬候鳥 Winter Visitor

1	2	3	4	5	6	7	8	9	10	11	12	常見月份

留鳥 Resident	迷鳥 Vagrant	偶見鳥 Occasional Visitor

東黃鶺鴒

[普] dōng huáng jí líng
[粵] 鶺鴒：音脊零

體長 length：16.5cm

Eastern Yellow Wagtail | *Motacilla tschutschensis*

有很多亞種，但全部都是下體黃色，背部橄欖綠色，嘴和腳黑色。第一年度冬的鳥有明顯白色眉紋，上體淡棕色，下體白色。*taivana* 亞種(香港最常見)頭頂及背部橄欖綠，眉紋及喉部鮮黃色。*simillima* 亞種頭頂藍灰色，眉白色，喉沾白，下體黃色較鮮。*macronyx* 亞種(香港甚少)，頭部深灰色，喉黃色，沒有眉紋。

Many sub-species, all with yellow underparts, olive-green back, black legs and bill. First winter bird has prominent white eyebrow, pale brown upperparts and white underparts. Race *taivana* (commonest in Hong Kong) has olive-green crown and mantle, with yellow eyebrow and throat. Race *simillima* has bluish grey head, white eyebrow, whitish throat and stronger yellow underparts. Race *macronyx* (least numerous in Hong Kong) has deep grey head, yellow throat and no eyebrows.

1 breeding, race *taivanna* 繁殖羽, *taivanna* 亞種：Fung Lok Wai 豐樂圍：Apr-06, 06 年 4 月：Jemi and John Holmes 孔思義、黃亞萍
2 1st winter, race *taivanna* 第一年冬天, *taivanna* 亞種：Long Valley 塱原：Nov-08, 08 年 11 月：Michelle and Peter Wong 江敏兒、黃理沛
3 non-breeding, race *taivanna* 非繁殖羽, *taivanna* 亞種：Long Valley 塱原：Jan-09, 09 年 1 月：Christine and Samuel Ma 馬志榮、蔡美蓮
4 1st winter, race *taivanna* 第一年冬天, *taivanna* 亞種：Long Valley 塱原：Feb-03, 03 年 2 月：Henry Lui 呂德恒
5 breeding, race *simillima* 繁殖羽, *simillima* 亞種：Lut Chau 甩洲：Apr-06, 06 年 4 月：Jemi and John Holmes 孔思義、黃亞萍
6 non-breeding, race *thunbergi* 非繁殖羽, *thunbergi* 亞種：Long Valley 塱原：Oct-04, 04 年 10 月：Cherry Wong 黃卓研
7 non-breeding, race *macronyx* 非繁殖羽, *macronyx* 亞種：Mai Po 米埔：Feb-08, 08 年 2 月：Danny Ho 何國海

春季過境遷徙鳥 Spring Passage Migrant	夏候鳥 Summer Visitor	秋季過境遷徙鳥 Autumn Passage Migrant	冬候鳥 Winter Visitor

常見月份 1 2 3 4 5 6 7 8 9 10 11 12

留鳥 Resident	迷鳥 Vagrant	偶見鳥 Occasional Visitor

黃頭鶺鴒

㊟ huáng tóu jí líng
㊟ 鶺鴒：音脊零

體長 length：16.5-20cm

Citrine Wagtail | *Motacilla citreola*

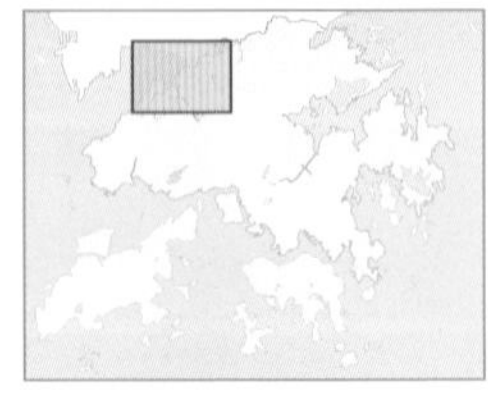

繁殖期上體灰色，頭及下體全黃。非繁殖期頭頂、面頰和下體沾灰，眉紋和頰下紋黃色，在耳羽後相連。靜立時，黑色飛羽上有兩道白色橫紋。幼鳥黃色較少，甚至沒有。

Breeding plumage has grey upperparts with yellow head and underparts. In non-breeding plumage, crown, cheeks and underparts turn grey. Yellow supercilium and submoustachial stripe meet behind ear coverts. Shows two white wing bars while at rest. Juvenile has little or no yellow.

1 1st winter 第一年冬天；Long Valley 塱原；Jan-08, 08 年 1 月；Chan Kai Wai 陳佳瑋
2 1st winter 第一年冬天；Long Valley 塱原；18-Nov-07, 07 年 11 月 18 日；Michelle and Peter Wong 江敏兒、黃理沛
3 1st winter 第一年冬天；Long Valley 塱原；Owen Chiang 深藍
4 1st winter 第一年冬天；Long Valley 塱原；18-Nov-07, 07 年 11 月 18 日；Michelle and Peter Wong 江敏兒、黃理沛

春季過境遷徙鳥 Spring Passage Migrant	夏候鳥 Summer Visitor	秋季過境遷徙鳥 Autumn Passage Migrant	冬候鳥 Winter Visitor

常見月份	1	2	3	4	5	6	7	8	9	10	11	12

留鳥 Resident	迷鳥 Vagrant	偶見鳥 Occasional Visitor

灰鶺鴒

[普] huī jí líng
[粵] 鶺鴒：音脊零

體長 length：17-20cm

Grey Wagtail | *Motacilla cinerea*

1

2

3

與黃鶺鴒相似，但頭至背部灰色，腰黃色。有明顯的長尾、白色眼眉和頰下紋。下體黃色，脇部有時沾白。繁殖期雄鳥喉部黑色(在香港不常見)，非繁殖期及幼鳥下體較淡黃。

Resembles Yellow Wagtail, but grey from head to mantle and rump is yellow. Obvious long tail, white supercilium and submoustrachial stripe. Underparts yellow, flanks whiter. Male in breeding plumage (rare in Hong Kong) has black throat. Pale yellow underparts in non-breeding plumage and juvenile.

1 breeding 繁殖羽：Mai Po 米埔：Apr-08, 08 年 4 月：James Lam 林文華
2 non-breeding adult 非繁殖羽成鳥：Sha Po 沙埔：Jan-04, 04 年 1 月：Jemi and John Holmes 孔思義、黃亞泙
3 non-breeding adult 非繁殖羽成鳥：Cheung Chau 長洲：Dec-04, 04 年 12 月：Henry Lui 呂德恆
4 non-breeding adult 非繁殖羽成鳥：Kowloon Hills Catchwater 九龍水塘引水道：Jan-04, 04 年 1 月：Henry Lui 呂德恆

4

春季過境遷徙鳥 Spring Passage Migrant	夏候鳥 Summer Visitor	秋季過境遷徙鳥 Autumn Passage Migrant	冬候鳥 Winter Visitor

1 2 3 4 5 6 7 8 9 10 11 12 常見月份

留鳥 Resident	迷鳥 Vagrant	偶見鳥 Occasional Visitor

白鶺鴒

[普] bái jí líng
[粵] 鶺鴒：音脊零

體長 length：16.5-18cm

White Wagtail | *Motacilla alba*

黑白灰三色，尾長，冬羽顏色較暗。有四個亞種，*leucopsis* 亞種(最常見)：臉部全白，無貫眼紋。*ocularis* 亞種有黑貫眼紋、黑喉和胸。*personata* 亞種(極少見)：頭部全黑，前額和眼週圍白色。*lugens* 亞種(又稱黑背鶺鴒)繁殖期上體和貫眼紋黑色，飛行時可見飛羽接近全白。

Plumage consists of black, white and grey. Long-tailed. Grey upperparts are more prominent in winter. Four subspecies in Hong Kong. Race *leucopsis* (commonest): completely white face with no eye-stripes. Race *ocularis*: black eye-stripes, black throat and breast. Race *personata* (very rare): black hood with white on forehead and around eyes. Race *lugens* (also known as Black-backed Wagtail): black upperparts and eye-stripes in breeding plumage, wings almost completely white noticeable in flight.

[1] non-breeding adult male, race *leucopsis* 非繁殖羽雄成鳥, *leucopsis* 亞種：Long Valley 塱原：Oct-03, 03 年 10 月：Michelle and Peter Wong 江敏兒、黃理沛
[2] 1st winter, race *leucopsis* 第一年冬天, *leucopsis* 亞種：Kam Tin 錦田：Dec-04, 04 年 12 月：Cherry Wong 黃卓研
[3] non-breeding adult male, race *leucopsis* 非繁殖羽雄成鳥, *leucopsis* 亞種：Long Valley 塱原：Nov-08, 08 年 11 月：Ng Lin Yau 吳璉宥
[4] non-breeding moulting into breeding adult male, race *leucopsis* 非繁殖羽轉換繁殖羽雄成鳥, *leucopsis* 亞種：Pui O 貝澳：Mar-08, 08 年 3 月：Isaac Chan 陳家強
[5] non-breeding adult male, race *ocularis* 非繁殖羽雌成鳥, *ocularis* 亞種：Ma On Shan 馬鞍山：Oct-08, 08 年 10 月：Eling Lee 李佩玲
[6] HK Wetland Park 香港濕地公園：Dec-04, 04年12月：Henry Lui 呂德恒
[7] non-breeding adult male, race *ocularis* 非繁殖羽雄成鳥, *ocularis* 亞種：Fo Tan 火炭：Dec-07, 07 年 12 月：Christine and Samuel Ma 馬志榮、蔡美蓮
[8] non-breeding adult female, race *leucopsis* 非繁殖羽雌成鳥, *leucopsis* 亞種：Long Valley 塱原：11-Nov-08, 08 年 11 月 11 日：Sung Yik Hei 宋亦希
[9] non-breeding adult female, race *leucopsis* 非繁殖羽雌成鳥, *leucopsis* 亞種：Long Valley 塱原：Oct-04, 04 年 10 月：Henry Lui 呂德恒

	春季過境遷徙鳥 Spring Passage Migrant			夏候鳥 Summer Visitor			秋季過境遷徙鳥 Autumn Passage Migrant			冬候鳥 Winter Visitor		
常見月份	1	2	3	4	5	6	7	8	9	10	11	12
	留鳥 Resident				迷鳥 Vagrant				偶見鳥 Occasional Visitor			

2
3
4
5
6
7
8

9

理氏鷚

[普] lǐ shì liù
[粵] 鷚：音溜

體長 length：17-18cm

Richard's Pipit | *Anthus richardi*

全身偏褐，下體較淡，胸前有深色細縱紋，嘴淡黃色，腳淡紅色。腳長，後爪亦較長，可以此和其他鷚區分。站立時姿勢挺直。*sinensis*亞種在香港的高山草地繁殖。叫聲為高音單節的「chich…」聲。

Overall tawny brown. Underparts pale, narrow and dark streaks on breast. Bill pale yellow, legs pale red. Distinctive long legs and long hind claws. Erect posture at rest. Race *sinensis* breeds in upland grassy areas. Call a high-pitched single note "chich…".

1 race *richardi*, *richardi* 亞種：Long Valley 塱原：Oct-04, 04 年 10 月：Cherry Wong 黃卓研
2 race *richardi*, *richardi* 亞種：Long Valley 塱原：Apr-07, 07 年 4 月：Sammy Sam and Winnie Wong 森美與雲妮
3 race *sinensis*, *sinensis* 亞種：Tai Mo Shan 大帽山：24-Jun-06, 06 年 6 月 24 日：Michelle and Peter Wong 江敏兒、黃理沛
4 race *sinensis*, *sinensis* 亞種：Tai Mo Shan 大帽山：Martin Hale 夏敖天
5 race *richardi*, *richardi* 亞種：Fung Lok Wai 豐樂圍：Apr-06, 06 年 4 月：Jemi and John Holmes 孔思義、黃亞萍
6 race *richardi*, *richardi* 亞種：Shek Kong 石崗：Feb-04, 04 年 2 月：Jemi and John Holmes 孔思義、黃亞萍
7 race *sinensis*, *sinensis* 亞種：Tai Mo Shan 大帽山：2-Jul-07, 07 年 7 月 2 日：Michelle and Peter Wong 江敏兒、黃理沛

	春季過境遷徙鳥 Spring Passage Migrant			夏候鳥 Summer Visitor			秋季過境遷徙鳥 Autumn Passage Migrant			冬候鳥 Winter Visitor		
常見月份	1	2	3	4	5	6	7	8	9	10	11	12
	留鳥 Resident				迷鳥 Vagrant				偶見鳥 Occasional Visitor			

2
3
4
5
6

7

樹鷚

㊢ shù liù
[粵] 鷚：音漏

體長 length：15-17cm

Olive-backed Pipit | *Anthus hodgsoni*

其他名稱 Other names：Indian Tree Pipit

1

上體橄欖綠色具有縱紋，下體較淡，胸前及脇有濃密深色縱紋，眉紋明顯，耳羽後有白點，上嘴深色，下嘴粉紅色，腳淡紅色。常上下擺尾，受驚時會飛到電線或樹枝上。叫聲為輕柔的「tseep」聲。

Upperparts olive with bold darkish streaks. Underparts paler, heavy dark streaks on breast and flanks. Prominent eyebrow and spot behind ear coverts, upper bill dark, lower bill pale pink. Wags tail up and down. When flushed, it flies up to wires or bare branches. Call a light and soft "tseep".

1 Long Valley 塱原：Dec-06, 06 年 12 月：Andy Kwok 郭匯昌
2 Shing Mun 城門水塘：Dec-08, 08 年 12 月：Ken Fung 馮漢城
3 Mai Po 米埔：Jan-09, 09 年 1 月：Cheng Nok Ming 鄭諾銘
4 Mai Po 米埔：Nov-04, 04 年 11 月：Henry Lui 呂德恒
5 Mai Po 米埔：Dec-05, 05 年 12 月：Henry Lui 呂德恒
6 Long Valley 塱原：Jan-09, 09 年 1 月：Aka Ho
7 Mai Po 米埔：Nov-06, 06 年 11 月：Law Kam Man 羅錦文
8 Shek Kong 石崗：4-Feb-07, 07 年 2 月 4 日：Michelle and Peter Wong 江敏兒、黃理沛

春季過境遷徙鳥 Spring Passage Migrant	夏候鳥 Summer Visitor	秋季過境遷徙鳥 Autumn Passage Migrant	冬候鳥 Winter Visitor

常見月份 1 2 3 4 5 6 7 8 9 10 11 12

留鳥 Resident	迷鳥 Vagrant	偶見鳥 Occasional Visitor

2
3
4
5
6
7

8

北鷚

㊟bĕi liù
㊟鷚：音漏

體長 length：14cm

Pechora Pipit | *Anthus gustavi*

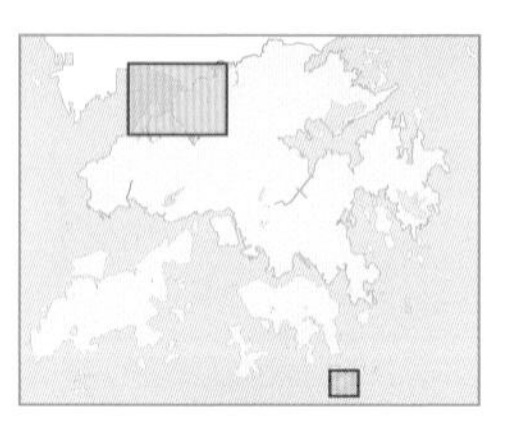

上體褐色，滿佈深色紋，下體偏白，胸至脇部有深色紋。和紅喉鷚相似，但翼摺合時初級飛羽較三級飛羽長得多，上體有較顯眼的淡色紋。

Upperparts brown scattered with dark stripes. White underparts. Breast to flanks with dark stripes. Similar to Red-throated Pipit but with longer primary projection and obvious pale stripes on upperparts.

1 Long Valley 塱原：Oct-07, 07 年 10 月；Tam Yiu Leung 譚耀良
2 Po Toi 蒲台：Oct-07, 07 年 10 月；Tam Yiu Leung 譚耀良

春季過境遷徙鳥 Spring Passage Migrant	夏候鳥 Summer Visitor	秋季過境遷徙鳥 Autumn Passage Migrant	冬候鳥 Winter Visitor

常見月份 1 2 3 **4** **5** 6 7 8 **9** **10** **11** 12

留鳥 Resident	迷鳥 Vagrant	偶見鳥 Occasional Visitor

水鷚

㊟shuǐ liù
㊟鷚：音溜

體長length：15-17cm

Water Pipit | *Anthus spinoletta*

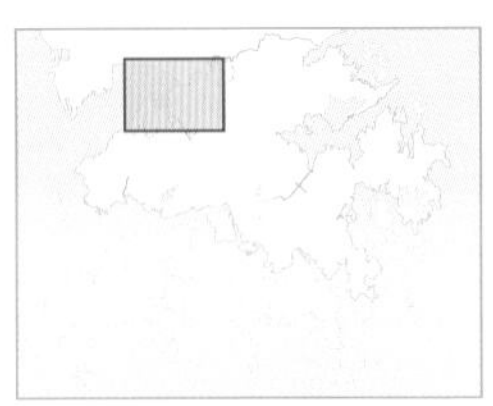

全身淡褐色，無明顯眉斑，嘴淡黃色，嘴端較深色，喉部及下體沾橙粉紅色，胸有淡褐色細縱紋，腳深色。

Pale brown, no obvious eyebrow. Bill yellow with dark tip. Throat to underparts pinkish orange, with pale brown narrow streaks on breast. Dark legs.

1 San Tin 新田：Paul Leader 利雅德

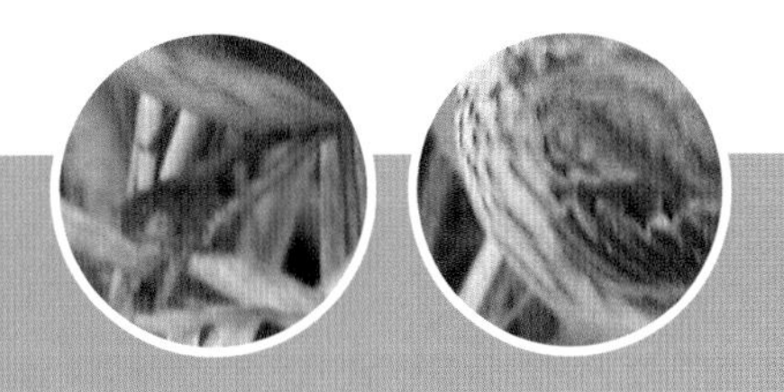

春季過境遷徙鳥 Spring Passage Migrant			夏候鳥 Summer Visitor			秋季過境遷徙鳥 Autumn Passage Migrant			冬候鳥 Winter Visitor			
1	2	3	4	5	6	7	8	9	10	11	12	常見月份

留鳥 Resident	迷鳥 Vagrant	偶見鳥 Occasional Visitor

粉紅胸鷚

㊟ fěn hóng xiōng liù
㊟ 鷚：音漏

體長 length：15-16.5cm

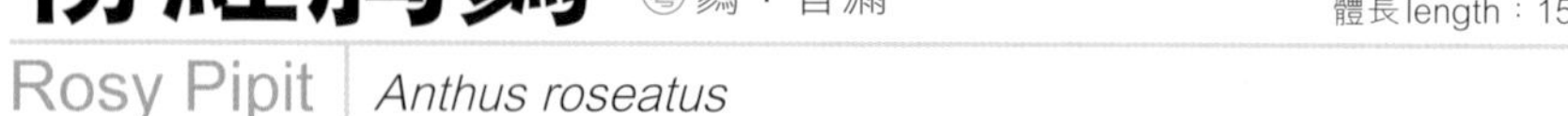

Rosy Pipit | *Anthus roseatus*

1

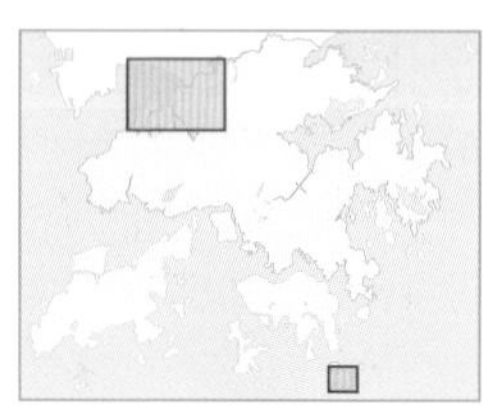

上體灰褐色，佈滿深色紋，下體偏白，胸至脇部淡褐色並有深色紋，繁殖期胸部粉紅色，沒有深色紋。嘴黑色，腳淡黃褐色，有淺色粗眼眉。

Greyish brown upperparts scattered with dark stripes. Whitish underparts. Breast to flanks are pale brown with dark stripes. Breeding bird has pale pink breast without dark stripes. Black bill and pale yellowish brown legs. Pale eyebrow.

1 breeding 繁殖羽；Long Valley 塱原；May-06, 06 年 5 月；Martin Hale 夏敖天
2 breeding 繁殖羽；Long Valley 塱原；May-06, 06 年 5 月；Martin Hale 夏敖天
3 breeding 繁殖羽；Long Valley 塱原；May-06, 06 年 5 月；Martin Hale 夏敖天
4 breeding 繁殖羽；Long Valley 塱原；May-06, 06 年 5 月；Jemi and John Holmes 孔思義、黃亞萍
5 breeding 繁殖羽；Long Valley 塱原；May-06, 06 年 5 月；Henry Lui 呂德恆
6 breeding 繁殖羽；Long Valley 塱原；14-May-06, 06 年 5 月 14 日；Michelle and Peter Wong 江敏兒、黃理沛

春季過境遷徙鳥 Spring Passage Migrant	夏候鳥 Summer Visitor	秋季過境遷徙鳥 Autumn Passage Migrant	冬候鳥 Winter Visitor

常見月份 1 2 3 4 5 6 7 8 9 10 11 12

留鳥 Resident	迷鳥 Vagrant	偶見鳥 Occasional Visitor

2
3
4
5
6

紅喉鷚

普 hóng hóu liù
粵 鷚：音漏

體長 length：14-15cm

Red-throated Pipit | *Anthus cervinus*

1

全身淡褐色，上體有粗黑色縱紋，下體較淡，胸前及脇有濃密深色縱紋，頭部顏色偏灰，有淡色眉紋，上嘴深色，下嘴淡黃色，腳淡紅色。繁殖羽面頰和喉明顯紅褐色。

Overall plumage pale brown, bold darkish streaks on upperparts, underparts paler, with heavy dark streaks on breast and flanks. Greyish head with pale eyebrow. Upper bill dark, lower bill pale yellow, legs pale red. In breeding plumage shows prominent reddish head and throat.

1 breeding 繁殖羽；Mai Po 米埔；Mar-08, 08 年 3 月；Michelle and Peter Wong 江敏兒、黃理沛
2 non-breeding moulting into breeding plumage 非繁殖羽轉換繁殖羽；Mai Po 米埔；Apr-06, 06 年 4 月；Martin Hale 夏敖天
3 non-breeding 非繁殖羽；Shek Kong 石崗；Feb-04, 04 年 2 月；Jemi and John Holmes 孔思義、黃亞萍
4 breeding 繁殖羽；Fung Lok Wai 豐樂圍；Apr-04, 04 年 4 月；Jemi and John Holmes 孔思義、黃亞萍
5 non-breeding moulting into breeding plumage 非繁殖羽轉換繁殖羽；Mai Po 米埔；Mar-08, 08 年 3 月；Kami Hui 許淑君
6 breeding 繁殖羽；Long Valley 塱原；13-Mar-05, 05 年 3 月 13 日；Doris Chu 朱詠兒
7 breeding 繁殖羽；Mai Po 米埔；Apr-07, 07 年 4 月；Pippen Ho 何志剛

春季過境遷徙鳥 Spring Passage Migrant			夏候鳥 Summer Visitor			秋季過境遷徙鳥 Autumn Passage Migrant			冬候鳥 Winter Visitor		
1	2	3	4	5	6	7	8	9	10	11	12
留鳥 Resident				迷鳥 Vagrant				偶見鳥 Occasional Visitor			

常見月份

2
3
4
5
6

7

黃腹鷚

㊟huáng fù liù
㊟鷚：音漏

體長length：14-17cm

Buff-bellied Pipit | *Anthus rubescens*

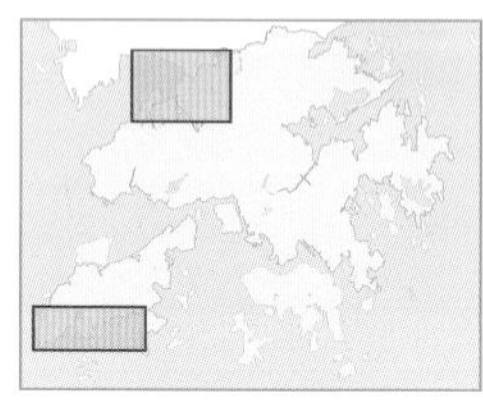

全身灰褐色，無明顯眉紋，嘴淡黃色，嘴端較深色，白色喉部兩側有明顯黑斑。下體較淡，胸前及脇有濃密深色縱紋，腳淡粉紅色。

Overall greyish brown, no obvious eyebrow. Bill is pale yellow with dark tip. Prominent blackish marks on both sides of the white throat. Heavily streaked breast and flanks, pinkish legs.

1 non-breeding 非繁殖羽：Mai Po 米埔：Feb-07, 07 年 2 月：Cherry Wong 黃卓研
2 non-breeding 非繁殖羽：Mai Po 米埔：2-Feb-06, 06 年 2 月 2 日：Michelle and Peter Wong 江敏兒、黃理沛
3 non-breeding 非繁殖羽：Mai Po 米埔：Feb-07, 07 年 2 月：Cherry Wong 黃卓研
4 non-breeding 非繁殖羽：Mai Po 米埔：Jan-06, 06 年 1 月：Martin Hale 夏敖天
5 non-breeding moulting into breeding plumage 非繁殖羽轉換繁殖羽：2-Feb-06, 06 年 2 月 2 日：Michelle and Peter Wong 江敏兒、黃理沛
6 non-breeding 非繁殖羽：Mai Po 米埔：2-Feb-06, 06 年 2 月 2 日：Michelle and Peter Wong 江敏兒、黃理沛
7 non-breeding 非繁殖羽：Mai Po 米埔：Jan-06, 06 年 1 月：Martin Hale 夏敖天

	春季過境遷徙鳥 Spring Passage Migrant			夏候鳥 Summer Visitor			秋季過境遷徙鳥 Autumn Passage Migrant			冬候鳥 Winter Visitor		
常見月份	1	2	3	4	5	6	7	8	9	10	11	12
	留鳥 Resident				迷鳥 Vagrant				偶見鳥 Occasional Visitor			

2
3
4
5
6

7

山鷚

㊟ shān liù
㊟ 鷚：音漏

體長 length：17cm

Upland Pipit | *Anthus sylvanus*

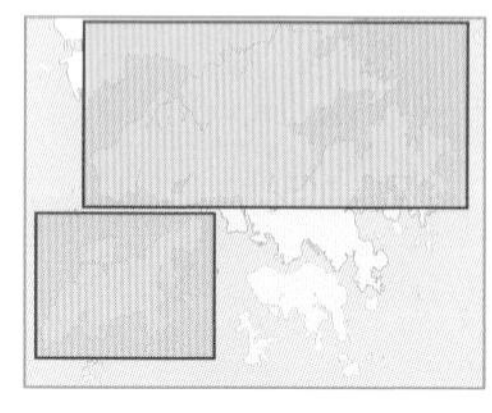

全身褐色，有明顯白眉紋。嘴粗短，上嘴深色，下嘴粉紅色。腳粉紅色。頭頂至上體有濃密斑點，下體偏白，胸和脇有淡色細紋，幼鳥細紋較多。站姿似理氏鷚，但沒有那麼挺直。叫聲較高較長久，通常在高地(500米以上)出沒。

Overall plumage brown. Obvious eyebrow. Bill short and thick; upper bill dark, lower bill pink. Pink legs. Heavily streaked from head to upperparts. Underparts whitish, pale narrow streaks on breast and flanks, heavier in juveniles. Posture resembles Richard's Pipit, but less erect. Long and high-pitched call. Usually found in upland areas above 500m.

1 adult 成鳥；Tai Mo Shan 大帽山；May-05, 05 年 5 月；Michelle and Peter Wong 江敏兒、黃理沛
2 Tai Mo Shan 大帽山；Cheung Ho Fai 張浩輝
3 adult 成鳥；Tai Mo Shan 大帽山；May-05, 05 年 5 月；Michelle and Peter Wong 江敏兒、黃理沛
4 Martin Hale 夏敖天
5 adult 成鳥；Tai Mo Shan 大帽山；May-05, 05 年 5 月；Michelle and Peter Wong 江敏兒、黃理沛

	春季過境遷徙鳥 Spring Passage Migrant			夏候鳥 Summer Visitor			秋季過境遷徙鳥 Autumn Passage Migrant			冬候鳥 Winter Visitor		
常見月份	1	2	3	4	5	6	7	8	9	10	11	12
	留鳥 Resident				迷鳥 Vagrant				偶見鳥 Occasional Visitor			

燕雀

㊂ yàn què

體長 length：13.5-16cm

Brambling | *Fringilla montifringilla*

頭和上身灰褐色，面頰、後枕和上背有明顯帶黑的斑紋，腰白色。喉、胸部和肩羽鮮橙色，腹部和尾下覆羽白色。嘴黃色，嘴尖黑色，腳淡粉紅色。雄鳥繁殖時頭明顯黑色。通常在地上活動。

Head and upperparts greyish brown, blackish marks on cheeks, nape and mantle, white rump. Throat and scapulars bright orange, belly and undertail coverts white. Yellow bill with black tip. Legs pale pink. Breeding male has prominent black head. Mostly found on the ground.

1 adult male 雄成鳥；Po Toi 蒲台；Apr-06, 06 年 4 月；Matthew and TH Kwan 關朗曦、關子凱
2 adult male moulting into breeding plumage 雄成鳥轉換繁殖羽；Po Toi 蒲台；8-Apr-06, 06 年 4 月 8 日；Doris Chu 朱詠兒
3 adult male 雄成鳥；Po Toi 蒲台；8-Apr-06, 06 年 4 月 8 日；Lo Kar Man 盧嘉孟
4 non-breeding adult male 非繁殖羽雄成鳥；Po Toi 蒲台；Nov-08, 08 年 11 月；Freeman Yue 余柏維
5 adult male 雄成鳥；Po Toi 蒲台；Apr-06, 06 年 4 月；Michelle and Peter Wong 江敏兒、黃理沛
6 non-breeding adult male 非繁殖羽雄成鳥；Po Toi 蒲台；Nov-08, 08 年 11 月；James Lam 林文華
7 non-breeding adult male 非繁殖羽雄成鳥；Po Toi 蒲台；Nov-08, 08 年 11 月；Freeman Yue 余柏維
8 adult male 雄成鳥；Po Toi 蒲台；Apr-06, 06 年 4 月；Michelle and Peter Wong 江敏兒、黃理沛

春季過境遷徙鳥 Spring Passage Migrant	夏候鳥 Summer Visitor	秋季過境遷徙鳥 Autumn Passage Migrant	冬候鳥 Winter Visitor

常見月份	1	2	3	4	5	6	7	8	9	10	11	12

留鳥 Resident	迷鳥 Vagrant	偶見鳥 Occasional Visitor

2
3
4
5
6
7
8

錫嘴雀

[普] xī zuǐ què

體長 length：16-18cm

Hawfinch | *Coccothraustes coccothraustes*

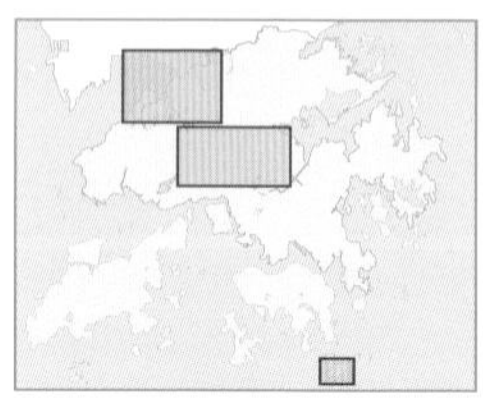

褐色的大型雀類，嘴下黑色小圓塊是特徵，粗厚的嘴灰色或偏黃。飛羽黑色，翼上有白色和深藍色帶，尾端白色。

Large brown finch with a small black bib in all plumages. The thick bill is grey or yellowish. Flying feathers are black, with white and dark blue bars on wings. Tail tipped white.

1 non-breeding adult female 非繁殖羽雌成鳥：Shek Kong 石崗：Jan-09, 09 年 1 月：Pippen Ho 何志剛
2 non-breeding adult female 非繁殖羽雌成鳥：Shek Kong 石崗：Jan-09, 09 年 1 月：James Lam 林文華
3 non-breeding adult female 非繁殖羽雌成鳥：Shek Kong 石崗：Jan-09, 09 年 1 月：Andy Kwok 郭匯昌
4 non-breeding adult female 非繁殖羽雌成鳥：Shek Kong 石崗：Jan-09, 09 年 1 月：Andy Kwok 郭匯昌
5 non-breeding adult female 非繁殖羽雌成鳥：Shek Kong 石崗：Jan-09, 09 年 1 月：Andy Kwok 郭匯昌
6 non-breeding adult female 非繁殖羽雌成鳥：Shek Kong 石崗：Jan-09, 09 年 1 月：Lee Kai Hong 李啟康

春季過境遷徙鳥 Spring Passage Migrant	夏候鳥 Summer Visitor	秋季過境遷徙鳥 Autumn Passage Migrant	冬候鳥 Winter Visitor

常見月份 1 2 3 4 5 6 7 8 9 10 11 12

留鳥 Resident	迷鳥 Vagrant	偶見鳥 Occasional Visitor

2
3
4
5

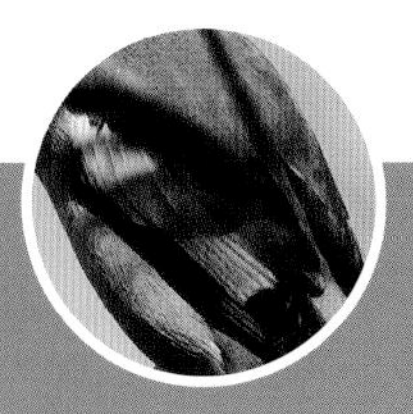

6

黑尾蠟嘴雀

㊟hēi wěi là zuǐ què

體長 length：15-18cm

Chinese Grosbeak | *Eophona migratoria*

1

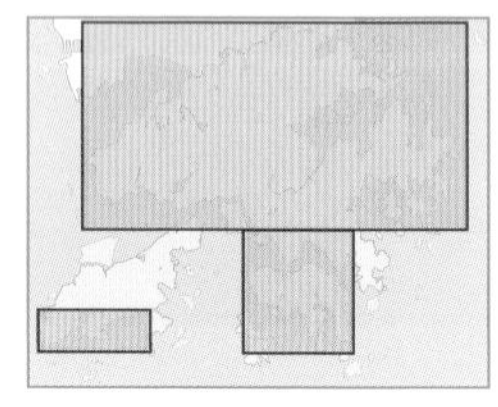

黃色的大嘴，頭至上體灰褐色，喉至下體淡灰褐，脇部沾橙色，尾下覆羽奶白色。雄鳥頭部黑色。飛行時黑色翼上有白斑，尾羽開叉。

Big yellow bill. Head to upperparts greyish brown, throat to underparts pale greyish brown. Flanks orange, undertail coverts milky white. Male has distinctive black head. In flight, it shows white marks on black flight feathers and deeply-notched tail.

1 adult male 雄成鳥：Fo Tan 火炭：Dec-07, 07 年 12 月：Christine and Samuel Ma 馬志榮、蔡美蓮
2 adult male 雄成鳥：Tsim Bei Tsui 尖鼻咀：Mar-08, 08 年 3 月：Ng Lin Yau 吳璉宥
3 adult female 雌成鳥：Shatin Central Park 沙田中央公園：16-Mar-08, 08 年 3 月 16 日：Christine and Samuel Ma 馬志榮、蔡美蓮
4 adult male 雄成鳥：Shatin Central Park 沙田中央公園：16-Mar-08, 08 年 3 月 16 日：Christine and Samuel Ma 馬志榮、蔡美蓮
5 adult female 雌成鳥：Shek Kong 石崗：Dec-05, 05 年 12 月：Jemi and John Holmes 孔思義、黃亞萍
6 adult male 雄成鳥：Shek Kong 石崗：Dec-05, 05 年 12 月：Jemi and John Holmes 孔思義、黃亞萍
7 adult male 雄成鳥：Mai Po 米埔：Jan-07, 07 年 1 月：Eling Lee 李佩玲
8 adult female 雌成鳥：Tsim Bei Tsui 尖鼻咀：Mar-08, 08 年 3 月：Ng Lin Yau 吳璉宥

	春季過境遷徙鳥 Spring Passage Migrant			夏候鳥 Summer Visitor			秋季過境遷徙鳥 Autumn Passage Migrant			冬候鳥 Winter Visitor		
常見月份	1	2	3	4	5	6	7	8	9	10	11	12
	留鳥 Resident				迷鳥 Vagrant				偶見鳥 Occasional Visitor			

2
3
4
5
6
7
8

黑頭蠟嘴雀

㊥ hēi tóu là zuǐ què

體長 length：18-23cm

Japanese Grosbeak | *Eophona personata*

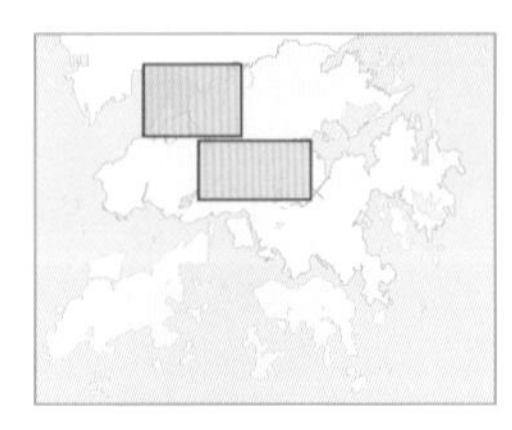

似較常見的黑尾蠟嘴雀，但體型較大及體色偏灰藍。頭前半、飛羽及尾黑色，飛羽中間有白斑，其餘全身羽色是帶灰藍的淺褐色。粗厚的嘴全黃色，腳淺褐色。

Resembles more common Chinse Grosbeak but bigger in size and greyish-blue in colour. Black at front half of head. Black flight feathers and tail. White wing spots on edge of wings. Mainly light brown with hint of greyish blue on the rest of body. Thick yellow bill. Light brown legs.

1 adult 成鳥：Shek Kong Airfield Road 石崗機場路：Jan-15, 15 年 1 月：Kinni Ho 何建業
2 adult 成鳥：Fung Yuen 鳳園：Jan-11, 11 年 1 月：Allen Chan 陳志雄
3 adult 成鳥：Lam Tsuen 林村; Mar-11, 11 年 3 月：jemi and John Holmes 孔思義、黃亞萍
4 adult 成鳥：Fung Yuen 鳳園：Feb-11, 11 年 2 月：Allen Chan 陳志雄
5 adult 成鳥：Shek Kong Airfield Road 石崗機場路：Jan-15, 15 年 1 月：Kinni Ho 何建業

春季過境遷徙鳥 Spring Passage Migrant	夏候鳥 Summer Visitor	秋季過境遷徙鳥 Autumn Passage Migrant	冬候鳥 Winter Visitor

常見月份 1 2 3 4 5 6 7 8 9 10 11 12

留鳥 Resident	迷鳥 Vagrant	偶見鳥 Occasional Visitor

2
3
4
5

普通朱雀

[普] pǔ tōng zhū què

體長 length：13-15cm

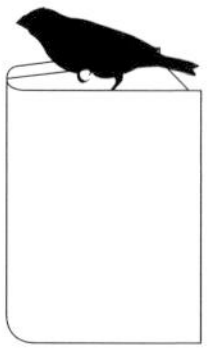

Common Rosefinch | *Carpodacus erythrinus*

其他名稱 Other names：朱雀

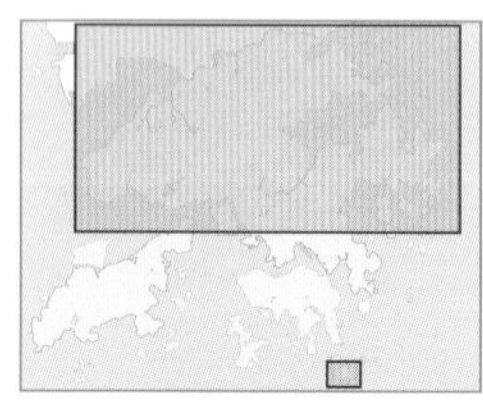

中型雀科鳥。嘴粗，尾部有凹陷缺位。雄性頭、胸和腰紅色，其他上體部分暗紅色。雌性有褐色縱紋和兩道淡黃色翼斑，但腰部沒有紅色。稀少冬候鳥，單隻或小群在開揚廣闊郊野進食漿果。

Medium-sized finch. Thick bill, notched tail. Male has red head, breast and rump, with duller red elsewhere on upperparts. Females, slightly streaked brown with two buff wingbars and no red rump. Winter visitor, getting rarer, singles or small parties feeding on tree berries in open countryside.

1 adult male 雄成鳥；Shek Kong 石崗；Feb-08, 08 年 2 月；Martin Hale 夏敖天
2 adult female 雌成鳥；28-Feb-08, 08 年 2 月 28 日；Martin Hale 夏敖天
3 adult female 雌成鳥；Shek Kong 石崗；Jan-07, 07 年 1 月；Andy Kwok 郭匯昌
4 adult male 雄成鳥；28-Feb-08, 08 年 2 月 28 日；Martin Hale 夏敖天
5 adult female 雌成鳥；Shek Kong 石崗；Jan-07, 07 年 1 月；Andy Kwok 郭匯昌
6 adult female 雌成鳥；3-Mar-08, 08 年 3 月 3 日；Martin Hale 夏敖天
7 adult male 雄成鳥；Shek Kong 石崗；13-Feb-05, 05 年 2 月 13 日；Doris Chu 朱詠兒

	春季過境遷徙鳥 Spring Passage Migrant			夏候鳥 Summer Visitor			秋季過境遷徙鳥 Autumn Passage Migrant			冬候鳥 Winter Visitor		
常見月份	1	2	3	4	5	6	7	8	9	10	11	12
	留鳥 Resident				迷鳥 Vagrant				偶見鳥 Occasional Visitor			

2
3
4
5
6

7

金翅雀

㊟ jīn chì què

體長 length：12.5-14cm

Grey-capped Greenfinch | *Chloris sinica*

其他名稱 Other names：Chinese Greenfinch

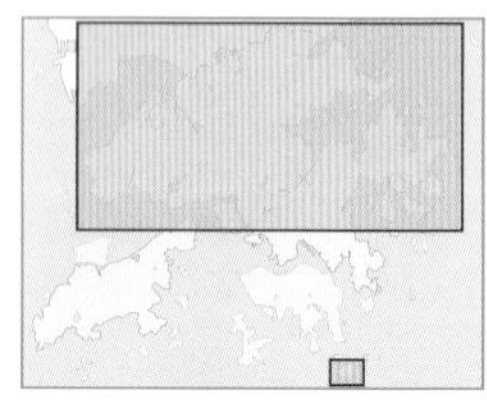

頭部灰黃色，上體深黃褐色，飛羽黑色，有顯眼黃斑，尾羽黑色。喉至下體黃褐色，尾下覆羽黃色。嘴和腳淡粉紅色。雌鳥羽色較淡。

Head yellowish grey, upperparts dark yellowish brown, flight feathers black with distinctive yellow marks, tail black. Throat to underparts yellowish-brown, undertail coverts yellow. Bill and legs pale pink. Female is paler.

1 adult male 雄成鳥；Tsing Yi 青衣；8-Feb-08, 08年2月8日；Owen Chiang 深藍
2 adult male / 1st winter 雄成鳥 / 第一年冬天雄鳥；Tsing Yi 青衣；Feb-08, 08 年 2 月；Isaac Chan 陳家強
3 adult male 雄成鳥；Tsing Yi 青衣；Jan-08, 08 年 1 月；Andy Kwok 郭匯昌
4 adult male / 1st winter 雄成鳥 / 第一年冬天雄鳥；Tsing Yi 青衣；Feb-08, 08 年 2 月；Eling Lee 李佩玲
5 adult male 雄成鳥；Tsing Yi 青衣；Feb-08, 08 年 2 月；Lee Kai Hong 李啟康
6 adult male 雄成鳥；Tsing Yi 青衣；Feb-08, 08 年 2 月；Lee Kai Hong 李啟康
7 adult male 雄成鳥；Tsing Yi 青衣；Feb-08, 08 年 2 月；Danny Ho 何國海

春季過境遷徙鳥 Spring Passage Migrant			夏候鳥 Summer Visitor			秋季過境遷徙鳥 Autumn Passage Migrant			冬候鳥 Winter Visitor		
常見月份 1	2	3	4	5	6	7	8	9	10	11	12
留鳥 Resident				迷鳥 Vagrant				偶見鳥 Occasional Visitor			

2
3
4
5
6

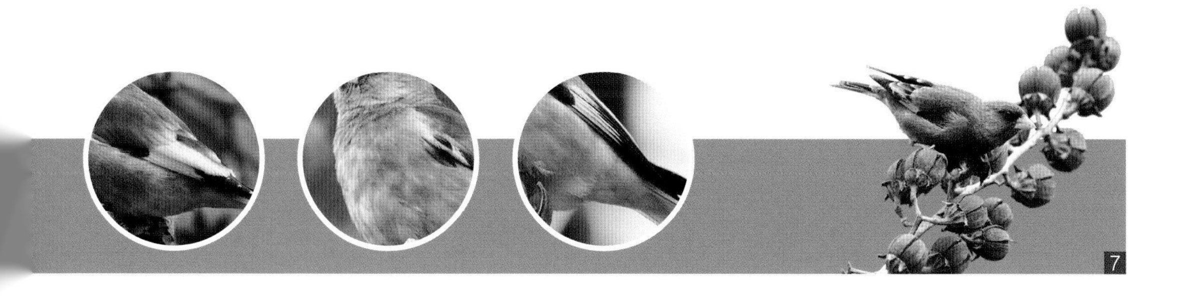
7

黃雀

㊟huáng què

體長length：11-12cm

Eurasian Siskin | *Spinus spinus*

其他名稱 Other names：Siskin

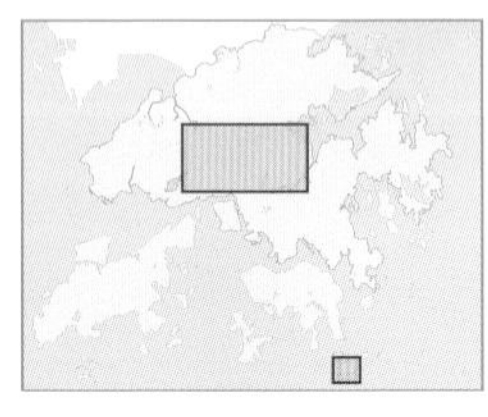

小型雀科，主要顏色有黃、綠和黑。雌雄均有兩道明顯黃色翼斑。間中出現的冬候鳥，大群可於馬尾松的頂部出現。來源可疑的單隻可能隨時出現。

Tiny finch, mainly yellow, green and black in colour. Both sexes have two bright yellow wing-bars on a dark flying feathers. Irruptive winter visitor, can occur in large flocks favouring the tops of trees such as Casuarina equisetifolia and Liquidambar formosana. Singles of dubious origin may turn up here and there.

1 adult male 雄成鳥；Po Toi 蒲台；Nov-07, 07 年 11 月；Michelle and Peter Wong 江敏兒、黃理沛
2 adult male 雄成鳥；Po Toi 蒲台；Nov-07, 07 年 11 月；Michelle and Peter Wong 江敏兒、黃理沛
3 adult female 雌成鳥；Po Toi 蒲台；Nov-07, 07 年 11 月；Michelle and Peter Wong 江敏兒、黃理沛
4 adult female 雌成鳥；Po Toi 蒲台；Nov-07, 07 年 11 月；Michelle and Peter Wong 江敏兒、黃理沛
5 adult male and female 雄成鳥和雌成鳥；Po Toi 蒲台；Nov-07, 07 年 11 月；Michelle and Peter Wong 江敏兒、黃理沛
6 adult female 雌成鳥；Po Toi 蒲台；Nov-07, 07 年 11 月；Michelle and Peter Wong 江敏兒、黃理沛

春季過境遷徙鳥 Spring Passage Migrant	夏候鳥 Summer Visitor	秋季過境遷徙鳥 Autumn Passage Migrant	冬候鳥 Winter Visitor

常見月份 1 2 3 4 5 6 7 8 9 10 11 12

留鳥 Resident	迷鳥 Vagrant	偶見鳥 Occasional Visitor

黃額絲雀

㊟huáng è sī què

體長 length：11-13cm

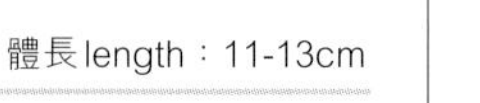

Yellow-fronted Canary | *Crithagra mozambica*

俗名「石燕」。頭部灰褐色，有明顯的黃色眉和頰紋，頰下紋黑色。上體褐灰色，有深色縱紋。腰鮮黃色，嘴和腳粉紅色。下體鮮黃色。幼鳥顏色較淡，下體沾有灰斑。所有記錄被判斷為逸鳥。

Greyish brown head with yellow supercilium and submoustachial stripe, malar stripe black. Upperparts brownish-grey with dark streaks. Rump light yellow, bill and legs pink. Underparts yellow. Juvenile is paler, with underparts tinted greyish. All records are considered as birds escaped from captivity.

1 Mai Po 米埔；7-Jun-06, 06年6月7日；Owen Chiang 深藍
2 Fung Lok Wai 豐樂圍；Jan-06, 06 年 1 月；Henry Lui 呂德恆
3 Mai Po 米埔；7-Jun-06，06年6月7日；Owen Chiang 深藍
4 Lai Chi Kok 荔枝角；Jan-08, 08 年 1 月；James Lam 林文華
5 Cheung Chau 長洲；Dec-05, 05 年 12 月；Henry Lui 呂德恆

春季過境遷徙鳥 Spring Passage Migrant			夏候鳥 Summer Visitor			秋季過境遷徙鳥 Autumn Passage Migrant			冬候鳥 Winter Visitor		
常見月份 1	2	3	4	5	6	7	8	9	10	11	12
留鳥 Resident				迷鳥 Vagrant				偶見鳥 Occasional Visitor			

2

3

4

5

鳳頭鵐

[普]fèng tóu wū
[粵]鵐：音無

體長 length：16cm

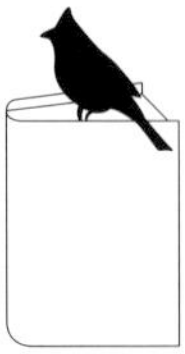

Crested Bunting | *Emberiza lathami*

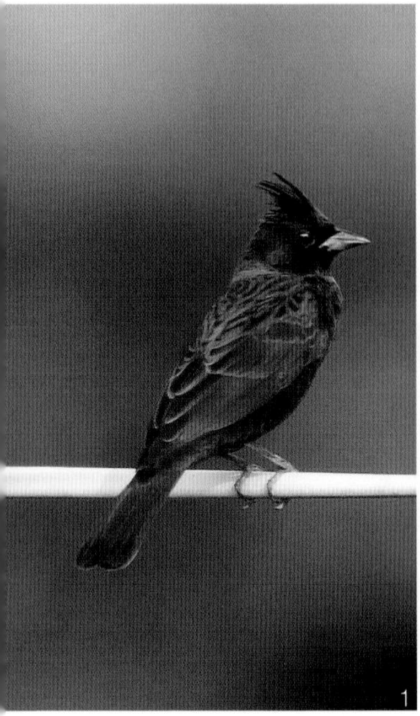
1

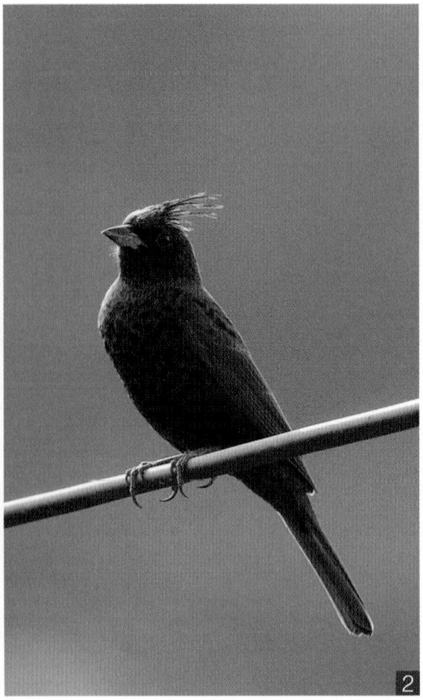
2

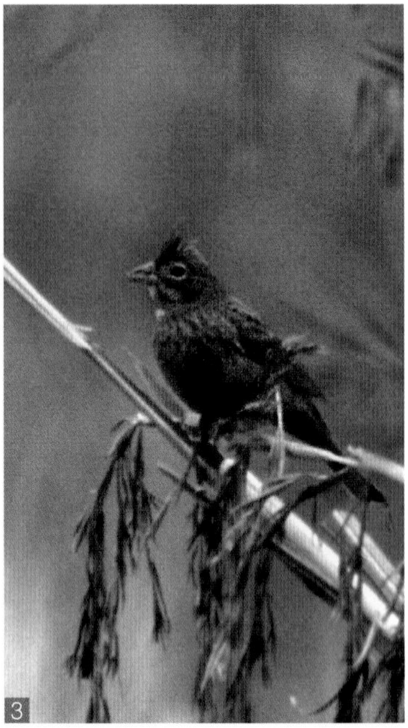
3

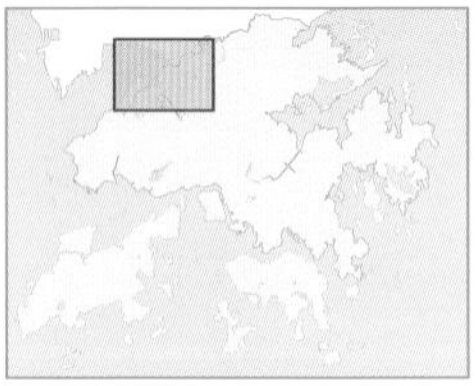

🌳

大型的鵐，頭帶細長冠羽。雄性黑色，翅膀及尾栗色。雌性深橄欖褐，上背及胸有深色縱紋，翼較深色，羽毛邊緣紅褐色。上嘴灰褐色，下嘴粉紅。以往常見於農田，現已非常罕見。

Large-sized bunting with long crest on head. Male dark in color with chestnut wing and tail. Female is dark olive brown, with dark strips at mantle and breast. Dark wings with rufous-red fringe. Upper bill greyish brown, lower bill pink. Occurs around cultivation but very rare in Hong Kong in recent years.

[1] adult male 雄成鳥：Long Valley塱原：26 Nov-14, 14年11月26日：Wong Shui Chi 黃瑞芝
[2] adult male 雄成鳥：Long Valley塱原：19 Jan-13, 13年1月19日：Beetle Cheng 鄭諾銘
[3] adult female 雌成鳥：Wu Kau Tang 烏蛟騰：20-Mar-05, 05 年 3 月 20 日：Doris Chu 朱詠兒
[4] adult female/immature 雌成鳥 / 未成年鳥：Long Valley 塱原：14 Nov-13, 13年11月14日：Wong Shui Chi 黃瑞芝

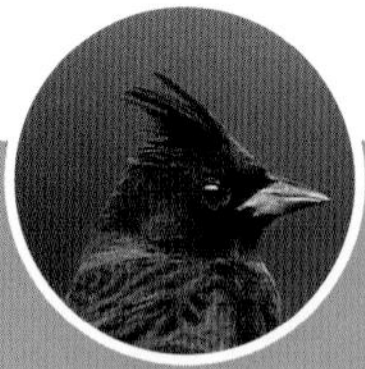

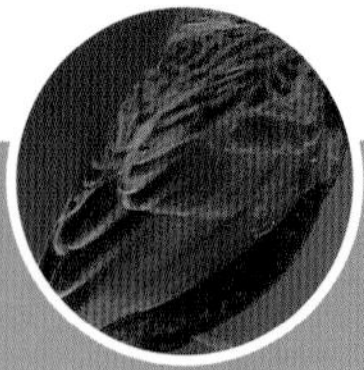

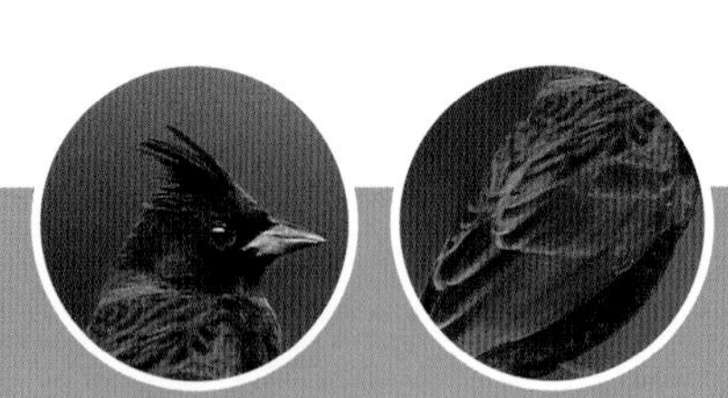

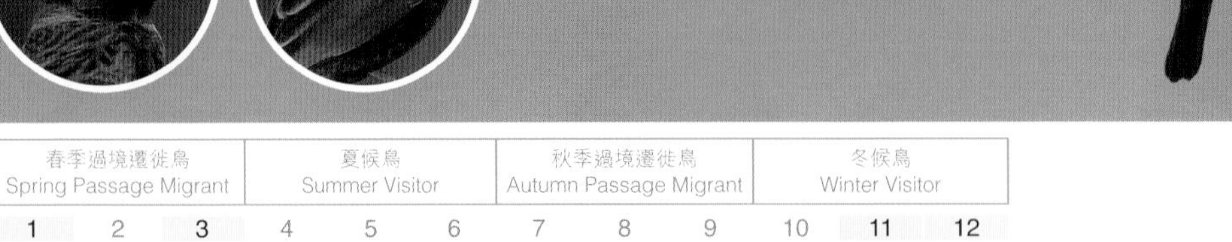

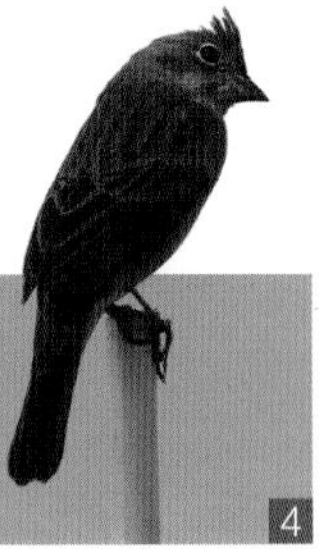
4

	春季過境遷徙鳥 Spring Passage Migrant			夏候鳥 Summer Visitor			秋季過境遷徙鳥 Autumn Passage Migrant			冬候鳥 Winter Visitor	
常見月份	1	2	3	4	5	6	7	8	9	10	11	12
	留鳥 Resident				迷鳥 Vagrant				偶見鳥 Occasional Visitor			

藍鵐

普 lán wū
粵 鵐：音無

體長 length：13cm

Slaty Bunting | *Emberiza siemsseni*

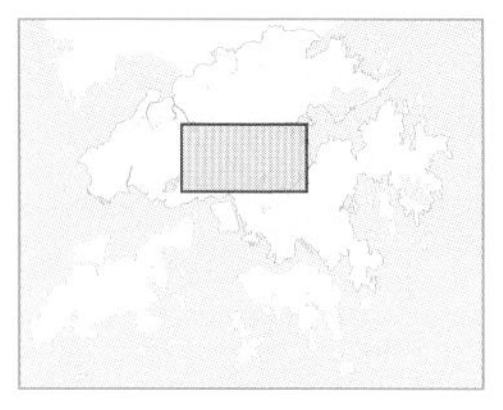

雄鳥除下腹至臀部及尾外緣白色外，全身羽色大致灰藍色。雌鳥除頭及胸部棕色、腰灰色外，全身大致暗褐，有兩條淺色翼斑。嘴黑色，腳淡粉紅色。

Male has mainly greyish-blue body. White from lower belly to vent. White outer tail feathers. Female has mainly dull brown body with two pale wing bars. Brown head and breast. Grey rump. Black bill. Pale pink legs.

1 adult female 雌成鳥；Tai Po Kau 大埔滘；Feb-13, 13 年 2 月；Wing Tang 鄧詠詩
2 adult female 雌成鳥；Tai Po Kau 大埔滘；Feb-13, 13 年 2 月；Gary Chow 周家禮

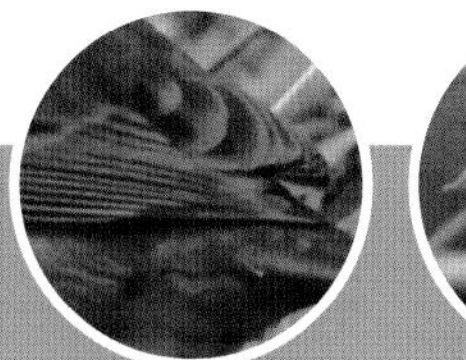

春季過境遷徙鳥 Spring Passage Migrant	夏候鳥 Summer Visitor	秋季過境遷徙鳥 Autumn Passage Migrant	冬候鳥 Winter Visitor

1 2 3 4 5 6 7 8 9 10 11 12 常見月份

留鳥 Resident	迷鳥 Vagrant	偶見鳥 Occasional Visitor

白頭鵐

㊟ bái tóu wū
㊣ 鵐：音無

體長 length：16-17.5cm

Pine Bunting | *Emberiza leucocephalos*

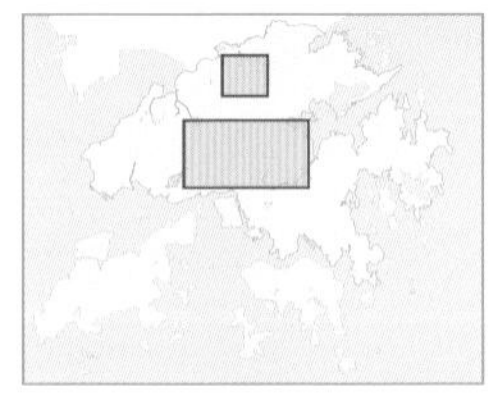

上體褐色具黑色縱紋，下體灰白色具深色幼縱紋。雄鳥頭部栗色，繁殖羽白色頂冠紋兩邊黑色，非繁殖羽白色頂冠紋轉為褐色，白色耳羽邊緣黑色，胸帶白色。嘴灰藍色而上嘴中線褐色，腳粉褐色。

Brown upperparts with black streaks. Greyish-white underparts with thin dark streaks. Male has chestnut head. Black lateral crown-stripes on sides of white median crown-stripe in breeding plumage. Median crown-stripe becomes brown in non-breeding plumage. Black edges on white ear-coverts. White breast band. Greyish-blue bill with brown culmen. Pale brown legs.

1 adult female 雌成鳥：Long Valley 塱原：Nov-14, 14 年 11 月：Allen Chan 陳志雄
2 adult female 雌成鳥：Long Valley 塱原：Nov-14, 14 年 11 月：Kinni Ho 何建業
3 adult female 雌成鳥：Long Valley 塱原：Nov-14, 14 年 11 月：Allen Chan 陳志雄

	春季過境遷徙鳥 Spring Passage Migrant			夏候鳥 Summer Visitor			秋季過境遷徙鳥 Autumn Passage Migrant			冬候鳥 Winter Visitor		
常見月份	1	2	3	4	5	6	7	8	9	10	11	12
	留鳥 Resident				迷鳥 Vagrant				偶見鳥 Occasional Visitor			

2
3

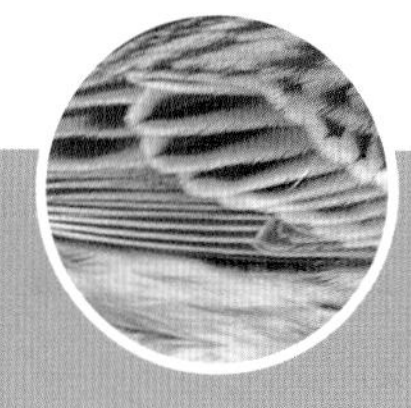

白眉鵐

普 bái méi wū
粵 鵐：音無

體長 length：15cm

Tristram's Bunting | *Emberiza tristrami*

1

有顯眼的白色粗眼眉、頰下紋和冠紋，面頰後部有一白點。上體灰褐色，有深色粗縱紋，腰部呈栗紅色，兩側尾羽白色，腳粉紅色。雄鳥的頭部黑色，喉部至尾下覆羽白色，胸和脇沾褐色。雌鳥頭部的顏色較淡，胸、脇褐色有深色縱紋。常在樹林底層覓食。

Distinctive thick white crown stripes, supercilium and submoustachial stripe, and a white spot behind the ear coverts. Upperparts greyish brown with dark streaks, rufous rump, white outertail feathers and pink legs. Male has black head, white throat and underparts, breast and flanks tinted brown. Female has paler head, breast and flanks brownish yellow with narrow brownish stripes. Forages at ground level in woodland.

1 breeding male 繁殖羽雄鳥：Wonderland Villas 華景山莊：Jan-07, 07 年 1 月：Pippen Ho 何志剛
2 non-breeding adult male moulting into breeding plumage 非繁殖羽雄成鳥轉換繁殖羽：Wonderland Villas 華景山莊：Feb-07, 07 年 2 月：Martin Hale 夏敖天
3 non-breeding adult male moulting into breeding plumage 非繁殖羽雄成鳥轉換繁殖羽：Kwai Chung 葵涌：Jan-07, 07 年 1 月：Allen Chan 陳志雄
4 non-breeding adult female 非繁殖羽雌成鳥：Kwai Chung 葵涌：Jan-07, 07 年 1 月：Allen Chan 陳志雄
5 adult female 雌成鳥：Wonderland Villas 華景山莊：Jan-07, 07 年 1 月：Pippen Ho 何志剛
6 breeding male 繁殖羽雄鳥：Wonderland Villas 華景山莊：Feb-07, 07 年 2 月：Martin Hale 夏敖天

春季過境遷徙鳥 Spring Passage Migrant	夏候鳥 Summer Visitor	秋季過境遷徙鳥 Autumn Passage Migrant	冬候鳥 Winter Visitor

常見月份 1 2 3 4 5 6 7 8 9 10 11 12

留鳥 Resident	迷鳥 Vagrant	偶見鳥 Occasional Visitor

2
3
4
5

6

栗耳鵐

㊟lì ěr wū
㊣鵐：音無

體長 length：16cm

Chestnut-eared Bunting | *Emberiza fucata*

其他名稱 Other names：赤胸鵐, Grey-headed Bunting

1

眼眉和頰下紋粗而淡黃，臉頰栗色，頰下紋黑色，伸延至胸部。上體淡黃褐色，有濃密黑色和褐色粗縱紋，肩羽栗色。下體淡黃褐色，胸部有深色縱紋。嘴灰色，腳粉紅色。雄鳥頭部灰色而有深色縱紋，胸部有栗色胸帶。

Bunting with thick pale yellow supercilium and submoustachial stripe, chestnut cheeks, black malar stripe extends to the breast. Upperparts pale yellowish brown, with heavy, thick black and brown streaks, scapulars chestnut. Underparts pale yellowish brown, with dark stripes forming a band on upper breast. Grey bill and pink legs. Male has grey head with dark stripes and brown breast band.

1 1st winter male 第一年冬天雄鳥：Long Valley 塱原：Nov-07, 07 年 11 月：Michelle and Peter Wong 江敏兒、黃理沛
2 1st winter female 第一年冬天雌鳥：Long Valley 塱原：Nov-08, 08 年 11 月：Ng Lin Yau 吳璉宥
3 1st winter male 第一年冬天雄鳥：Long Valley 塱原：Nov-07, 07 年 11 月：Michelle and Peter Wong 江敏兒、黃理沛
4 1st winter female 第一年冬天雌鳥：Long Valley 塱原：Nov-04, 04 年 11 月：Jemi and John Holmes 孔思義、黃亞萍
5 1st winter female 第一年冬天雌鳥：Long Valley 塱原：Nov-07, 07 年 11 月：Pippen Ho 何志剛

	春季過境遷徙鳥 Spring Passage Migrant			夏候鳥 Summer Visitor			秋季過境遷徙鳥 Autumn Passage Migrant			冬候鳥 Winter Visitor		
常見月份	1	2	3	4	5	6	7	8	9	10	11	12
	留鳥 Resident				迷鳥 Vagrant				偶見鳥 Occasional Visitor			

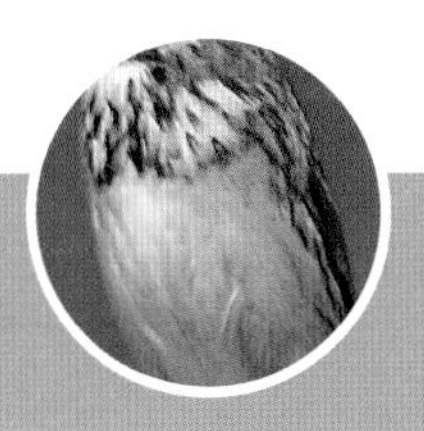

小鵐

[普] xiǎo wū
[粵] 鵐：音無

體長 length：12-13.5cm

Little Bunting | *Emberiza pusilla*

1

小型的鵐。冠紋、眉紋和頰下紋粗而淡黃色，貫眼紋和側冠紋深褐色，面頰栗色。上體紅褐色，有深色的粗縱紋。下體白色，胸部和脇部有黑色縱紋。嘴灰色，腳淡紛紅色。

Small bunting with thick buffy crown stripe, supercilium and submoustachial stripe. Eyestripe and lateral crown stripes dark brown. Cheeks chestnut. Upperparts rufous brown, with thick and dark streaks. Underparts white with with black stripes on the breast and flanks. Grey bill and pink legs.

1 breeding 繁殖羽：Long Valley 塱原：Feb-06, 06 年 2 月：Michelle and Peter Wong 江敏兒、黃理沛
2 non-breeding 非繁殖羽：Long Valley 塱原：Dec-05, 05 年 12 月：Henry Lui 呂德恆
3 breeding 繁殖羽：Po Toi 蒲台：Apr-08, 08 年 4 月：James Lam 林文華
4 breeding 繁殖羽：Po Toi 蒲台：Apr-07, 07 年 4 月：Allen Chan 陳志雄
5 breeding 繁殖羽：Po Toi 蒲台：Apr-08, 08 年 4 月：Joyce Tang 鄧玉蓮
6 breeding 繁殖羽：Long Valley 塱原：Feb-06, 06 年 2 月：Michelle and Peter Wong 江敏兒、黃理沛
7 breeding 繁殖羽：Long Valley 塱原：Feb-06, 06 年 2 月：Michelle and Peter Wong 江敏兒、黃理沛

春季過境遷徙鳥 Spring Passage Migrant	夏候鳥 Summer Visitor	秋季過境遷徙鳥 Autumn Passage Migrant	冬候鳥 Winter Visitor

常見月份	1	2	3	4	5	6	7	8	9	10	11	12

留鳥 Resident	迷鳥 Vagrant	偶見鳥 Occasional Visitor

2
3
4
5
6

7

黃眉鵐

㊟ huáng méi wū
㊟ 鵐：音無

體長 length：13-15cm

Yellow-browed Bunting | *Emberiza chrysophrys*

1

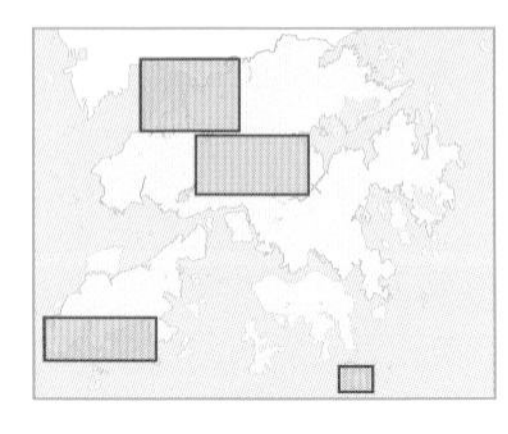

有顯眼的粗眼眉、頰下紋和冠紋，白眼眉前端黃色。上體灰褐色，有深色粗縱紋，兩側尾羽白色，腳粉紅色。首次越冬和雌鳥頭部的顏色較淡。

Bunting with distinctive crown stripes, supercilium and submoustachial stripe. White supercilium yellow at the front. Upperparts greyish brown with dark streaks, outertail feathers white, pink legs. First winter birds and females are paler.

1 adult female 雌成鳥；Po Toi 蒲台；Apr-07, 07 年 4 月；Wallace Tse 謝鑑超
2 adult female 雌成鳥；Po Toi 蒲台；Apr-07, 07 年 4 月；Law Kam Man 羅錦文
3 female / non-breeding male 雌鳥 / 非繁殖羽雄鳥；Mai Po 米埔；Oct-07, 07 年 10 月；Cheng Nok Ming 鄭諾銘
4 adult female 雌成鳥；Po Toi 蒲台；Apr-07, 07 年 4 月；Michelle and Peter Wong 江敏兒、黃理沛
5 adult female 雌成鳥；Po Toi 蒲台；28-Apr-07, 07 年 4 月 28 日；Owen Chiang 深藍
6 adult female 雌成鳥；Po Toi 蒲台；Apr-07, 07 年 4 月；Sammy Sam and Winnie Wong 森美與雲妮
7 adult female 雌成鳥；Po Toi 蒲台；Apr-07, 07 年 4 月；Wallace Tse 謝鑑超

春季過境遷徙鳥 Spring Passage Migrant			夏候鳥 Summer Visitor			秋季過境遷徙鳥 Autumn Passage Migrant			冬候鳥 Winter Visitor		
1	2	3	4	5	6	7	8	9	10	11	12

常見月份

留鳥 Resident	迷鳥 Vagrant	偶見鳥 Occasional Visitor

2
3
4
5
6

7

田鵐

[普] tián wū
[粵] 鵐：音無

體長 length：13-14.5cm

Rustic Bunting | *Emberiza rustica*

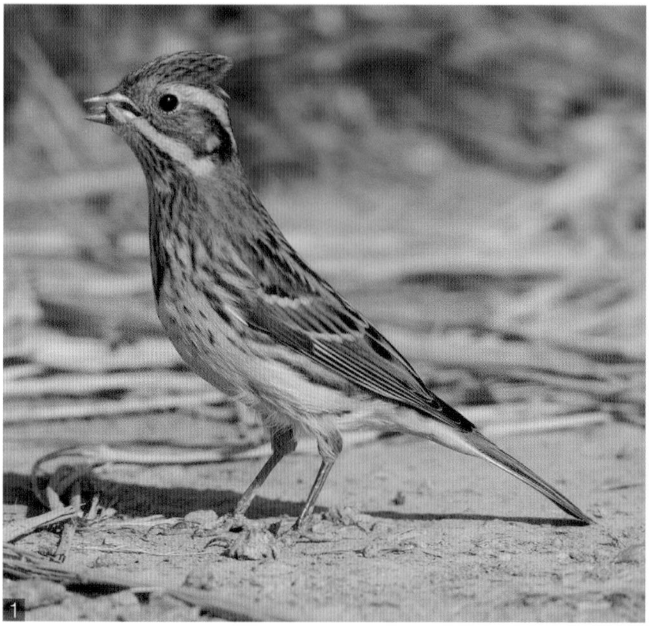
1

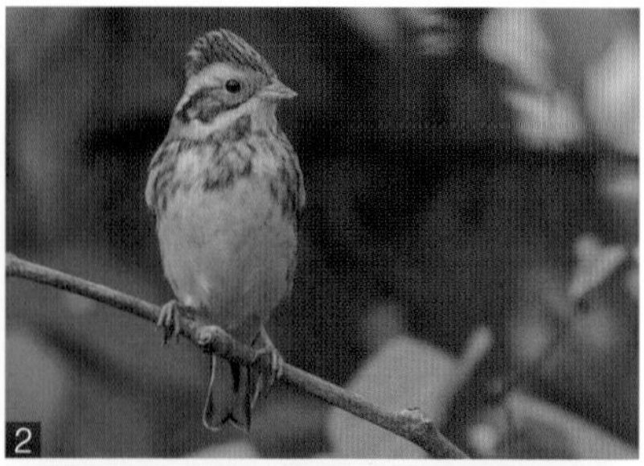
2

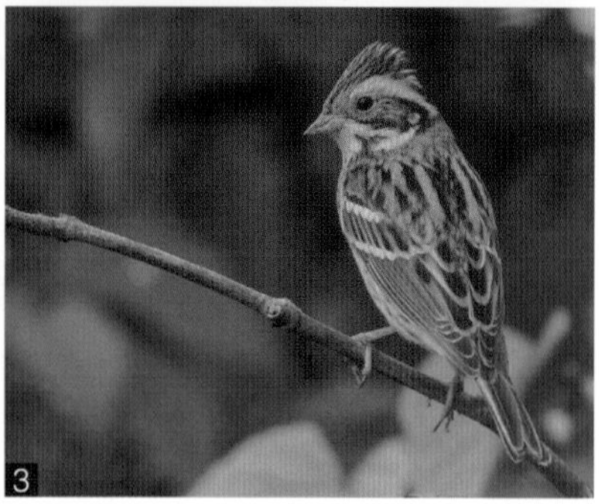
3

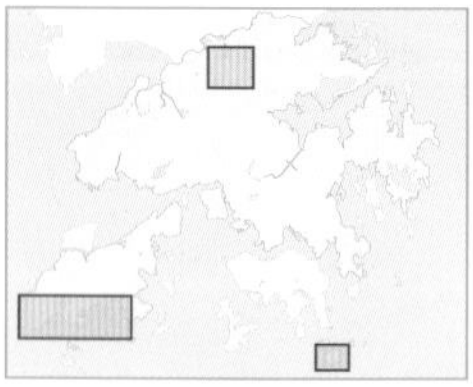

頭部具黑白條紋，上體栗黑斑駁且羽邊白色，略具羽冠，耳羽有一白點，腰棕色。下體白色，胸及脇部具棕色幼縱紋。嘴深灰色，基部粉褐色，腳粉褐色。

Black and white stripes on head. Mix of chestnut and black on upperparts with white fringes on feathers. Short crest. White spots on ear-coverts. Brown rump. White underparts. Thin brown streaks on breast and flanks. Dark grey bill with pink at base. Pale brown legs.

[1] adult female 雌成鳥：Long Valley 塱原：Nov-14, 14 年 11 月：Allen Chan 陳志雄
[2] adult female 雌成鳥：Long Valley 塱原：Nov-14, 14 年 11 月：Kwok Tsz Ki 郭子祈
[3] adult female 雌成鳥：Long Valley 塱原：Nov-14, 14 年 11 月：Kwok Tsz Ki 郭子祈
[4] adult female 雌成鳥：Long Valley 塱原：Nov-14, 14 年 11 月：Kwok Tsz Ki 郭子祈

4

春季過境遷徙鳥 Spring Passage Migrant	夏候鳥 Summer Visitor	秋季過境遷徙鳥 Autumn Passage Migrant	冬候鳥 Winter Visitor

常見月份	1	2	3	4	5	6	7	8	9	10	11	12

留鳥 Resident	迷鳥 Vagrant	偶見鳥 Occasional Visitor

黑頭鵐

[普] hēi tóu wū
[粵] 鵐：音無

體長 length：15.5-17.5cm

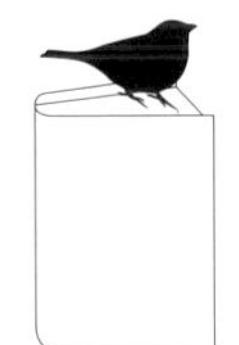

Black-headed Bunting | *Emberiza melanocephala*

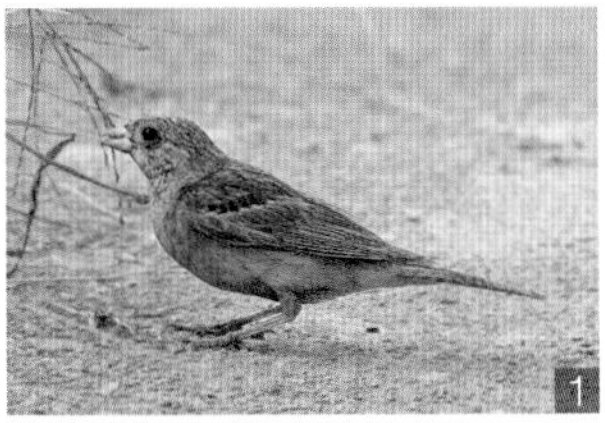
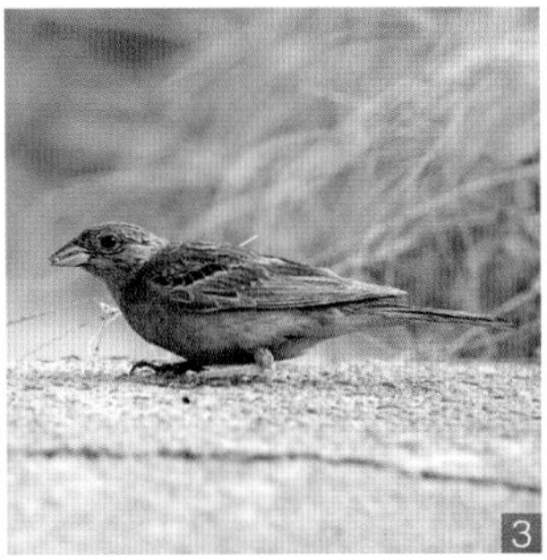
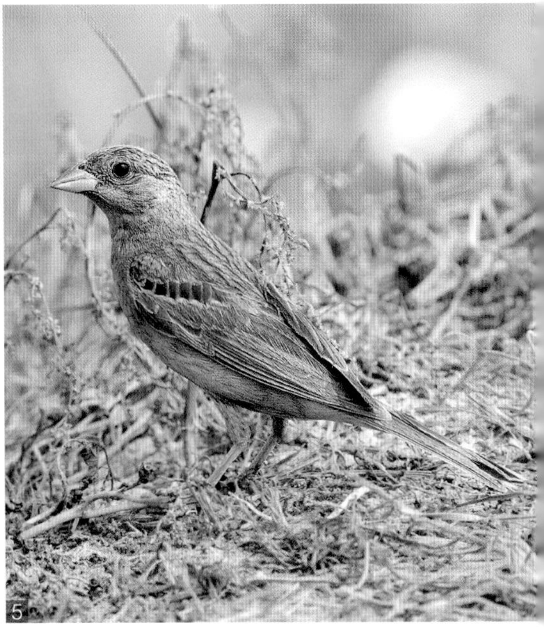
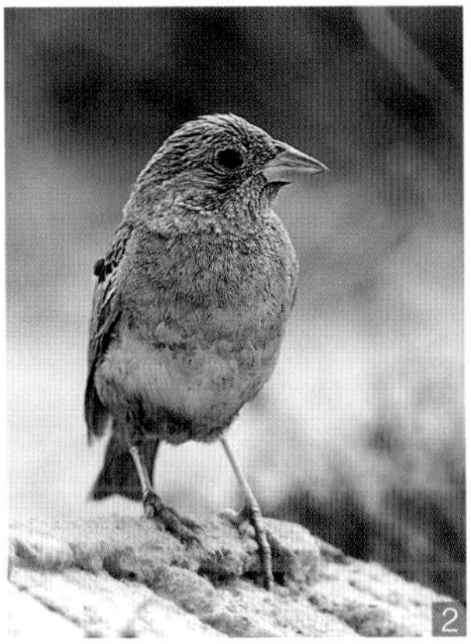
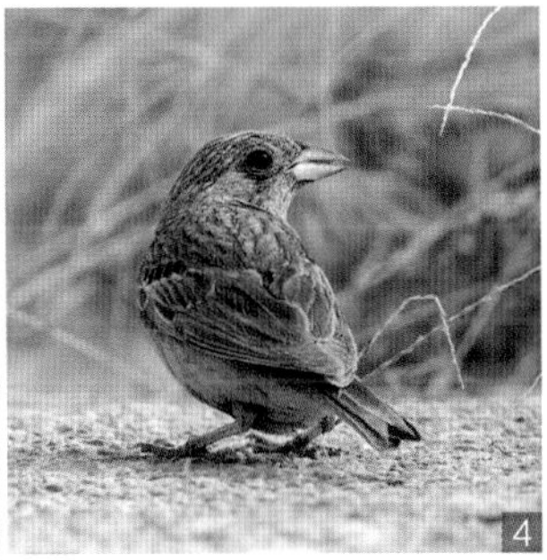

嘴部粗大的大型鵐。繁殖期雄鳥頭枕黑色背栗色，喉至胸部鮮黃色。雌鳥上體淺褐色、下體淡黃色。外尾羽沒有白邊。飛羽羽端白色在翅膀摺合時看似翼斑。香港大部分記錄均在秋季錄得，大都屬於非繁殖期或雌性，很易與褐頭鵐混淆。

Large bunting with thick bill. Male in breeding plumage shows black cap and face, bright yellow breast and rufous upperparts. Female has yellow-buff underparts and pale brown upperparts. Lacks white outer tail feathers. Fine white tips at flight feathers give folded wing a barred appearance. Most Hong Kong records were in autumn are non-breeding or female birds, which are easily confused with Red-headed Bunting *E. bruniceps*.

[1] 1st winter male 第一年冬天雄鳥：Long Valley 塱原：Nov-09, 09 年 11 月：Pang Chun Chiu 彭俊超
[2] 1st winter male 第一年冬天雄鳥：Long Valley 塱原：18-Nov-08, 08 年 11 月 18 日：James Lam 林文華
[3] 1st winter male 第一年冬天雄鳥：Long Valley 塱原：Nov-08, 08 年 11 月：Danny Ho 何國海
[4] 1st winter male 第一年冬天雄鳥：Long Valley 塱原：Nov-08, 08 年 11 月：Danny Ho 何國海
[5] 1st winter male 第一年冬天雄鳥：Long Valley 塱原：Nov-08, 08 年 11 月：Cheng Nok Ming 鄭諾銘

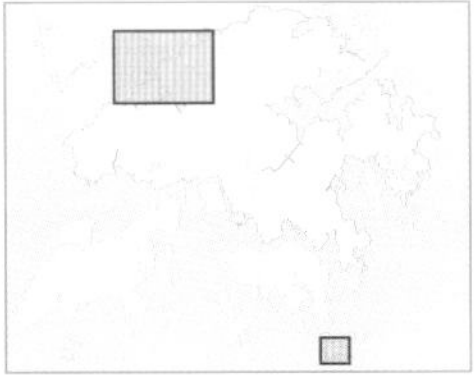

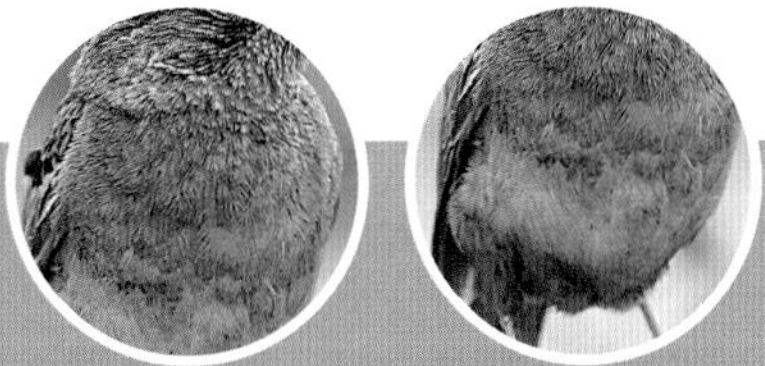

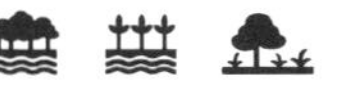

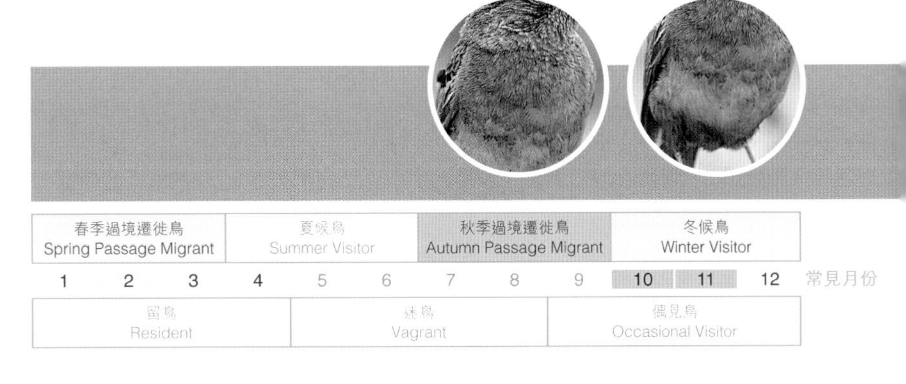

春季過境遷徙鳥 Spring Passage Migrant	夏候鳥 Summer Visitor	秋季過境遷徙鳥 Autumn Passage Migrant	冬候鳥 Winter Visitor

1 2 3 4 5 6 7 8 9 **10** **11** 12 常見月份

留鳥 Resident	迷鳥 Vagrant	偶見鳥 Occasional Visitor

黃喉鵐

[普]huáng hóu wū
[粵]鵐：音無

體長 length：14-22cm

Yellow-throated Bunting | *Emberiza elegans*

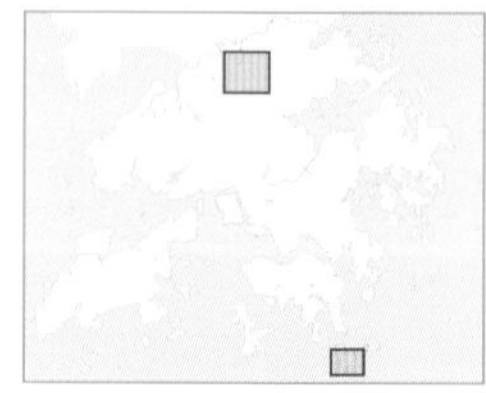

繁殖期雄鳥枕和喉有獨特鮮黃色與黑褐色的眼，面頰和胸帶斑紋搶眼易認。非繁殖期鳥體黑色部位轉為褐色，而黃色部位變得暗淡。罕有冬候鳥，在中國中部繁殖，沿海低地越冬。

Breeding male shows distinctive bright yellow throat and hindcrown with black eyemask. Distinctive black patch on cheek and breast. For Female and male non-breeding, black on head is replaced by brown, yellow parts duller. Breeds in Central China and winters on the coastal lowlands.

1 Po Toi 蒲台：Nov-09, 09 年 11 月：Michelle and Peter Wong 江敏兒、黃理沛
2 Po Toi 蒲台：Nov-09, 09 年 11 月：Michelle and Peter Wong 江敏兒、黃理沛
3 Po Toi 蒲台：Apr-10, 10 年 4 月：Michelle and Peter Wong 江敏兒、黃理沛
4 Po Toi 蒲台：Apr-10, 10 年 4 月：Michelle and Peter Wong 江敏兒、黃理沛
5 Po Toi 蒲台：Apr-10, 10 年 4 月：Michelle and Peter Wong 江敏兒、黃理沛
6 adult female 雌成鳥：Long Valley 塱原：Nov-16, 16 年 11 月：Kwok Tsz Ki 郭子祈
7 adult female 雌成鳥：Long Valley 塱原：Nov-16, 16 年 11 月：Kwok Tsz Ki 郭子祈
8 adult female 雌成鳥：Po Toi 蒲台：24-Apr-06, 06 年 4 月 24 日：Geoff Welch

春季過境遷徙鳥 Spring Passage Migrant	夏候鳥 Summer Visitor	秋季過境遷徙鳥 Autumn Passage Migrant	冬候鳥 Winter Visitor

常見月份 1 2 3 4 5 6 7 8 9 10 11 12

留鳥 Resident	迷鳥 Vagrant	偶見鳥 Occasional Visitor

2
5
3
6
4
7

8

黃胸鵐

㊥huáng xiōng wū
㊕鵐：音無

體長 length：14-15.5cm

Yellow-breasted Bunting | *Emberiza aureola*

俗名「禾花雀」。上體褐色而有粗濃密縱紋，有明顯白肩及多於一條翼帶，下體黃色，脇部有縱紋。雄鳥頭黑色，後枕栗色與栗色胸帶相連。幼鳥眼眉、頰下紋和喉部黃色，沒有胸帶。雌性肩部沒有白色。

Known as “Rice Flower Bird” in Chinese. Upperparts brown with heavy thick stripes and distinctive white scapulars. Yellow underparts with stripes on flanks. Male has black head and a chestnut nape that connects with a chestnut breast band. Juvenile has yellow supercilium, submoustachial stripes and throat, but no breast band. Female lacks white scapulars.

1 adult male 雄成鳥：Long Valley 塱原：Apr-05, 05 年 4 月：Michelle and Peter Wong 江敏兒、黃理沛
2 female 雌鳥：Long Valley 塱原：18-Dec-05, 05 年 12 月 18 日：Doris Chu 朱詠兒
3 1st winter male 第一年冬天雄鳥：Lok Ma Chau 落馬洲：Nov-06, 06 年 11 月：Martin Hale 夏敖天
4 female 雌鳥：Fung Lok Wai 豐樂圍：Apr-04, 04 年 4 月：Jemi and John Holmes 孔思義、黃亞萍
5 1st summer male 第一年夏天雄鳥：Mai Po 米埔：28-Apr-08, 08 年 4 月 28 日：Martin Hale 夏敖天
6 1st summer male 第一年夏天雄鳥：Mai Po 米埔：May-08, 08 年 5 月：Andy Kwok 郭匯昌
7 1st summer male 第一年夏天雄鳥：Mai Po 米埔：May-08, 08 年 5 月：Andy Kwok 郭匯昌
8 1st winter male 第一年冬天雄鳥：Long Valley 塱原：Apr-07, 07 年 4 月：Matthew and TH Kwan 關朗曦、關子凱

春季過境遷徙鳥 Spring Passage Migrant	夏候鳥 Summer Visitor	秋季過境遷徙鳥 Autumn Passage Migrant	冬候鳥 Winter Visitor

常見月份 1 2 3 4 5 6 7 8 9 10 11 12

留鳥 Resident	迷鳥 Vagrant	偶見鳥 Occasional Visitor

2
3
4
5
6
7

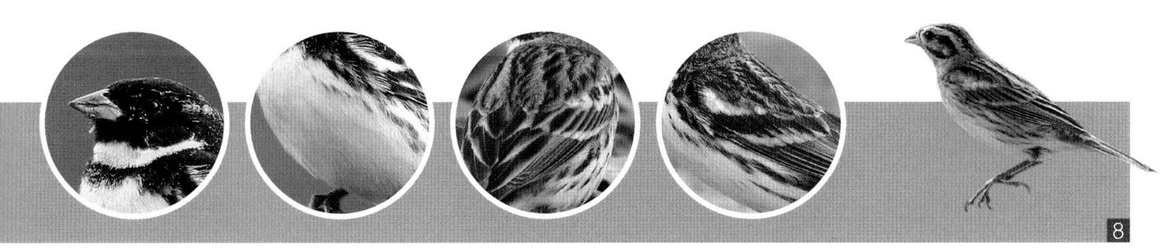
8

栗鵐

[普]lì wū
[粵]鵐：音無

體長length：14-15cm

Chestnut Bunting | *Emberiza rutila*

雄鳥頭部和上體栗色，喉至胸部栗色，腹部黃色，脇部有深色幼縱紋。越冬時栗色羽毛邊緣淡黃褐色，個別有栗色胸帶。雌鳥和幼鳥顏色較淡，縱紋較多。

Male has chestnut-brown head, upperparts and from throat to breast. Yellow belly and striped flanks. Wintering birds have chestnut-brown feathers fringed with yellowish brown. Some individuals have chestnut-brown breast band. Female and juvenile look paler and more heavily streaked.

1 adult male 雄成鳥：Po Toi 蒲台：26-Apr-08, 08 年 4 月 26 日：Lee Yat Ming 李逸明
2 1st summer male 第一年夏天雄鳥：Po Toi 蒲台：Apr-08, 08 年 4 月：Michelle and Peter Wong 江敏兒、黃理沛
3 adult male 雄成鳥：Po Toi 蒲台：Apr-08, 08 年 4 月：Kitty Koo 古愛婉
4 female 雌鳥：Po Toi 蒲台：Nov-08, 08 年 11 月：Michelle and Peter Wong 江敏兒、黃理沛
5 adult male 雄成鳥：Po Toi 蒲台：Apr-07, 07 年 4 月：Michelle and Peter Wong 江敏兒、黃理沛
6 1st winter male 第一年冬天雄鳥：Po Toi 蒲台：Nov-08, 08 年 11 月：Wallace Tse 謝鑑超
7 adult male 雄成鳥：Po Toi 蒲台：Apr-07, 07 年 4 月：Michelle and Peter Wong 江敏兒、黃理沛

	春季過境遷徙鳥 Spring Passage Migrant			夏候鳥 Summer Visitor			秋季過境遷徙鳥 Autumn Passage Migrant			冬候鳥 Winter Visitor		
常見月份	1	2	3	4	5	6	7	8	9	10	11	12
	留鳥 Resident				迷鳥 Vagrant				偶見鳥 Occasional Visitor			

2
3
4
5
6

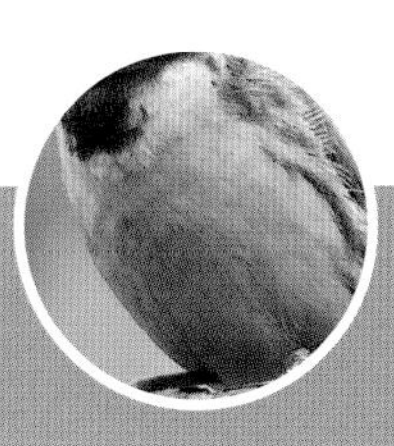

7

硫磺鵐

[普] liú huáng wū
[粵] 鵐：音無

體長 length：13-14cm

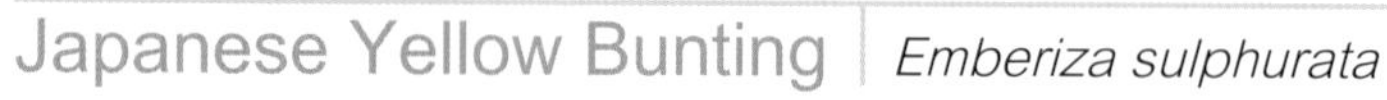

Japanese Yellow Bunting | *Emberiza sulphurata*

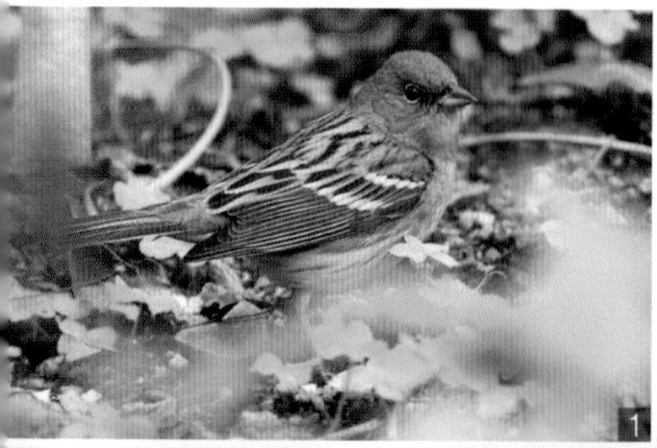
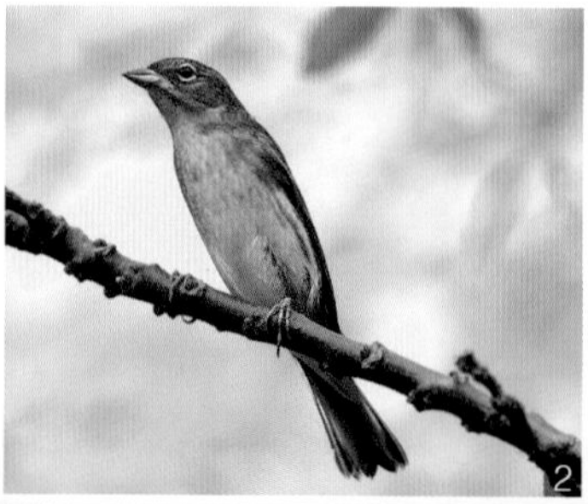
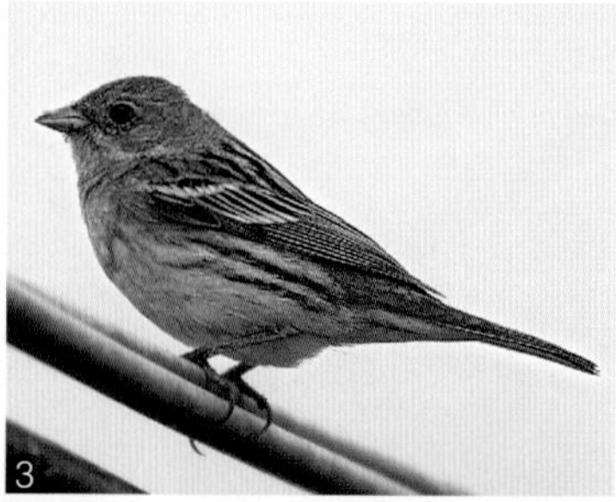
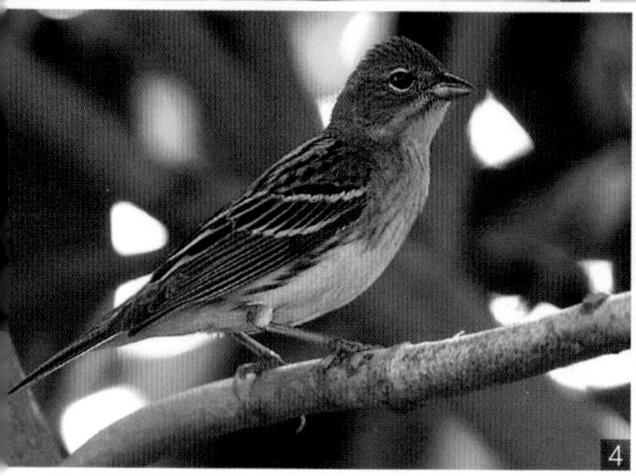
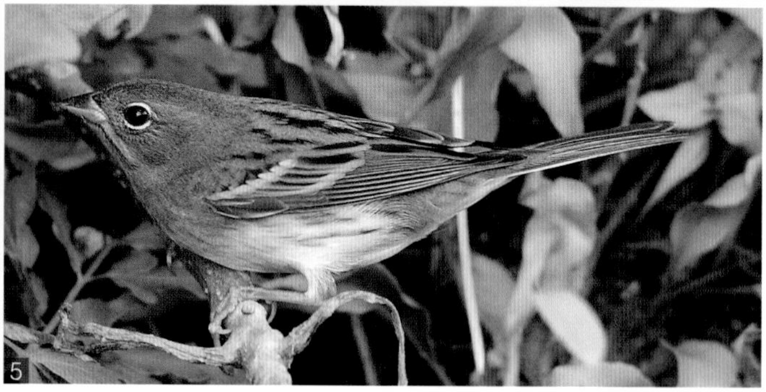

🌳

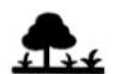

上體主要為灰綠色，背部有深色粗縱紋及兩道白色翼斑。下體主要為黃綠色，脇部有少量深色縱紋。頭部偏綠及有明顯不完整的白色眼圈。雄鳥有黑色眼先及較黃色的下體，雌鳥顏色較暗。

Upperparts are overall greenish-grey with some dark streaks on mantle and double white wing bars. Underparts are mainly greenish-yellow. Flanks have a few dark streaks. Head is plain and more greenish with conspicuous broken white eye-ring. Male has black lores and more yellowish underparts. Female is duller in colour.

[1] adult male 雄成鳥；Po Toi 蒲台；Apr-07, 07 年 4 月；Pippen Ho 何志剛
[2] 1st winter male 第一年冬天雄鳥；Po Toi 蒲台；Oct-08, 08 年 10 月；Isaac Chan 陳家強
[3] adult female 雌成鳥；Long Valley 塱原；Apr-07, 07 z年 4 月；Michelle and Peter Wong 江敏兒、黃理沛
[4] adult female 雌成鳥；Long Valley 塱原；Apr-06, 06 年 4 月；Michelle and Peter Wong 江敏兒、黃理沛
[5] female 雌鳥；Po Toi 蒲台；Nov-08, 08 年 11 月；Michelle and Peter Wong 江敏兒、黃理沛
[6] female 雌鳥；Po Toi 蒲台；Nov-08, 08 年 11 月；Wallace Tse 謝鑑超

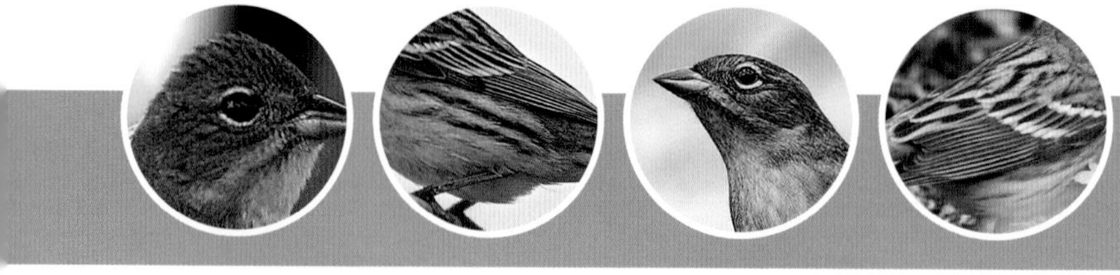
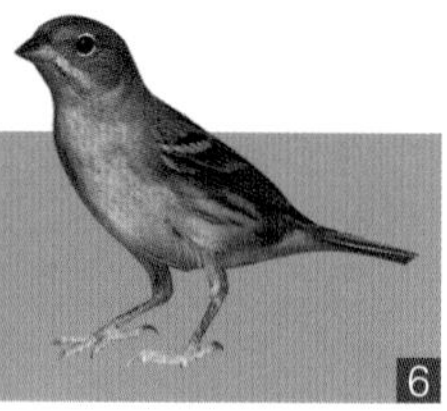

春季過境遷徙鳥 Spring Passage Migrant	夏候鳥 Summer Visitor	秋季過境遷徙鳥 Autumn Passage Migrant	冬候鳥 Winter Visitor

常見月份	1	2	3	4	5	6	7	8	9	10	11	12

留鳥 Resident	迷鳥 Vagrant	偶見鳥 Occasional Visitor

葦鵐

[普] wěi wū
[粵] 鵐：音無

體長 length：12-13.5cm

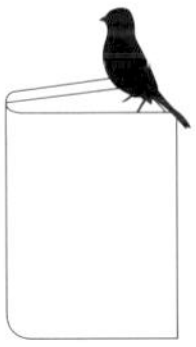

Pallas's Reed Bunting | *Emberiza pallasi*

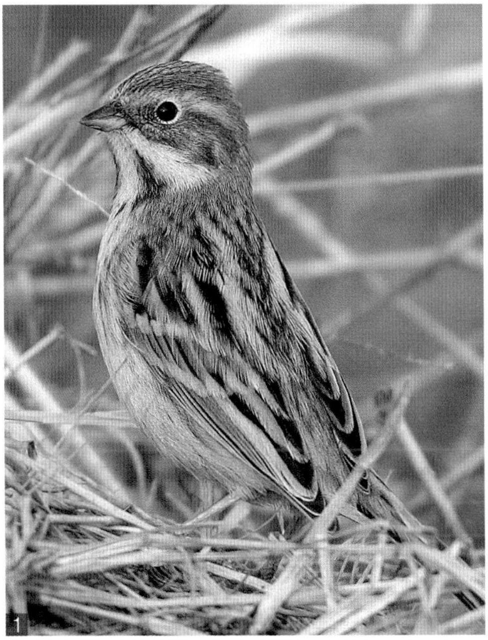

1

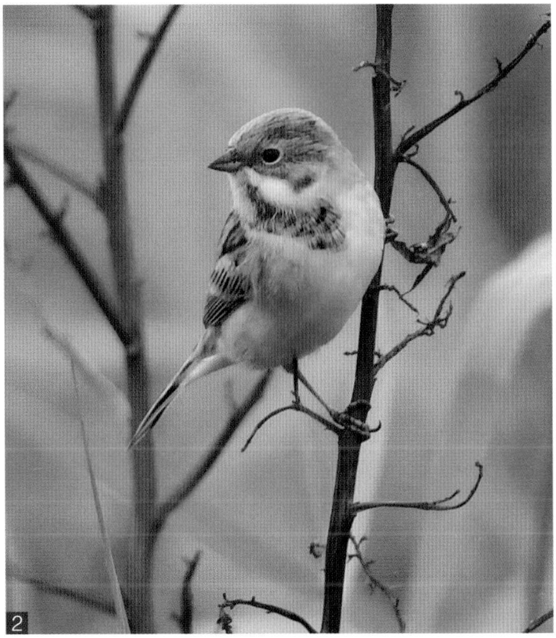

2

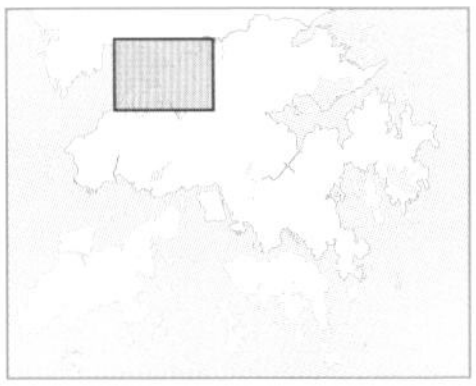

體型較蘆鵐小，顏色較淡。繁殖期雄性頭與喉黑色，白色頦紋與白領相連。淡褐色上體有少量橫紋。雌鳥和非繁殖雄性頭部淺褐色，眉和頰紋白色。較罕有，香港曾在深秋錄得非繁殖期的雄鳥和雌鳥。

Smaller and plainer than Common Reed Bunting. Breeding male has black head and throat separated by white submoustachial strip joining the white collar. Buffy upperparts, lightly streaked. Females and non-breeders have buffy head with white supercilium and submoustachial strip. Rare birds. All Hong Kong records are in late autumn of females or non-breeding males.

[1] female 雌鳥：Lok Ma Chau 落馬洲：Nov-06, 06 年 11 月：Martin Hale 夏敖天
[2] adult female / i mmature 雌成鳥 / 未成年鳥：San Tin新田：3 Nov-18, 18年11月3日：Beetle Cheng 鄭諾銘

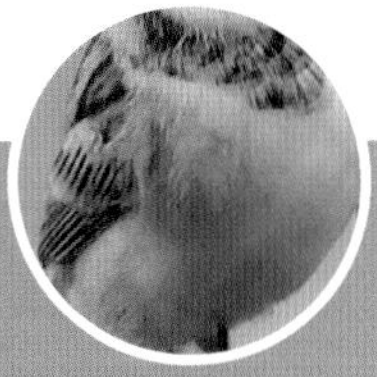

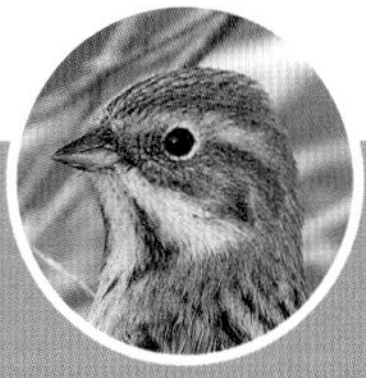

春季過境遷徙鳥 Spring Passage Migrant			夏候鳥 Summer Visitor			秋季過境遷徙鳥 Autumn Passage Migrant			冬候鳥 Winter Visitor			
1	2	3	4	5	6	7	8	**9**	**10**	**11**	**12**	常見月份
留鳥 Resident				迷鳥 Vagrant				偶見鳥 Occasional Visitor				

灰頭鵐

㊢ huī tóu wū
[粵] 鵐：音無

體長 length：13.5-16cm

Black-faced Bunting | *Emberiza spodocephala*

其他名稱 Other names：Masked Bunting

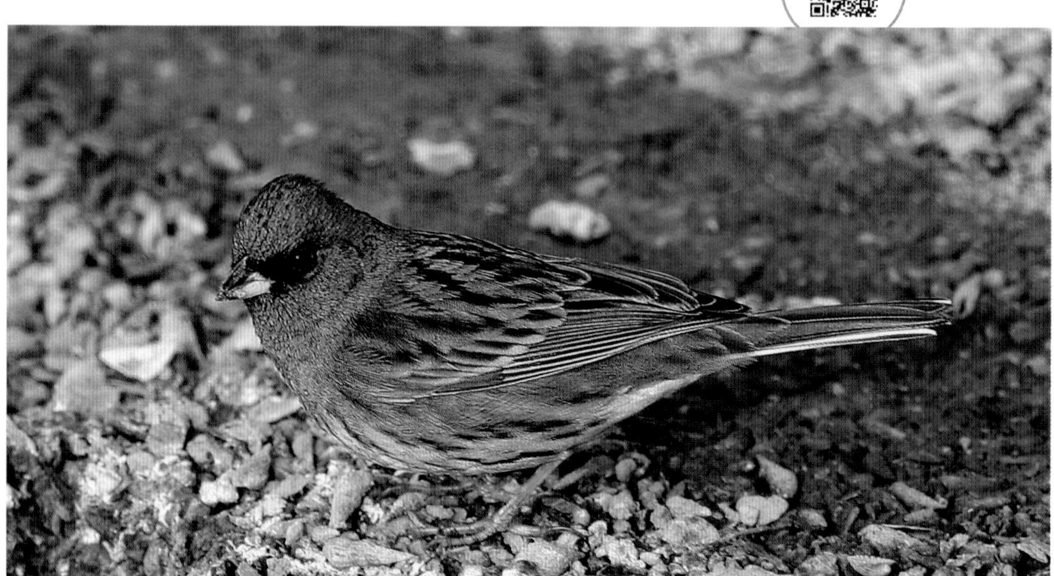

1

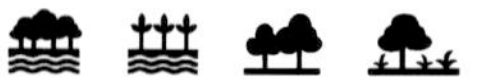

頭部深灰綠色，上體褐色，有濃密深色縱紋。下體淡黃色，有褐色縱紋。香港有幾個亞種，最常見的是*personata*亞種，有淡白色頰下紋，胸前有灰褐色縱紋。*spodocephala*亞種雄鳥頭為灰色，淡黃褐色下體，縱紋較少，雌鳥的顏色較淡，頭部沒有灰色。*sordida*亞種有黃色下體。

Head greenish grey, upperparts brown with heavy dark streaks, underparts yellow with brownish stripes. There are several sub-species. The most common, *personata* has a pale submoustachial stripe and greenish grey stripes on the breast, *spodocephala* male has grey head, buffy underparts that are slightly striped, female is paler and head is not grey. *sordida* has yellow underparts.

1 adult male, race *sordida* 雄成鳥, *sordida* 亞種：Po Toi 蒲台：Jan-05, 05 年 1 月：Michelle and Peter Wong 江敏兒、黃理沛
2 adult male, race *spodocephala* 雄成鳥, *spodocephala* 亞種：Po Toi 蒲台：Apr-07, 07 年 4 月：Jemi and John Holmes 孔思義、黃亞萍
3 1st winter male, race *sordida* 第一年冬天雄鳥, *sordida* 亞種：Shek Pik 石壁：7-Feb-08, 08 年 2 月 7 日：Sung Yik Hei 宋亦希
4 1st winter male, race *sordida* 第一年冬天雄鳥, *sordida* 亞種：Po Toi 蒲台：Jan-05, 05 年 1 月：Michelle and Peter Wong 江敏兒、黃理沛
5 female, race *sordida* 雌鳥, *sordida* 亞種：Long Valley 塱原：Feb-07, 07 年 2 月：James Lam 林文華
6 female, race *spodocephala* 雌鳥, *spodocephala* 亞種：Po Toi 蒲台：Apr-08, 08 年 4 月：Michelle and Peter Wong 江敏兒、黃理沛
7 adult female, race *spodocephala* 雌成鳥, *spodocephala* 亞種：Mai Po 米埔：Apr-04, 04 年 4 月：Martin Hale 夏敖天
8 male, race *spodocephala* 雄鳥, *spodocephala* 亞種：Mai Po 米埔：Feb-05, 05 年 2 月：Cherry Wong 黃卓研

	春季過境遷徙鳥 Spring Passage Migrant			夏候鳥 Summer Visitor			秋季過境遷徙鳥 Autumn Passage Migrant			冬候鳥 Winter Visitor		
常見月份	1	2	3	4	5	6	7	8	9	10	11	12
	留鳥 Resident				迷鳥 Vagrant				偶見鳥 Occasional Visitor			

蘆鵐

㊟ lú wū
㊟ 鵐：音無

體長 length：14-16.5cm

Common Reed Bunting | *Emberiza schoeniclus*

其他名稱 Other names：Reed Bunting

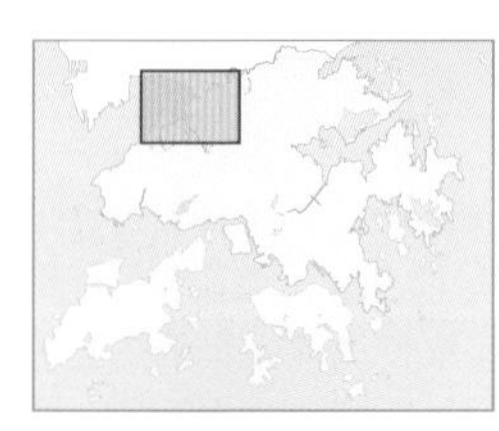

體型比葦鵐大，縱紋較多，翼上有紅褐色塊。雄性背紅褐色，雌性背和腰淺褐色。沒有田鵐耳後的淡色斑塊。香港在冬季和初春曾在蘆葦叢有小群記錄。

Bigger than Pallas's Reed Bunting, more streaked and all plumages have some rufous-red on wings. Male bird also has rufous back, while female has more buffy colour on back, concolourous with rump. Unlike Rustic Bunting, it has no dark patch on ear covert. In Hong Kong, small parties in phragmites reedbeds have been recorded in winter and early spring.

1 non-breeding 非繁殖羽；Mai Po 米埔；Dec-06, 06 年 12 月；Cherry Wong 黃卓研
2 adult 成鳥；Long Valley 塱原；Nov-16, 16 年 11 月；Kwok Tsz Ki 郭子祈

春季過境遷徙鳥 Spring Passage Migrant				夏候鳥 Summer Visitor			秋季過境遷徙鳥 Autumn Passage Migrant			冬候鳥 Winter Visitor	
1	2	3	4	5	6	7	8	9	10	11	12
留鳥 Resident				迷鳥 Vagrant				偶見鳥 Occasional Visitor			

常見月份：1, 2, 3, 4, 12

參考資料
References

Sonobe, K. and Usui, S. (ed.) 1993. *A Field Guide to the Waterbirds of Asia*. Wild Bird Society of Japan, Tokyo.

Wang, S, Zheng G. and Wang, Q. (ed.) 1998. *China Red Data Book of Endangered Animals: Aves.* Science Press, Beijing.

Carey, G. J., Chalmers, M. L., Diskin, D. A., Kennerley, P. R., Leader, P. J., Leven, M. R., Lewthwaite, R. W., Melville, D. S., Turnbull, M., and Young, L. 2001. *The Avifauna of Hong Kong.* Hong Kong Bird Watching Society, Hong Kong.

Hayman P., Marchant J. and Prater T. 1995. *Shorebirds. An identification Guide to the Waders of the World.* Christopher Helm (Publishers) Ltd.

Hong Kong Bird Watching Society. 1957-2004. *Hong Kong Bird Reports (1957-2004)* – annual report publish by Hong Kong Bird Watching Society.

Viney, C., Phillips, K. and Lam, C.Y., 2005. *Birds of Hong Kong and South China.* Government Printer. Hong Kong.

MacKinnon, M., and Phillipps, K. 2000. *A Field Guide to the Birds of China.* Oxford University Press, UK

Hong Kong Bird Watching Society 2005. *A Photographic Guide to the Birds of Hong Kong* Wan Li Book Company Limited

King, B.F., Woodcock, M.W. and Dickinaon E.C. 1975 Collins Field Guide Birds of South-East Asia Harper Collins Publishers

聶延秋 2019 中國鳥類識別手冊(第二版) 中國林業出版社

林超英 2000 大埔滘觀鳥樂 天地圖書有限公司、郊野公園之友、香港大學地理及地質學系

馬嘉慧、馮寶基 1999 觀鳥一從后海灣開始 長春社

馬嘉慧、馮寶基、蘇毅雄 2001 觀鳥一從城市開始 香港觀鳥會

約翰・馬敬能、卡倫・菲利普斯、何芬奇 2000 中國鳥類野外手冊 湖南教育出版社

王嘉雄等 1991 台灣野鳥圖鑑 台灣野鳥資訊社

許維樞等 1996 中國野鳥圖鑑 翠鳥文化出版社

鄭光美 2002 世界鳥類分類與分布名錄 科學出版社

尹璉、費嘉倫、林超英 2006 香港及華南鳥類 政府新聞處

香港觀鳥會 2005 香港鳥類攝影圖鑑 萬里機構

常用網頁
Websites

Hong Kong Bird Watching Society 香港觀鳥會

www.hkbws.org.hk

Hong Kong Observatory 香港天文台網頁

Hong Kong Tide table 香港潮汐表

http://www.hko.gov.hk/tc/tide/marine/realtide.htm?s=TBT&t=CHART（中文 Chinese）

http://www.hko.gov.hk/en/tide/marine/realtide.htm?s=TBT&t=CHART（英文 English）

中文鳥名索引
Index by Chinese Names

英文鳥名索引
Index by English Common Names

學名索引
Index by Scientific Names

觀鳥系列02：香港觀鳥全圖鑑 A Photographic Guide to the Birds of Hong Kong

著者 Author
香港觀鳥會 The Hong Kong Bird Watching Society

策劃 Project Co-ordinator
謝妙華 Pheona Tse

責任編輯 Project Editor
簡詠怡、陳芷欣 Karen Kan, Kitty Chan

裝幀設計 Design
鍾啟善 Nora Chung

排版 Typography
劉葉青 Rosemary Liu

出版者 Publisher
萬里機構出版有限公司 Wan Li Book Company Limited
香港北角英皇道499號北角工業大廈20樓 20/F, North Point Industrial Building, 499 King's Road, North Point, Hong Kong
電話 Tel: 2564 7511
傳真 Fax: 2565 5539
電郵 Email: info@wanlibk.com
網址 http://www.wanlibk.com
http://www.facebook.com/wanlibk

發行者 Distributor
香港聯合書刊物流有限公司 SUP Publishing Logistics (HK) Ltd.
香港荃灣德士古道220-248號荃灣工業中心16樓 16/F, Tsuen Wan Industrial Centre, 220-248 Texaco Road, Tsuen Wan, NT., Hong Kong
電話 Tel: 2150 2100
傳真 Fax: 2407 3062
電郵 E-mail: info@suplogistics.com.hk
網址 http://www.suplogistics.com.hk

承印者 Printer
美雅印刷製本有限公司 Elegance Printing & Book Binding Co., Ltd.
香港九龍觀塘榮業街6號海濱工業大廈4樓A室 Block A, 4/F, Hoi Bun Industrial Building, 6 Wing Yip Street, Kwun Tong, Kln., Hong Kong

出版日期 Publishing Date
二〇二〇年六月第一次印刷 First print in June 2020
二〇二一年二月第二次印刷 Second print in February 2021

規格 Specifications
特 16 開（240mm × 171mm） 16K (240mm × 171mm)

ISBN 978-962-14-7239-7

雀鳥視頻 Video of Birds

創作 Creator
文緝明 Man Chup Ming

製作 Production
香港觀鳥會 The Hong Kong Bird Watching Society

鳴謝 Acknowledgements
文緝明先生 Mr Man Chup Ming
黃亞萍小姐 Ms Jemi AP Wong
孔思義先生 Mr John Holmes